TAKING SIDES

Clashing Views on

Environmental Issues

TWELFTH EDITION, EXPANDED

TAKING SIDES

Clashing Views on

Environmental Issues

TWELFTH EDITION, EXPANDED

Selected, Edited, and with Introductions by

Thomas Easton
Thomas College

Mc Graw Hill **Contemporary Learning Series**

A Division of The McGraw-Hill Companies

Photo Acknowledgment
Cover image: U.S. Navy photo

Cover Acknowledgment
Maggie Lytle

Compositor: ICC Macmillan Inc.

Manufactured in the United States of America

Twelfth Edition, Expanded

123456789DOCDOC987

Library of Congress Cataloging-in-Publication Data
Main entry under title:
Taking sides: clashing views on controversial issues in environmental issues/selected, edited, and
with introductions by Thomas A. Easton—12th ed.
Includes bibliographical references and index.
1. Environmental policy. 2. Environmental protection. I. Easton, Thomas A. *comp.*

363.7

MHID: 0-07-351443-8
ISBN: 978-0-07-351443-7
ISSN: 1091-8825

Printed on Recycled Paper

Preface

Most fields of academic study evolve over time. Some evolve in turmoil, for they deal in issues of political, social, and economic concern. That is, they involve controversy.

It is the mission of the *Taking Sides* series to capture current, ongoing controversies and make the opposing sides available to students. This book focuses on environmental issues, from the philosophical to the practical. It does not pretend to cover all such issues, for not all provoke controversy or provoke it in suitable fashion. But there is never any shortage of issues that can be expressed as pairs of opposing essays that make their positions clearly and understandably.

The basic technique—presenting an issue as a pair of opposing essays—has risks. Students often display a tendency to remember best those essays that agree with the attitudes they bring to the discussion. They also want to know what the "right" answers are, and it can be difficult for teachers to refrain from taking a side, or from revealing their own attitudes. Should teachers so refrain? Some do not, but of course they must still cover the spectrum of opinion if they wish to do justice to the scientific method and the complexity of an issue. Some do, though rarely so successfully that students cannot see through the attempt.

For any *Taking Sides* volume, the issues are always phrased as yes/no questions. Which answer—yes or no—is the correct answer? Perhaps neither. Perhaps both. Perhaps we will not be able to tell for another hundred years. Students should read, think about, and discuss the readings and then come to their own conclusions without letting my or their instructor's opinions dictate theirs. The additional readings mentioned in the introductions and postscripts should prove helpful.

This edition of *Taking Sides: Clashing Views on Environmental Issues* Expanded contains 44 readings arranged in pro and con pairs to form 19 issues. For each issue, an *introduction* provides historical background and a brief description of the debate. The *postscript* after each pair of readings offers recent contributions to the debate, additional references, and sometimes a hint of future directions. Each part is preceded by an *On the Internet* page that lists several links that are appropriate for further pursuing the issues in that part.

Changes to this Edition   Over half of this book consists of new material. Two issues, *Should the Military Be Exempt from Environmental Regulations?* (Issue 7) and *Is Additional Federal Oversight Needed for the Construction of LNG Import Facilities?* (Issue 11) were added for the 2006 partial revision. The global warming issue has been recast as *Should the U.S. Be Doing More to Combat Global Warming?* (Issue 9). The population issue has become *Do Falling Birth Rates Pose a Threat to Human Welfare?* (Issue 13). The genetic engineering issue is now *Is Genetic Engineering the Answer to Hunger?* (Issue 14). The nuclear waste issue is now *Should the United States Reprocess Spent Nuclear Fuel?* (Issue 19).

In addition, for nine of the issues retained from the previous edition, one or both of the readings have been replaced. In all, 24 of the readings in this edition were not in the 11th edition.

A word to the instructor   An *Instructor's Manual With Test Questions* (multiple-choice and essay) is available through the publisher for the instructor using *Taking Sides* in the classroom. Also available is a general guidebook, *Using Taking Sides in the Classroom*, which offers suggestions for adapting the pro-con approach in any classroom setting. An online version of *Using Taking Sides in the Classroom* and a correspondence service for Taking Sides adopters can be found at http://www.mhcls.com/usingts/.

Taking Sides: Clashing Views on Environmental Issues is only one title in the Taking Sides series. If you are interested in seeing the table of contents for any of the other titles, please visit the Taking Sides Web site at http://www.mhcls.com/takingsides/.

Thomas A. Easton
Thomas College

Contents In Brief

Contents

Nancy Myers, communications director for the Science and Environmental Health Network, argues that because the precautionary principle "makes sense of uncertainty," it has gained broad international recognition as being crucial to environmental policy. John D. Graham, dean of the Frederick S. Pardee RAND Graduate School, argues that the precautionary principle is so subjective that it permits "precaution without principle" and threatens innovation and public and environmental health. It must therefore be used cautiously.

Jeremy Rifkin, president of the Foundation on Economic Trends, argues that Europeans pride themselves on their quality of life, and their emphasis on sustainable development promises to maintain that quality of life into the future. Environmental journalist Ronald Bailey states that sustainable development results in economic stagnation and threatens both the environment and the world's poor.

Jim Morrison argues that ecosystem services such as cleaning water, controlling floods, and pollinating crops have sufficient economic value to make it profitable to spend millions of dollars to protect natural systems. Professors of applied ecology Marino Gatto and Giulio A. De Leo contend that the pricing approach to valuing nature's services is misleading because it falsely implies that only economic values matter.

Professor of economics David N. Laband argues that the public demands excessive amounts of biodiversity largely because decision makers and voters do not have to bear the costs of producing it. Wildlife conservation researcher and writer Howard Youth argues that the actions needed to protect biodiversity not only have economic benefits, but also are the same actions needed to ensure a sustainable future for humanity.

Professor Julian Agyeman argues that although there is much debate over whether sustainable development means addressing environmental issues or environmental justice, equity, human rights, and poverty reduction, the two can and must be integrated. Writer and social analyst David Friedman denies the existence of environmental racism. He argues that the environmental justice movement is a government-sanctioned political ploy that will hurt urban minorities by driving away industrial jobs.

Freelance science writer Charles W. Schmidt argues that economic incentives such as emissions rights trading offer the most useful approaches to reducing pollution. Author, college teacher, and environmental activist Brian Tokar maintains that pollution credits and other market-oriented environmental protection policies do nothing to

drastic action. The Bush administration's plan for dealing with global warming insists that short-term economic health must come before reducing emissions of greenhouse gases. It is more useful to reduce "greenhouse gas intensity" or emissions per dollar of economic activity than to reduce total emissions.

David L. Bodde argues that there is no question whether hydrogen can satisfy the nation's energy needs. The real issue is how to handle the transition from the current energy system to the hydrogen system. Michael Behar argues that the public has been misled about the prospects of the "hydrogen economy." We must overcome major technological, financial, and political obstacles before hydrogen can be a viable alternative to fossil fuels.

Edward J. Markey argues that the risks—including those associated with terrorist attack—associated with LNG (liquefied natural gas) tankers and terminals are so great that additional federal regulation is essential in order to protect the public. Donald F. Santa, Jr., argues that meeting demand for energy requires public policies that "do not unreasonably limit resource and infrastructure development." The permitting process for LNG import facilities should be governed by existing Federal Regulatory Commission procedures without additional regulatory impediments.

Michael J. Wallace argues that because the benefits of nuclear power include energy supply and price stability, air pollution control, and greenhouse gas reduction, new nuclear power plant construction—with federal support—is essential. *Public Citizen* argues that nuclear power is too unreliable and risky to count on. We must "embrace safe, clean, sustainable energy sources."

although the use of marine protected areas can be beneficial, limiting fishing effort is a more effective way of achieving sustainable fisheries.

Phillip J. Finck argues that by reprocessing spent nuclear fuel, the United States can enable nuclear power to expand its contribution to the nation's energy needs while reducing carbon emissions, nuclear waste, and the need for waste repositories such as Yucca Mountain. Matthew Bunn argues that there is no near-term need to embrace nuclear spent fuel reprocessing, costs are highly uncertain, and there is a worrisome risk that the increased availability of bomb-grade nuclear materials will increase the risk of nuclear war and terrorism.

Brian Halweil, senior researcher at the Worldwatch Institute, argues that organic agriculture is potentially so productive that it could sustainably increase world food supply, although the future may be more likely to see a mix of organic and nonorganic techniques. John J. Miller argues that organic farming is not productive enough to feed today's population, much less larger future populations, it is prone to dangerous biological contamination, and it is not sustainable.

Representing the National Wildlife Foundation, John Kostyack argues that the Endangered Species Act has been so successful that it should not be weakened but strengthened. Speaking for the National Endangered Species Act Reform Coalition, a group that represents those affected by the Endangered Species Act, Monita Fontaine argues that federal regulation under the ESA should be replaced by a system that relies more on voluntary and state species conservation efforts.

Introduction

Environmental Issues: The Never-Ending Debate

Thomas A. Easton

One of the courses I teach is "Environmentalism: Philosophy, Ethics, and History." I begin by explaining the roots of the word "ecology," from the Greek *oikos* (house or household), and assigning the students to write a brief paper about their own household. How much, I ask them, do you need to know about the place where you live? And why?

The answers vary. Some of the resulting papers focus on people—roommates if the "household" is a dorm room, spouses and children if the students are older, parents and siblings if they live at home—and the needs to cooperate and get along, and perhaps the need not to overcrowd. Some pay attention to houseplants and pets, and occasionally even bugs and mice. Some focus on economics—possessions, services, and their costs, where the checkbook is kept, where the bills accumulate, the importance of paying those bills, and of course the importance of earning money to pay those bills. Some focus on maintenance—cleaning, cleaning supplies, repairs, whom to call if something major breaks. For some the emphasis is operation—garbage disposal, grocery shopping, how to work the lights, stove, fridge, and so on. A very few recognize the presence of toxic chemicals under the sink and in the medicine cabinet and the need for precautions in their handling. Sadly, a few seem to be oblivious to anything that does not have something to do with entertainment.

Not surprisingly, some students object initially that the exercise seems trivial. "What does this have to do with environmentalism?" they ask. Yet the course is rarely very old before most are saying, "Ah! I get it!" That nice, homey microcosm has a great many of the features of the macrocosmic environment, and the multiple ways people can look at the microcosm mirror the ways people look at the macrocosm. It's all there, as is the question of priorities: What is important? People or fellow creatures or economics or maintenance or operation or waste disposal or food supply or toxics control or entertainment? Or all of the above?

And how do you decide? I try to illuminate this question by describing a parent trying to teach a teenager not to sit on a woodstove. In July, the kid answers, "Why?" and continues to perch. In August, likewise. And still in September. But in October or November, the kid yells "Ouch!" and jumps off in a hurry.

That is, people seem to learn best when they get burned.

This is surely true in our homely *oikos*, where we may not realize our fellow creatures deserve attention until houseplants die of neglect or cockroaches invade the cupboards. Economics comes to the fore when the phone gets cut off, repairs when a pipe ruptures, air quality when the air conditioner breaks or strange fumes rise from the basement, waste disposal when the bags pile up and begin to stink or the toilet backs up. Toxics control suddenly matters when a child or pet gets into the rat poison.

In the larger *oikos* of environmentalism, such events are paralleled by the loss of a species, or an infestation by another, by floods and droughts, by lakes turned into cesspits by raw sewage, by air turned foul by industrial smokestacks, by groundwater contaminated by toxic chemicals, by the death of industries and the loss of jobs, by famine and plague and even war.

If nothing is going wrong, we are not very likely to realize there is something we should be paying attention to. And this too has its parallel in the larger world. Indeed, the history of environmentalism is, . . . important part, a history of people carrying on with business as usual until something goes obviously awry. Then, if they can agree on the nature of the problem (Did the floor cave in because the joists were rotten or because there were too many people at the party?), they may learn something about how to prevent recurrences.

The Question of Priorities

There is of course a crucial "if" in that last sentence: *If people can agree . . .* It is a truism to say that agreement is difficult. In environmental matters, people argue endlessly over whether anything is actually wrong, what its eventual impact will be, what if anything can or should be done to repair the damage, and how to prevent recurrence. Not to mention who's to blame and who should take responsibility for fixing the problem! Part of the reason is simple: Different things matter most to different people. Individual citizens may want clean air or water or cheap food or a convenient commute. Politicians may favor sovereignty over international cooperation. Economists and industrialists may think a few coughs (or cases of lung cancer, or shortened lifespans) a cheap price to pay for wealth or jobs.

No one now seems to think that protecting the environment is not important. But different groups—even different environmentalists—have different ideas of what "environmental responsibility" means. To a paper company cutting trees for pulp, it may mean leaving a screen of trees (a "beauty strip") beside the road and minimizing erosion. To hikers following trails through or within view of the same tract of land, that is not enough; they want the trees left alone. The hikers may also object to seeing the users of trail bikes and all-terrain-vehicles on the trails. They may even object to hunters and anglers, whose activities they see as diminishing the wilderness experience. They may therefore push for protecting the land as limited-access wilderness. The hunters and anglers object to that, of course, for they want to be able to use their vehicles to bring their game home, or to bring their boats to their favorite rivers and lakes. They also argue, with some justification, that their license fees support a great deal of environmental protection work.

To a corporation, dumping industrial waste into a river may make perfect sense, for alternative ways of disposing of waste are likely to cost more and diminish profits. Of course, the waste renders the water less useful to wildlife or downstream humans, who may well object. Yet, telling the corporation it cannot dump may be seen as depriving it of property. A similar problem arises when regulations prevent people and corporations from using land—and making money—as they had planned. Conservatives have claimed that environmental regulations thus violate the Fifth Amendment to the U.S. Constitution, which says "No person shall . . . be deprived of . . . property, without due process of law; nor shall private property be taken for public use, without just compensation."

One might think the dangers of such things as dumping industrial waste in rivers are obvious. But scientists can and do disagree, even given the same evidence. For instance, a chemical in waste may clearly cause cancer in laboratory animals. Is it therefore a danger to humans? A scientist working for the company dumping that chemical in a river may insist that no such danger has been proven. Yet, a scientist working for an environmental group such as Greenpeace may insist that the danger is obvious because carcinogens do generally affect more than one species.

Scientists are human. They have not only employers but also values, rooted in political ideology and religion. They may feel that the individual matters more than corporations or society, or vice versa. They may favor short-term benefits over long-term benefits, or vice versa.

And scientists, citizens, corporations, and government all reflect prevailing social attitudes. When America was expanding westward, the focus was on building industries, farms, and towns. If problems arose, there was vacant land waiting to be moved to. But when the expansion was done, problems became more visible and less avoidable. People could see that there were "trade-offs" involved in human activity: more industry meant more jobs and more wealth, but there was a price in air and water pollution and human health (among other things).

Nowhere, perhaps, are these trade-offs more obvious than in Eastern Europe. The former Soviet Union was infamous for refusing to admit that industrial activity was anything but desirable. Anyone who spoke up about environmental problems risked jail. The result, which became visible to Western nations after the fall of the Iron Curtain in 1990, was industrial zones where rivers had no fish, children were sickly, and life expectancies were reduced. The fate of the Aral Sea, a vast inland body of water once home to a thriving fishery and a major regional transportation route, is emblematic: Because the Soviet Union wanted to increase its cotton production, it diverted for irrigation the rivers that delivered most of the Aral Sea's fresh water supply. The Sea then began to lose more water to evaporation than it gained, and it rapidly shrank, exposing sea-bottom so contaminated by industrial wastes and pesticides that wind-borne dust is now responsible for a great deal of human illness. The fisheries are dead, and freighters lie rusting on bare ground where once waves lapped.

The Environmental Movement

The twentieth century saw immense changes in the conditions of human life and in the environment that surrounds and supports human life. According to historian J. R. McNeill, in *Something New Under the Sun: An Environmental History of the Twentieth-Century World* (W. W. Norton, 2000), the environmental impacts that resulted from the interactions of burgeoning population, technological development, shifts in energy use, politics, and economics in that period are unprecedented in both degree and kind. Yet, a worse impact may be that we have come to accept as "normal" a very temporary situation that "is an extreme deviation from any of the durable, more 'normal,' states of the world over the span of human history, indeed over the span of earth history." We are thus not prepared for the inevitable and perhaps drastic changes ahead.

Environmental factors cannot be denied their role in human affairs. Nor can human affairs be denied their place in any effort to understand environmental change. As McNeill says, "Both history and ecology are, as fields of knowledge go, supremely integrative. They merely need to integrate with each other."

The environmental movement, which grew during the twentieth century in response to increasing awareness of human impacts, is a step in that direction. Yet, environmental awareness reaches back long before the modern environmental movement. When John James Audubon (1785–1851), famous for his bird paintings, was young, he was an enthusiastic slaughterer of birds (a few of which he used as models for the paintings). Later in life, he came to appreciate that birds were diminishing in numbers, as were the American bison, and he called for conservation measures. His was a minority voice, however. It was not until later in the century that John Muir (1838–1914; founder of the Sierra Club) began to call for the preservation of natural wilderness, untouched by human activities. In 1890, Gifford Pinchot (1865–1946) found "the nation . . . obsessed by a fury of development. The American Colossus was fiercely intent on appropriating and exploiting the riches of the richest of all continents." Under President Theodore Roosevelt, Pinchot became the first head of the U.S. Forest Service and a strong voice for conservation (not to be confused with preservation; Gifford's conservation meant using nature but in such a way that it was not destroyed; his aim was "the greatest good of the greatest number in the long run"). By the 1930s, Aldo Leopold (1887–1948), best known for his concept of the "land ethic" and his book, *A Sand County Almanac*, could argue that we had a responsibility not only to maintain the environment but also to repair damage done in the past.

The modern environmental movement was kick-started by Rachel Carson's *Silent Spring* (Houghton Mifflin, 1962). In the 1950s, Carson realized that the use of pesticides was having unintended consequences—the death of nonpest insects, food-chain accumulation of poisons and the consequent loss of birds, and even human illness—and meticulously documented the case. When her book was published, she and it were immediately vilified by pesticide proponents in government, academia, and industry (most notably, the pesticides industry). There was no problem, the critics said; the negative effects, if any, were worth it, and she—a *woman* and a nonscientist—could not possibly know what she was talking about. But the facts won out. A decade later, DDT was banned and other pesticides were regulated in ways unheard of before Carson spoke out.

Other issues have followed or are following a similar course.

The situation before Rachel Carson and *Silent Spring* is nicely captured by Judge Richard Cudahy, who in "Coming of Age in the Environment," *Environmental Law* (Winter 2000), writes, "It doesn't seem possible that before 1960 there was no 'environment'—or at least no environmentalism. I can even remember the Thirties, when we all heedlessly threw our trash out of car windows, burned coal in the home furnace (if we could afford to buy any), and used a lot of lead for everything from fishing sinkers and paint to no-knock gasoline. Those were the days when belching black smoke meant a welcome end to the Depression and little else."

Historically, humans have felt that their own well-being mattered more than anything else. The environment existed to be used. Unused, it was only wilderness or wasteland, awaiting the human hand to "improve" it and make it valuable. This is not surprising at all, for the natural tendency of the human mind is to appraise all things in relation to the self, the family, and the tribe. An important aspect of human progress has lain in enlarging our sense of "tribe" to encompass nations and groups of nations. Some now take it as far as the human species. Some include other animals. Some embrace plants as well, and bacteria, and even landscapes.

The more limited standard of value remains common. Add to that a sense that wealth is not just desirable but a sign of virtue (the Puritans brought an explicit version of this with them when they colonized North America; see Lynn White, Jr., "The Historical Roots of Our Ecological Crisis," *Science*, 1967), and it is hardly surprising that humans have used and still use the environment intensely. People also tend to resist any suggestion that they should restrain their use out of regard for other living things. Human needs, many insist, come first.

The unfortunate consequences include the loss of other species. Lions vanished from Europe about 2,000 years ago. The dodo of Mauritius was extinguished in the 1600s (see the American Museum of Natural History's account at http://www.amnh.org/exhibitions/expeditions/ treasure_fossil/Treasures/Dodo/dodo.html?dinos). The last of North America's passenger pigeons died in a Cincinnati zoo in 1914 (see http://www.amnh.org/exhibitions/expeditions/treasure_ fossil/Treasures/Passenger_Pigeons/pigeons.html?dinos). Concern for such species was at first limited to those of obvious value to humans. In 1871, the U.S. Commission on Fish and Fisheries was created and charged with finding solutions to the decline in food fishes and promoting aquaculture. The first federal legislation designed to protect game animals was the Lacey Act of 1900. It was not until 1973 that the U.S. Endangered Species Act was adopted to shield all species from the worst human impacts.

Other unfortunate consequences of human activities include dramatic erosion, air and water pollution, oil spills, accumulations of hazardous (including nuclear) waste, famine, and disease. Among the many "hot stove" incidents that have caught public attention are the following:

- The Dust Bowl—in 1934, wind blew soil from drought-stricken farms in Oklahoma all the way to Washington, D.C.;
- Cleveland's Cuyahoga River caught fire in the 1960s;
- The Donora, Pennsylvania, smog crisis—in one week of October 1948, 20 died and over 7000 were sickened;
- The London smog crisis in December 1952—4,000 dead,
- The Torrey Canyon and Exxon Valdez oil spills, which fouled shores and killed seabirds, seals, and fish;
- Love Canal, where industrial wastes seeped from their burial site into homes and contaminated groundwater;
- Union Carbide's toxics release at Bhopal, India—3,800 dead and up to 100,000 ill, according to Union Carbide; others claim a higher toll;
- The Three Mile Island and Chornobyl nuclear accidents;
- The decimation of elephants and rhinoceroses to satisfy a market for tusks and horns;
- The loss of forests—in 1997, fires set to clear Southeast Asian forest lands produced so much smoke that regional airports had to close;
- Ebola, a virus that kills nine-tenths of those it infects, apparently first struck humans because growing populations reached into its native habitat;
- West Nile Fever, a mosquito-borne virus with a much less deadly record, was brought to North America by travelers or immigrants from Egypt;
- Acid rain, global climate change, and ozone depletion, all caused by substances released into the air by human activities.

The alarms have been raised by many people in addition to Rachel Carson. For instance, in 1968 (when world population was only a little over half of what it is today), Paul Ehrlich's *The Population Bomb* (Ballantine Books) described the ecological threats of a rapidly growing population and Garrett Hardin's influential essay, "The Tragedy of the Commons," *Science* (December 13, 1968) described the consequences of using self-interest alone to guide the exploitation of publicly owned resources (such as air and water). (In 1974, Hardin introduced the unpleasant concept of "lifeboat ethics," which says that if there are not enough resources to go around, some people must do without). In 1972, a group of economists, scientists, and business leaders calling themselves "The Club of Rome" published *The Limits to Growth* (Universe Books), an analysis of population, resource use, and pollution trends that predicted difficult times within a century; the study was redone as *Beyond the Limits to Growth: Confronting Global Collapse, Envisioning a Sustainable Future* (Chelsea Green, 1992), using more powerful computer models, and came to very similar conclusions. Among the most recent books is Jared Diamond's *Collapse: How Societies Choose to Fail or Succeed* (Viking, 2005), which uses historical cases to illuminate the roles of human biases and choices in dealing with environmental problems. Among Diamond's important points is the idea that in order to cope successfully with such problems, a society may have to surrender cherished traditions.

The following list of selected U.S. and U.N. laws, treaties, conferences, and reports illustrates the national and international responses to the various cries of alarm:

1967   The U.S. Air Quality Act set standards for air pollution.

1968   The U.N. Biosphere Conference discussed global environmental problems.

1969   The U.S. Congress passed the National Environmental Policy Act, which (among other things) required federal agencies to prepare environmental impact statements for their projects.

1970   The first Earth Day demonstrated so much public concern that the Environmental Protection Agency (EPA) was created; the Endangered Species Act, Clean Air Act, and Safe Drinking Water Act soon followed.

1971   The U.S. Environmental Pesticide Control Act gave the EPA authority to regulate pesticides.

1972   The U.N. Conference on the Human Environment, held in Stockholm, Sweden, recommended government action and led to the U.N. Environment Programme.

1973   The Convention on International Trade in Endangered Species of Wild Fauna and Flora (CITES) restricted trade in threatened species; because enforcement was weak, however, a black market flourished.

1976   The U.S. Resource Conservation and Recovery Act and the Toxic Substances Control Act established control over hazardous wastes and other toxic substances.

1979   The Convention on Long-Range Transboundary Air Pollution addressed problems such as acid rain (recognized as crossing national borders in 1972).

1982   The Law of the Sea addressed marine pollution and conservation.

1982   The second U.N. Conference on the Human Environment (the Stockholm +10 Conference) renewed concerns and set up a commission to prepare a "global agenda for change," leading to the 1987 Brundtland Report (*Our Common Future*).

1983   The U.S. Environmental Protection Agency and the U.S. National Academy of Science issued reports calling attention to the prospect of global warming as a consequence of the release of greenhouse gases such as carbon dioxide.

1987   The Montreal Protocol (strengthened in 1992) required nations to phase out use of chlorofluorocarbons (CFCs), the chemicals responsible for stratospheric ozone depletion (the "ozone hole").

1987   The Basel Convention controlled cross-border movement of hazardous wastes.

1988   The U.N. assembled the Intergovernmental Panel on Climate Change, which would report in 1995, 1998, and 2001 that the dangers of global warming were real, large, and increasingly ominous.

1992   The U.N. Convention on Biological Diversity required nations to act to protect species diversity.

1992   The U.N. Conference on Environment and Development (also known as the Earth Summit), held in Rio de Janeiro, Brazil, issued a broad call for environmental protections.

1992   The U.N. Convention on Climate Change urged restrictions on carbon dioxide release to avoid climate change.

1994   The U.N. Conference on Population and Development, held in Cairo, Egypt, called for stabilization and reduction of global population growth, largely by improving women's access to education and health care.

1997   The Kyoto Protocol attempted to strengthen the 1992 Convention on Climate Change by requiring reductions in carbon dioxide emissions, but U.S. resistance limited success.

2001   The U.N. Stockholm Convention on Persistent Organic Pollutants required nations to phase out use of many pesticides and other chemicals. It took effect May 17, 2004, after ratification by over 50 nations (not including the United States and the European Union).

2002   The U.N. World Summit on Sustainable Development, held in Johannesburg, South Africa, brought together representatives of governments, nongovernmental organizations, businesses, and other groups to examine "difficult challenges, including improving people's lives and conserving our natural resources in a world that is growing in population, with ever-increasing demands for food, water, shelter, sanitation, energy, health services and economic security."

2003   The World Climate Change Conference held in Moscow, Russia, concluded that global climate is changing, very possibly because of human activities, and the overall issue must be viewed as one of intergenerational justice. "Mitigating global climate change will be possible only with the coordinated actions of all sectors of society."

2005   The U.N. Millennium Project Task Force on Environmental Sustainability released its report, *Environment and Human Well-Being A Practical Strategy* (http://www.unep.org/Documents.Multilingual/Default.asp?DocumentID=392& ArticleID=4747&l=en).

2005   The U.N. Millennium Ecosystem Assessment released its report, *Ecosystems and Human Well-Being: Synthesis* (http://www.millenniumassessment.org/en/index.aspx) (Island Press).

2005   The U.N. Climate Change Conference held in Montreal, Canada, marked the taking effect of the Kyoto Protocol, ratified in 2004 by 141 nations (not including the U.S. and Australia).

Rachel Carson would surely have been pleased by these responses, for they suggest both concern over the problems identified and determination to solve those problems. But she would just as surely have been frustrated, for a simple listing of laws, treaties, and reports does nothing to reveal the endless wrangling and the way political and business forces try to block progress whenever it is seen as interfering with their interests. Agreement on banning chlorofluorocarbons was relatively easy to achieve because CFCs were not seen as essential to civilization and there were substitutes available. Restraining greenhouse gas emissions is harder because we see fossil fuels as essential and though substitutes may exist, they are so far more expensive.

The Globalization of the Environment

Years ago, it was possible to see environmental problems as local. A smokestack belched smoke and made the air foul. A city sulked beneath a layer of smog. Bison or passenger pigeons declined in numbers and even vanished. Rats flourished in a dump where burning garbage produced clouds of smoke and runoff contaminated streams and groundwater and made wells unusable. Sewage, chemical wastes, and oil killed the fish in streams, lakes, rivers, and harbors. Toxic chemicals such as lead and mercury entered the food chain and affected the health of both wildlife and people.

By the 1960s, it was becoming clear that environmental problems did not respect borders. Smoke blows with the wind, carrying one locality's contamination to others. Water flows to the sea, carrying sewage and other wastes with it. Birds migrate, carrying with them whatever toxins they have absorbed with their food. In 1972, researchers were able to report that most of the acid rain falling on Sweden came from other countries. Other researchers have shown that the rise and fall of the Roman Empire can be tracked in Greenland, where glaciers preserve lead-containing dust deposited over the millennia—the amount rises as Rome flourished, falls with the Dark Ages, and rises again with the Renaissance and Industrial Revolution. Today we know that pesticides and other chemicals can show up in places (such as the Arctic) where they have never been used, even years after their use has been discontinued. The 1979 Convention on Long-Range Transboundary Air Pollution has been strengthened several times with amendments to address persistent organic pollutants, heavy metals, and other pollutants.

We are also aware of new environmental problems that exist only in a global sense. Ozone depletion, first identified in the stratosphere over Antarctica, threatens to increase the amount of ultraviolet light reaching the ground, and thereby increase the incidence of skin cancer and cataracts, among other things. The cause is the use by the industrialized world of chlorofluorocarbons (CFCs) in refrigeration, air conditioning, aerosol cans, and electronics (for cleaning grease off circuit boards). The effect is global. Worse yet, the cause is rooted in northern lands such as the United States and Europe, but the worst effects may be felt where the sun shines brightest—in the tropics, which are dominated by developing nations. A serious issue of justice or equity is therefore involved.

A similar problem arises with global warming, which is also rooted in the industrialized world and its use of fossil fuels. The expected climate effects will hurt worst the poorer nations of the tropics, and perhaps worst of all those on low-lying South Pacific islands, which are expecting to be wholly inundated by rising seas.

Both the developed and the developing world are aware of the difficulties posed by environmental issues. In Europe, "green" political parties play a major and growing part in government. In Japan, some environmental regulations are more demanding than those of the United States. Developing nations understandably place dealing with their growing populations high on their list of priorities, but they play an important role in UN conferences on environmental issues, often demanding more responsible behavior from developed nations such as the United States (which often resists these demands; it has refused to ratify international agreements such as the Kyoto Protocol, for example).

Western scholars have been known to suggest that developing nations should forgo industrial development because if their huge populations ever attain the same per-capita environmental impact as the populations of wealthier lands, the world will be laid waste. It is not hard to understand why the developing nations object to such suggestions; they too want a better standard of living. Nor do they think it fair that they suffer for the environmental sins of others.

Are global environmental problems so threatening that nations must surrender their sovereignty to international bodies? Should the U.S. or Europe have to change energy supplies to protect South Pacific nations from inundation by rising seas? Should developing nations be obliged to reduce birth rates or forgo development because their population growth is seen as exacerbating pollution or threatening biodiversity?

Questions such as these play an important part in global debates today. They are not easy to answer, but their very existence says something important about the general field of environmental studies. This field is based in the science of ecology. Ecology focuses on living things and their interactions with each other and their surroundings. It deals with resources and limits and coexistence. It can see problems, their causes, and even potential solutions. And it can turn its attention to human beings as easily as it can to deer mice.

Yet, human beings are not mice. We have economies and political systems, vested interests, and conflicting priorities and values. Ecology is only one part of environmental studies. Other sciences—chemistry, physics, climatology, epidemiology, geology, and more—are involved. So are economics, history, law, and politics. Even religion can play a part.

Unfortunately, no one field sees enough of the whole to predict problems (the chemists who developed CFCs could hardly have been expected to realize what would happen if these chemicals reached the stratosphere). Environmental studies is a field for teams. That is, it is a holistic, multidisciplinary field.

This gives us an important basic principle to use when evaluating arguments on either side of any environmental issue: Arguments that fail to recognize the complexity of the issue are necessarily suspect. On the other hand, arguments that endeavor to convey the full complexity of an issue may be impossible to understand; a middle ground is essential for clarity, but any reader or student must realize that something important might be left out.

Current Environmental Issues

In 2001, the National Research Council's Committee on Grand Challenges in Environmental Sciences published *Grand Challenges in Environmental Sciences* (National Academy Press, 2001) in an effort to reach "a judgment regarding the most important environmental research challenges of the next generation—the areas most likely to yield results of major scientific and practical importance if pursued vigorously now." These areas include the following:

- Biogeochemical cycles (the cycling of plant nutrients, the ways human activities affect them, and the consequences for ecosystem functioning, atmospheric chemistry, and human activities)
- Biological diversity
- Climate variability
- Hydrologic forecasting (groundwater, droughts, floods, etc.)
- Infectious diseases
- Resource use
- Land use
- Reinventing the use of materials (e.g., recycling)

Similar themes appeared when *Issues in Science and Technology* celebrated its twentieth anniversary with its Summer 2003 issue. The editors noted that over the life of the magazine to date, some problems have hardly changed, nor has our sense of what must be done to solve them. Others have been affected, sometimes drastically, by changes in scientific knowledge, technological capability, and political trends. In the environmental area, the magazine paid special attention to:

- Biodiversity
- Overfishing
- Climate change
- The Superfund program
- The potential revival of nuclear power
- Sustainability

Many of the same basic themes were reiterated when *Science* magazine (published weekly by the American Association for the Advancement of Science) published in November and December 2003 a four-week series on the "State of the Planet," followed by a special issue on "The Tragedy of the Commons." In the introduction to the series, H. Jesse Smith began with these words: "Once in a while, in our headlong rush toward greater prosperity, it is wise to ask ourselves whether or not we can get there from here. As global population increases, and the demands we make on our natural resources grow even faster, it becomes ever more clear that the well-being we seek is imperiled by what we do."

Among the topics covered in the series were:

- Human population
- Biodiversity
- Tropical soils and food security
- The future of fisheries
- Freshwater resources
- Energy resources
- Air quality and pollution
- Climate change
- Sustainability
- The burden of chronic disease

Many of the topics on these lists are covered in this book. There are, of course, a great many other environmental issues—many more than can be covered in any one book such as this one. I have not tried to deal here with invasive species, the Endangered Species Act, the removal of dams to restore populations of anadromous fishes such as salmon, the depletion of aquifers, floodplain development, urban planning, and many others. My sample of the variety available begins with the more philosophical issues. For instance, there is considerable debate over the "precautionary principle," which says in essence that even if we are not sure that our actions will have unfortunate consequences, we should take precautions just in case (see Issue 1). This principle plays an important part in many environmental debates, from those over the value of preserving biodiversity (Issue 4) or the wisdom of opening the Arctic National Wildlife Refuge to oil drilling (Issue 8) to the folly (or wisdom) of reprocessing nuclear waste (Issue 19).

I said earlier that many people believed (and still believe) that nature has value only when it is turned to human benefit. One consequence of this belief is that it may be easier to convince people that nature is worth protecting if one can somehow calculate a cash value for nature "in the raw." Some environmentalists object to even trying to do this, on the grounds that economic value is not the only value, or even the value that should matter (see Issue 3).

What other values might we consider? Past editions of this book have considered whether nature has a value all its own, or a right to exist unmolested, and whether human property rights should take precedence. Here we discuss whether we should strive for social justice (Issue 5) and whether environmental regulations should be set aside in favor of certain human activities (such as military actions; see Issue 7).

Should we be concerned about the environmental impacts of specific human actions or products? Here too we can consider opening the Arctic National Wildlife Refuge to oil drilling (Issue 8), as well as the conflict between the value of DDT for preventing malaria and its impact on ecosystems (Issue 16), the hormone-like effects of some pesticides and other chemicals on both wildlife and humans (Issue 17), and the hazards of energy technologies (Issue 11) and global warming (Issue 9). Genetic engineering promises to do wonders for food production, but some worry about effects on ecosystems (Issue 14).

Waste disposal is a problem area all its own. It encompasses both nuclear waste (Issue 19) and hazardous waste (Issue 18). A new angle on hazardous waste comes from the popularity of the personal computer—or more specifically, from the huge numbers of PCs that are discarded each year.

What solutions are available? Some are specific to particular issues, as the hydrogen car (Issue 10) or a revival of nuclear power (Issue 12) may be to the problems associated with fossil fuels, or as marine reserves may be to declining fish stocks (Issue 15). Some are more general, as we might expect as soon as we hear someone speak of population growth as a primary cause of environmental problems (Issue 13) (there is some truth to this, for if the human population were small enough, its environmental impact—no matter how sloppy people were—would also be small).

Some analysts argue that whatever solutions we need, government need not impose them all. Private industry may be able to do the job if government can find a way to motivate industry, as with the idea of tradable pollution rights (Issue 6).

The overall aim, of course, is to avoid disaster and enable human life and civilization to continue prosperously into the future. The term for this is "sustainable development" (see Issue 2), and it was the chief concern of the U.N. World Summit on Sustainable Development, held in Johannesburg, South Africa, in August 2002. Exactly how to avoid disaster and continue prosperously into the future are the themes of the U.N. Millennium Ecosystem Assessment report, *Ecosystems and Human Well-Being: Synthesis* (http://www.millenniumassessment.org/en/index.aspx) (Island Press, 2005). The main findings of this report are that over the past half century, meeting human needs for food, fresh water, fuel, and other resources has had major negative effects on the world's ecosystems; those effects are likely to grow worse over the next half century and will pose serious obstacles to reducing global hunger, poverty, and disease; and although "significant changes in policies, institutions, and practices can mitigate many of the negative consequences of growing pressures on ecosystems, . . . the changes required are large and not currently under way." Also essential will be improvements in knowledge about the environment, the ways humans affect it, and the ways humans depend upon it, as well as improvements in technology, both for assessing environmental damage and for repairing and preventing damage, as emphasized by Bruce Sterling in "Can Technology Save the Planet?" *Sierra* (July/August 2005). Sterling concludes, perhaps optimistically, that "When we see our historical predicament in its full, majestic scope, we will stir ourselves to great and direly necessary actions. It's not beyond us to think and act in a better way. Yesterday's short-sighted habits are leaving us, the way gloom lifts with the dawn." George Musser, introducing the special September 2005 "Crossroads for Planet Earth" issue of *Scientific American* in "The Climax of Humanity," notes that the next few decades will determine our future. Sterling's optimism may be fulfilled, but if we do not make the right choices, the future may be very bleak indeed.

On the Internet . . .

The Earth Day Network

The Earth Day Network (EDN) promotes environmental citizenship and helps activists in their efforts to change local, national, and global environmental policies. It also makes the Ecological Footprint quiz (http://www.myfootprint.org/) available.

http://www.earthday.net/

The Natural Resources Defense Council

The Natural Resources Defense Council is one of the most active environmental research and advocacy organizations. Its home page lists its concerns as clean air and water, energy, global warming, toxic chemicals, nuclear waste, and much more.

http://www.nrdc.org/

The Earth Council

The Earth Council is an international nongovernmental organization (NGO) created to promote and advance the implementation of the Earth Summit agreements. Its mission is support and empower people in building a more secure, equitable, and sustainable future.

http://www.ecouncil.net/

The United Nations Environment Programme

The United Nation's Environment Programme "works to encourage sustainable development through sound environmental practices everywhere. Its activities cover a wide range of issues, from atmosphere and terrestrial ecosystems, the promotion of environmental science and information, to an early warning and emergency response capacity to deal with environmental disasters and emergencies."

http://www.unep.org/

The International Institute for Sustainable Development

The International Institute for Sustainable Development advances sustainable development policy and research by providing information and engaging in partnerships worldwide. It says it "promotes the transition toward a sustainable future. We seek to demonstrate how human ingenuity can be applied to improve the well-being of the environment, economy and society."

http://www.llsd.org/default.asp

PART 1

Environmental Philosophy

*E*nvironmental debates are rooted in questions of values—what is right? what is just?—and inevitably political in nature. It is worth stressing that people who consider themselves to be environmentalists can be found on both sides of most of the issues in this book. They differ in what they see as their own self-interest and even in what they see as humanity's long-term interest.

Understanding the general issues raised in this section is useful preparation for examining the more specific controversies that follow in later sections.

- Is the Precautionary Principle a Sound Basis for International Policy?
- Is Sustainable Development Compatible with Human Welfare?
- Should a Price Be Put on the Goods and Services Provided by the World's Ecosystems?

ISSUE 1

Is the Precautionary Principle a Sound Basis for International Policy?

YES: Nancy Myers, from "The Rise of the Precautionary Principle: A Social Movement Gathers Strength," *Multinational Monitor* (September 2004)

NO: John D. Graham, from "The Perils of the Precautionary Principle: Lessons from the American and European Experience," Heritage Lecture #818 (January 15, 2004)

ISSUE SUMMARY

YES: Nancy Myers, communications director for the Science and Environmental Health Network, argues that because the precautionary principle "makes sense of uncertainty," it has gained broad international recognition as being crucial to environmental policy.

NO: John D. Graham, dean of the Frederick S. Pardee RAND Graduate School, argues that the precautionary principle is so subjective that it permits "precaution without principle" and threatens innovation and public and environmental health. It must therefore be used cautiously.

The traditional approach to environmental problems has been reactive. That is, first the problem becomes apparent—wildlife or people sicken and die, or drinking water or air tastes foul. Then researchers seek the cause of the problem, and regulators seek to eliminate or reduce that cause. The burden is on society to demonstrate that harm is being done and that a particular cause is to blame.

An alternative approach is to presume that *all* human activities—construction projects, new chemicals, new technologies, etc.—have the potential to cause environmental harm. Therefore, those responsible for these activities should prove in advance that they will not do harm and should take suitable steps to prevent any harm from happening. A middle ground is occupied by the "precautionary principle," which has played an increasingly important part in environmental law ever since it first appeared in Germany in the mid-1960s. On the international scene, it has been applied to climate change, hazardous waste management, ozone depletion, biodiversity, and fisheries management.

In 1992 the Rio Declaration on Environment and Development, listing it as Principle 15, codified it thus:

> In order to protect the environment, the precautionary approach shall be widely applied by States according to their capabilities. When there are threats of serious or irreversible damage, lack of full scientific certainty shall not be used as a reason for postponing cost-effective measures to prevent environmental degradation.

Other versions of the principle also exist, but all agree that when there is reason to think—not absolute proof—that some human activity is or might be harming the environment, precautions should be taken. Furthermore, the burden of proof should be on those responsible for the activity, not on those who may be harmed. This has come to be broadly accepted as a basic tenet of ecologically or environmentally sustainable development. See Paul L. Stein, "Are Decision-Makers Too Cautious With the Precautionary Principle?" *Environmental and Planning Law Journal* (February 2000), and Marco Martuzzi and Roberto Bertollini, "The Precautionary Principle, Science and Human Health Protection," *International Journal of Occupational Medicine and Environmental Health* (January 2004).

The precautionary principle also contributes to thinking in the areas of risk assessment and risk management in general. Human activities—the manufacture of chemicals and other products; the use of pesticides, drugs, and fossil fuels; the construction of airports and shopping malls; and even agriculture—can damage health and the environment. Some people insist that action need not be taken against any particular activity until and unless there is solid, scientific proof that it is doing harm, and even then risks must be weighed against each other. Others insist that mere suspicion should be grounds enough for action.

Since solid, scientific proof can be very difficult to obtain, the question of just how much proof is needed to justify action is vital. Not surprisingly, if action threatens an industry, that industry's advocates will argue against taking precautions, generally saying that more proof is needed. A good example can be found in Stuart Pape, "Watch Out for the Precautionary Principle," *Prepared Foods* (October 1999): "In recent months, U.S. food manufacturers have experienced a rude introduction to the 'Precautionary Principle.' ... European regulators have begun to adopt extreme definitions of the Principle in order to protect domestic industries and place severe restrictions on the use of both old and new materials without justifying their action upon sound science."

In the following selections, Nancy Myers, communications director for the Science and Environmental Health Network, argues that the precautionary principle has gained broad international recognition as crucial to environmental policy for good reason. "The essence of the Precautionary Principle is that when lives and the future of the planet are at stake, people must act on ... clues and prevent as much harm as possible, despite imperfect knowledge and even ignorance." John D. Graham, now dean of the Frederick S. Pardee RAND Graduate School, but until recently a top official at the Office of Management and Budget, argues that the precautionary principle is so subjective that it permits "precaution without principle" and threatens innovation and public and environmental health. It must therefore be used cautiously.

Nancy Myers

The Rise of the Precautionary Principle: A Social Movement Gathers Strength

Ed Soph is a jazz musician and professor at the University of North Texas in Denton, a growing town of about 100,000 just outside Dallas, Texas. In 1997, Ed and his wife Carol founded Citizens for Healthy Growth, a Denton group concerned about the environment and future of their town. The Sophs and their colleagues—the group now numbers about 400—are among the innovative pioneers who are implementing the Precautionary Principle in the United States.

The Sophs first came across the Precautionary Principle in 1998, in the early days of the group's campaign to prevent a local copper wire manufacturer, United Copper Industries, from obtaining an air permit that would have allowed lead emissions. Ed remembers the discovery of the Wingspread Statement on the Precautionary Principle—a 1998 environmental health declaration holding that "When an activity raises threats of harm to human health or the environment, precautionary measures should be taken even if some cause and effect relationships are not fully established scientifically"—as "truly a life-changing experience." Using the Precautionary Principle as a guide, the citizens refused to be drawn into debates on what levels of lead, a known toxicant, might constitute a danger to people's health. Instead, they pointed out that a safer process was available and insisted that the wise course was not to issue the permit. The citizens prevailed.

The principle helped again in 2001, when a citizen learned that the pesticides 2,4-D, simazine, Dicamba and MCPP were being sprayed in the city parks. "The question was, given the 'suspected' dangers of these chemicals, should the city regard those suspicions as a reassurance of the chemicals' safety or as a warning of their potential dangers?" Ed recalls. "Should the city act out of ignorance or out of common sense and precaution?"

Soph learned that the Greater Los Angeles School District had written the Precautionary Principle into its policy on pesticide use and had turned to Integrated Pest Management (IPM), a system aimed at controlling pests without the use of toxic chemicals. The Denton group decided to advocate for a similar policy. They persuaded the city's park district to form a focus group of park users and organic gardening experts. The city stopped spraying the four problem chemicals and initiated a pilot IPM program.

The campaign brought an unexpected economic bonus to the city. In the course of their research, parks department staff discovered that corn gluten was a good turf builder and natural broadleaf herbicide. But the nearest supplier of corn gluten was in the Midwest, and that meant high shipping costs for the city. Meanwhile, a corn processing facility in Denton was throwing away the corn gluten it produced as a byproduct. The parks department made the link, and everyone was pleased. The local corn company was happy to add a new product line; the city was happy about the expanded local business and the lower price for a local product; and the environmental group chalked up another success.

The citizens of Denton, Texas, did not stop there. They began an effort to improve the community's air pollution standards. They got arsenic-treated wood products removed from school playgrounds and parks and replaced with nontoxic facilities. "The Precautionary Principle helped us define the problems and find the solutions," Ed says.

But, as he wrote in an editorial for the local paper, "The piecemeal approach is slow, costly and often more concerned with mitigation than prevention." Taking a cue from Precautionary Principle pioneers in San Francisco, they also began lobbying for a comprehensive new environmental code for the community, based on the Precautionary Principle.

In June 2003, San Francisco's board of supervisors had become the first government in the United States to embrace the Precautionary Principle. A new environmental code drafted by the city's environment commission put the Precautionary Principle at the top, as Article One. Step one in implementing the code was a new set of guidelines for city purchasing, pointing the way toward "environmentally preferable" purchases by careful analysis and choice of the best alternatives. The White Paper accompanying the ordinance pointed out that most of the city's progressive environmental policies were already in line with the Precautionary Principle, and that the new code provided unity and focus to the policies rather than a radically new direction.

That focus is important; too often, environmental matters seem like a long, miscellaneous and confusing list of problems and solutions.

Likewise in Denton, the Precautionary Principle has not been a magic wand for transforming policy, but it has put backbone into efforts to enact truly protective and far-sighted environmental policies. Ed Soph points out that, in his community as in others, growth had often been dictated by special interests in the name of economic development, and the environment got short shrift.

"Environmental protection and pollution prevention in our city have been a matter, not of proactive policy, but of reaction to federal and state mandates, to the threat of citizens' lawsuits, and to civic embarrassment. Little thought is given to future environmental impacts," he told the city council when he argued for a new environmental code.

He added, "The toxic chemical pollution emitted by area industries has been ignored or accepted for all the ill-informed or selfish reasons that we are too familiar with. The Precautionary Principle dispels that ignorance and empowers concerned citizens with the means to ensure a healthier future."

The Precautionary Principle has leavened the discussion of environmental and human health policy on many fronts—in international treaty negotiations and global trade forums, in city resolutions and national policies, among conservationists and toxicologists, and even in corporate decision making.

Two treaties negotiated in 2000 incorporated the principle for the first time as an enforceable measure. The Cartagena Protocol on Biosafety allows countries to invoke the Precautionary Principle in decisions on admitting imports of genetically modified organisms. It became operative in June 2003. The Stockholm Convention on Persistent Organic Pollutants prescribes the Precautionary Principle as a standard for adding chemicals to the original list of 12 that are banned by the treaty. This treaty went into force in February 2004.

Making Sense of Uncertainty

Understanding the need for the Precautionary Principle requires some scientific sophistication. Ecologists say that changes in ecological systems may be incremental and gradual, or surprisingly large and sudden. When change is large enough to cause a system to cross a threshold, it creates a new dynamic equilibrium that has its own stability and does not change back easily. These new interactions become the norm and create new realities.

Something of this new reality is evident in recently observed changes in patterns of human disease:

- Chronic diseases and conditions affect more than 100 million men, women, and children in the United States—more than a third of the population. Cancer, asthma, Alzheimer's disease, autism, birth defects, developmental disabilities, diabetes, endometriosis, infertility, multiple sclerosis and Parkinson's disease are becoming increasingly common.
- Nearly 12 million children in the United States (17 percent) suffer from one or more developmental disabilities. Learning disabilities alone affect at least 5 to 10 percent of children in public schools, and these numbers are increasing. Attention deficit hyperactivity disorder conservatively affects 3 to 6 percent of all school children. The incidence of autism appears to be increasing.
- Asthma prevalence has doubled in the last 20 years.
- Incidence of certain types of cancer has increased. The age-adjusted incidence of melanoma, non-Hodgkins lymphoma, and cancers of the prostate, liver, testis, thyroid, kidney, breast, brain, esophagus and bladder has risen over the past 25 years. Breast cancer, for example, now strikes more women worldwide than any other type of cancer, with rates increasing 50 percent during the past half century. In the

1940s, the lifetime risk of breast cancer was one in 22. Today's risk is one in eight and rising.

- In the United States, the incidence of some birth defects, including male genital disorders, some forms of congenital heart disease and obstructive disorders of the urinary tract, is increasing. Sperm density is declining in some parts of the United States and elsewhere in the world.

These changes in human health are well documented. But proving direct links with environmental causative factors is more complicated.

Here is how the scientific reasoning might go: Smoking and diet explain few of the health trends listed above. Genetic factors explain up to half the population variance for several of these conditions—but far less for the majority of them—and in any case do not explain the changes in disease incidence rates. This suggests that other environmental factors play a role. Emerging science suggests this as well. In laboratory animals, wildlife and humans, considerable evidence documents a link between environmental contamination and malignancies, birth defects, reproductive disorders, impaired behavior and immune system dysfunction. Scientists' growing understanding of how biological systems develop and function leads to similar conclusions.

But serious, evident effects such as these can seldom be linked decisively to a single cause. Scientific standards of certainty (or "proof") about cause and effect are high. These standards may never be satisfied when many different factors are working together, producing many different results. Sometimes the period of time between particular causes and particular results is so long, with so many intervening factors, that it is impossible to make a definitive link. Sometimes the timing of exposure is crucial—a trace of the wrong chemical at the wrong time in pregnancy, for example, may trigger problems in the child's brain or endocrine system, but the child's mother might never know she was exposed.

In the real world, there is no way of knowing for sure how much healthier people might be if they did not live in the modern chemical stew, because the chemicals are everywhere—in babies' first bowel movement, in the blood of U.S. teenagers and in the breastmilk of Inuit mothers. No unexposed "control" population exists. But clearly, significant numbers of birth defects, cancers and learning disabilities are preventable.

Scientific uncertainty is a fact of life even when it comes to the most obvious environmental problems, such as the disappearance of species, and the most potentially devastating trends, such as climate change. Scientists seldom know for sure what will happen until it happens, and seldom have all the answers about causes until well after the fact, if ever. Nevertheless, scientific knowledge, as incomplete as it may be, provides important clues to all of these conditions and what to do about them.

The essence of the Precautionary Principle is that when lives and the future of the planet are at stake, people must act on these clues and prevent as much harm as possible, despite imperfect knowledge and even ignorance.

Environmental Failures

A premise of Precautionary Principle advocates is that environmental policies to date have largely not met this challenge. Part of the explanation for why they have not is that the dimensions of the emerging problems are only now becoming apparent. The limits of the earth's assimilative capacity are much clearer now than they were when the first modern environmental legislation was enacted 30 years ago.

Another part of the explanation is that, although some environmental policies are preventive, most have focused on cleaning up messes after the fact—what environmentalists call "end of pipe" solutions. Scrubbers on power plant stacks, catalytic converters on tailpipes, recycling and super-sized funds dedicated to detoxifying the worst dumps have not been enough. The Precautionary Principle holds that earlier, more comprehensive and preventive approaches are necessary. Nor is it enough to address problems only after they have become so obvious that they cannot be ignored—often, literally waiting for the dead bodies to appear or for coastlines to disappear under rising tides.

The third factor in the failure of environmental policies is political, say Precautionary Principle proponents. After responding to the initial burst of concern for the environment, the U.S. regulatory system and others like it were subverted by commercial interests, with the encouragement of political leaders and, increasingly, the complicity of the court system. Environmental laws have been subjected to an onslaught of challenges since the 1980s; many have been modified or gutted, and all are enforced by regulators who have been chastened by increasing challenges to their authority by industry and the courts.

The courts, and now increasingly international trade organizations and agreements like the World Trade Organization (WTO) and the North American Free Trade Agreement (NAFTA), have institutionalized an anti-precautionary approach to environmental controls. They have demanded the kinds of proof and certainty of harms and efficacy of regulation that science often cannot provide.

False Certainties

Ironically, one tool that has proved highly effective in the battle against environmental regulations was one that was meant to strengthen the enforcement of such laws: quantitative risk assessment. Risk assessment was developed in the 1970s and 1980s as a systematic way to evaluate the degree and likelihood of harmful side effects from products and technologies. With precise, quantitative risk assessments in hand, regulators could more convincingly demonstrate the need for action. Risk assessments would stand up in court. Risk assessments could "prove" that a product was dangerous, would cause a certain number of deaths per million, and should be taken off the market.

Or not. Quantitative risk assessment, which became standard practice in the United States in the mid-1980s and was institutionalized in the global trade agreements of the 1990s, turned out to be most useful in "proving" that a product or technology was not inordinately dangerous. More precisely, risk assessments presented sets of numbers that purported to state definitively how much harm might occur. The next question for policymakers then became: How much harm is acceptable? Quantitative risk assessment not only provided the answers; it dictated the questions.

As quantitative risk assessment became the norm, commercial and industrial interests were increasingly able to insist that harm must be proven "scientifically"—in the form of a quantitative risk assessment demonstrating harm in excess of acceptable limits—before action was taken to stop a process or product. These exercises were often linked with cost-benefit assessments that heavily weighted the immediate monetary costs of regulations and gave little, if any, weight to costs to the environment or future generations.

Although risk assessments tried to account for uncertainties, those projections were necessarily subject to assumptions and simplifications. Quantitative risk assessments usually addressed a limited number of potential harms, often missing social, cultural or broader environmental factors. These risk assessments have consumed enormous resources in strapped regulatory agencies and have slowed the regulatory process. They have diverted attention from questions that could be answered: Do better alternatives exist? Can harm be prevented?

The slow pace of regulation, the insistence on "scientific certainty," and the weighting toward immediate monetary costs often give the benefit of doubt to products and technologies, even when harmful side effects are suspected. One result is that neither international environmental agreements nor national regulatory systems have kept up with the increasing pace and cumulative effects of environmental damage.

A report by the European Environment Agency in 2001 tallied the great costs to society of some of the most egregious failures to heed early warnings of harm. Radiation, ozone depletion, asbestos, Mad Cow disease and other case studies show a familiar pattern: "Misplaced 'certainty' about the absence of harm played a key role in delaying preventive actions," the authors conclude.

They add, "The costs of preventive actions are usually tangible, clearly allocated and often short term, whereas the costs of failing to act are less tangible, less clearly distributed and usually longer term, posing particular problems of governance. Weighing up the overall pros and cons of action, or inaction, is therefore very difficult, involving ethical as well as economic considerations."

The Precautionary Approach

As environmentalists looked at looming problems such as global warming, they were appalled at the inadequacy of policies based on quantitative risk assessment. Although evidence was piling up rapidly that human activities were having an unprecedented effect on global climate, for example, it was difficult to say when the threshold of scientific certainty would be crossed. Good science demanded caution about drawing hard and fast conclusions. Yet, the longer humanity waited to take action, the harder it would be to reverse any effect. Perhaps it was already too late. Moreover, action would have to take the form of widespread changes not only in human behavior but also in technological development. The massive shift away from fossil fuels that might yet mitigate the effects of global warming would require rethinking the way humans produce and use energy. Nothing in the risk-assessment-based approach to policy prepared society to do that.

The global meetings called to address the coming calamity were not helping much. Politicians fiddled with blame and with protecting national economic interests while the globe heated up. Hard-won and heavily compromised agreements such as the 1997 Kyoto agreement on climate change were quickly mired in national politics, especially in the United States, the heaviest fossil-fuel user of all.

In the United States and around the globe, a different kind of struggle had been going on for decades: the fight for attention to industrial pollution in communities. From childhood lead poisoning in the 1930s to Love Canal in the 1970s, communities had always faced an uphill battle in proving that pollution and toxic products were making them sick. Risk assessments often made the case that particular hazardous waste dumps were safe, or that a single polluting industry could not possibly have caused the rash of illnesses a community claimed. But these risk assessments missed the obvious fact that many communities suffered multiple environmental assaults, compounded by other effects of poverty. A landmark 1987 report by the United Church of Christ coined the term "environmental racism" and confirmed that the worst environmental abuses were visited on communities of color. This growing awareness generated the international environmental justice movement.

In early 1998, a small conference at Wingspread, the Johnson Foundation's conference center in Racine, Wisconsin, addressed these dilemmas head-on. Participants groped for a better approach to protecting the environment and human health. At that time, the Precautionary Principle, which had been named in Germany in the 1970s, was an emerging precept of international law. It had begun to appear in international environmental agreements, gaining reference in a series of protocols, starting in 1984, to reduce pollution in the North Sea; the 1987 Ozone Layer Protocol; and the Second World Climate Conference in 1990.

At the Rio Earth Summit in 1992, precaution was enshrined as Principle 15 in the Rio Declaration on Environment and Development: "In order to protect the environment, the precautionary approach shall be widely applied by states according to their capabilities. Where there are threats of serious or irreversible damage, lack of full scientific certainty shall not be used as a reason for postponing cost-effective measures to prevent environmental degradation."

In the decade after Rio, the Precautionary Principle began to appear in national constitutions and environmental policies worldwide and was occasionally invoked in legal battles. For example:

- The Maastricht Treaty of 1994, establishing the European Union, named the Precautionary Principle as a guide to EU environment and health policy.

THE U.S. CHAMBER OF COMMERCE: POLICY BRIEF AGAINST THE PRECAUTIONARY PRINCIPLE

Objective

To ensure that regulatory decisions are based on scientifically sound and technically rigorous risk assessments, and to oppose the adoption of the Precautionary Principle as the basis for regulation.

Summary of the Issue

The U.S. Chamber supports a science-based approach to risk management; where risk is assessed based on scientifically sound and technically rigorous analysis. Under this approach, regulatory actions are justified where there are legitimate, scientifically ascertainable risks to human health, safety, or the environment. That is, the greater the risk, the greater the degree of regulatory scrutiny. This standard has served the nation well, and has led to astounding breakthroughs in the fields of science, health care, medicine, biotechnology, agriculture, and many other fields. There is, however, a relatively new theory, known as the Precautionary Principle that is gaining popularity among environmentalists and other groups. The Precautionary Principle says that when the risks of a particular activity are unclear or unknown, assume the worst and avoid the activity. It is essentially a policy of risk avoidance.

The regulatory implications of the Precautionary Principle are huge. For instance, the Precautionary Principle holds that since the existence and extent of global warming and climate change are not known one should assume the worst and immediately restrict the use of carbon-based fuels. However the nature and extent of key environmental, health, and safety concerns requires careful scientific and technical analysis. That is why the U.S. Chamber has long supported the use of sound science, cost-benefit analysis, risk assessment and a full understanding of uncertainty when assessing a particular regulatory issue.

The Precautionary Principle has been explicitly incorporated into various laws and regulations in the European Union and various international bodies. In the United States, radical environmentalists are pushing for its adoption as a basis for regulating biotechnology, food and drug safety, environmental protection, and pesticide use.

U.S. Chamber Strategy

- Support a science-based approach to risk management, where risk is assessed based on scientifically sound and technically rigorous standards.
- Oppose the domestic and international adoption of the Precautionary Principle as a basis for regulatory decision making.
- Educate consumers, businesses, and federal policymakers about the implications of the Precautionary Principle.

- The Precautionary Principle was the basis for arguments in a 1995 International Court of Justice case on French nuclear testing. Judges cited the "consensus flowing from Rio" and the fact that the Precautionary Principle was "gaining increasing support as part of the international law of the environment."
- At the World Trade Organization in the mid-1990s, the European Union invoked the Precautionary Principle in a case involving a ban on imports of hormone-fed beef.

The Wingspread participants believed the Precautionary Principle was not just another weak and limited fix for environmental problems. They believed it could bring far-reaching changes to the way those policies were formed and implemented. But action to prevent harm in the face of scientific uncertainty alone did not translate into sound policies protective of the environment and human health. Other norms would have to be honored simultaneously and as an integral part of a precautionary decision-making process. Several other principles had often been linked with the Precautionary Principle in various statements of the principle or in connection with precautionary policies operating in Northern European countries. The statement released at the end of the meeting, the Wingspread Statement on the Precautionary Principle, was the first to put four of these primary elements on the same page—acting upon early evidence of harm, shifting the burden of proof, exercising democracy and transparency, and assessing alternatives. These standards form the basis of what has come to be known as the overarching or comprehensive Precautionary Principle or approach:

> When an activity raises threats of harm to human health or the environment, precautionary measures should be taken even if some cause and effect relationships are not fully established scientifically.
>
> In this context the proponent of an activity, rather than the public, should bear the burden of proof.
>
> The process of applying the Precautionary Principle must be open, informed and democratic and must include potentially affected parties. It must also involve an examination of the full range of alternatives, including no action.

The conference generated widespread enthusiasm for the principle among U.S. environmentalists and academics as well as among some policymakers. That was complemented by continuing and growing support for the principle among Europeans as well as ready adoption of the concept in much of the developing world. And in the years following Wingspread, the Precautionary Principle has gained new international status.

John D. Graham

 **NO**

The Perils of the Precautionary Principle: Lessons from the American and European Experience

The concept of a universal precautionary principle apparently has its origins in early German and Swedish thinking about environmental policy, particularly the need for policymakers to practice foresight in order to prevent long-range environmental problems. The concept was included in the Amsterdam Treaty—an important step toward establishment of the European Union—but the concept was left undefined and was applied only to environmental policy. In the past 20 years, there have been numerous references to precaution in various international treaties, statements of advocacy groups, and academic writings, but the significance of the principle in international law remains uncertain.

In recent years there has been growing international interest in the subject of precaution. Reacting to criticism that the principle was too ambiguous, the European Commission in 2000 issued a formal "Communication" about the precautionary principle. This Communication extended the applicability of the principle to public health and consumer protection as well as environmental policy. For several years, the German Marshall Fund has been working with Duke University to sponsor several informal dialogue sessions involving governmental officials and academics from Europe and the USA. Several months ago, the Canadian government released a "Framework" document for the application of precaution in science-based decisions about risk.

The United States government believes it is important to understand that, notwithstanding the rhetoric of our European colleagues, there is no such thing as the precautionary principle. Indeed, the Swedish philosopher Sandin has documented 19 versions of the precautionary principle in various treaties, laws, and academic writings. Although these versions are similar in some respects, they have major differences in terms of how uncertain science is evaluated, how the severity of consequences is considered, and how the costs and risks of precautionary measures are considered. The United States government believes that precaution is a sensible idea, but there are multiple approaches to implementing precaution in risk management.

Heritage Lecture #818, Heritage Foundation, January 15, 2004.

Defining the Principle

Given the ambiguity about the precautionary principle, it may be useful to start with a dictionary definition. Webster's 2nd Edition of the New World Dictionary defines precaution as "care taken beforehand" or "a measure taken beforehand against possible danger." Understood in this way, precaution is a well-respected notion that is practiced daily in the stock market, in medicine, on the highway, and in the workplace. In both business and politics, decision makers seek the right balance between taking risks and behaving in a precautionary manner.

Before joining the Office of Management and Budget (OMB), I served for 17 years on the faculty of the Harvard School of Public Health. In that capacity I learned that public health historians have documented the preventable pain and suffering that can occur from insufficient consideration of the need for precaution. In the United States we felt that pain as a result of how we handled emerging science about tobacco, lead, and asbestos. Historians teach us that the major health problems from these substances could have been reduced or prevented altogether if decision makers had reacted to early scientific indications of harm in a precautionary manner.

We should not belittle the scientific complexities in each of these examples. Take the link between smoking and lung cancer. Although this link now seems obvious, in the middle of the previous century the link was not obvious to many competent and thoughtful physicians. They knew that many lifetime smokers never developed lung cancer; they also knew that some lung cancer patients had never been smokers. Compounding the problem was the inability of laboratory scientists to produce lung tumors in laboratory animals exposed by inhalation. In the final analysis, it took large-scale statistical studies of smokers to resolve the issue. In fact, there was a large scale study of the health of British physicians that played an important role in building the medical consensus against smoking.

In each of these examples (tobacco, lead, and asbestos), it was epidemiology rather than the experimental sciences that played the most pivotal role in identifying health risks. Ironically, it is epidemiology that is now one of the more controversial contributors to public health science.

Exaggerated Claims of Hazard

There is no question that postulated hazards sometimes prove more serious and/or widespread than originally anticipated. Ralph Nader has previously argued that this is the norm in regulatory science, while the European Commission recently issued a report of case studies where hazards appear to have been underestimated. However, the dynamics of science are not so easily predicted. Sometimes claims of hazard prove to be exaggerated, and in fact there are cases of predictions of doom that have simply not materialized.

Consider the "dismal theorem" of the Reverend Thomas Malthus (1798). He hypothesized that population would grow exponentially while sources of sustenance would only grow arithmetically. The result, he predicted, would be that living standards would fail to rise beyond subsistence levels. However, history has shown this theorem to be incorrect. Malthus did not foresee the technological advances that have allowed both population and standard of living to rise steadily and substantially.

A more recent example in the USA concerns the popular artificial sweetener saccharine. The Food and Drug Administration (FDA) declared the regulatory equivalent of war against this product on the basis of experimental laboratory test results. The finding was that huge doses of saccharine cause bladder cancer in rodents. While the FDA attempted to ban saccharine based on this evidence, the U.S. Congress overturned the FDA's action. With the benefit of hindsight, it now appears that the FDA's attempted ban may have been poorly grounded in science. Just recently, the federal government in the USA removed saccharine from the official list of "carcinogens" for two reasons: experimental biologists have found that saccharin causes bladder tumors in rodents through a mechanism (cell proliferation) that is unlikely to be relevant to low-dose human exposures; and large-scale epidemiological studies of saccharine users have found no evidence that the product is linked to excess rates of bladder cancer in people.

Students of risk science are aware that the number of alleged hazards far exceeds the number that are ever proven based on sound science. Consider the following scares: electric power lines and childhood leukemia, silicone breast implants and auto-immune disorders, cell phones and brain cancer, and disruption of the endocrine system of the body from multiple, low-dose exposures to industrial chemicals. In each of these cases, early studies that suggested dangers were not replicated in subsequent studies performed by qualified scientists. Efforts at replication or verification were simply not successful. At the same time, when early studies are replicated by independent work, such as occurred with the acute mortality events following exposure to fine particles in the air, it is important for public health regulators to take this information seriously in their regulatory deliberations.

Given that the dynamics of science are not predictable, it is important to consider the dangers of excessive precaution. One of those is the threat to technological innovation. Imagine it is 1850 and the following version of the precautionary principle is adopted: No innovation shall be approved for use until it is proven safe, with the burden of proving safety placed on the technologist. Under this system, what would have happened to electricity, the internal combustion engine, plastics, pharmaceuticals, the Internet, the cell phone, and so forth? By its very nature, technological innovation occurs through a process of trial-and-error and refinement, and this process could be disrupted by an inflexible version of the precautionary principle.

Many risk specialists in the USA regret some of the prior policy steps we have taken on the basis of precaution. In U.S. energy policy, for example, the

Three Mile Island incident had a large policy impact, though even today there is no evidence of significant public health harm caused by the accident at Three Mile Island. In fact, there has been a de facto moratorium on the construction of new nuclear power plants in the USA. We have become more deeply dependent on fossil fuels for energy, and now precaution is being invoked as a reason to enact stricter rules on use of fossil fuels. Part of the answer may rest with clean coal technologies and renewable energy, but we should not foreclose the advanced nuclear option.

Recent Progress in Europe

In comparing the actions of different countries and regions, it is important to avoid the fallacy that Europe is precautionary while the USA is not. The late Aaron Wildavsky, in his studies of risk regulation, observed that cultures engage in risk selection. Some have argued that the USA is more tolerant than Europe of the possible risks of bioengineered foods, global climate change, and industrial chemical exposures. However, a fair analysis would also show that Europe has been less precautionary than the USA on diesel engine exhaust, environmental tobacco smoke, and lead in gasoline. In fact, the recent comparative research by Professor Jonathan Wiener of Duke University has found no evidence to support the popular myth that Europe is generally more precautionary than the USA.

A subjective concept such as "the precautionary principle" is itself dangerous because it permits what conservative scholars have called "precaution without principle." In particular, the principle may be easily manipulated by commercial interests for rent-seeking purposes. According to Conko and Miller, students of biotech policy, the EU policy on genetically modified organisms "creates a bizarre bureaucratic distinction that favors certain classes of products widely made in Europe." This practice is hardly new. That is precisely what the World Trade Organization found in its earlier decision against the EU ban on hormone-treated beef, a ban that had no grounding in public health science.

Although there are many reasons to be skeptical about Europe's stance on precaution, there are recent signs of progress from Europe. Take the response of Brussels to "mad cow's disease." Once the British government and industry had taken all reasonable steps to address this problem, Brussels instructed member states of the EU to lift their bans on beef imports from the UK. All member states complied except France, which argued that French beef might still be safer than British beef and that France has the right to invoke the precautionary principle. Brussels took France to the European Court of Justice, where the Court ruled against France, indicating that speculative appeals to the precautionary principle must have some grounding in science.

Much more recently, the European Commission has rejected an unauthorized use of the precautionary principle by the provincial government of Upper Austria. In March of this year Austria notified Brussels of its proposed ban of genetically modified seeds that the EC had approved for cultivation under the EC Directive 90/220. Upper Austria appealed to the precautionary principle but Brussels overruled them: "Recourse to the precautionary principle presupposes that potentially dangerous effects . . . have been identified, and that scientific evaluation does not allow the risk to be determined with sufficient certainty." The EC noted that Upper Austria had not made this case and there was certainly nothing unique about the safety of genetically modified seeds in Upper Austria.

While it is fashionable to criticize Europe on the subject of precaution, and much of that criticism is deserved, it should also be noted that the EC's official views on precaution are becoming more nuanced. In the February 2000 Communication, for example, we found the following views that are similar to the perspective of the U.S. government:

1. Precaution is a necessary and useful concept but it is subjective and susceptible to abuse by policymakers for trade purposes.
2. Scientific and procedural safeguards need to be applied to risk management decisions based on precaution.
3. Adoption of precautionary measures should be preceded by objective scientific evaluations, including risk assessment and benefit-cost analysis of alternative measures.
4. There are a broad range of precautionary measures, including bans, product restrictions, education, warning labels, and market-based approaches. Even targeted research programs to better understand a hazard are a precautionary measure.
5. Opportunities for public participation—to discuss efficiency, fairness and other public values—are critical to sound risk management.

In OMB's 2003 Report to Congress on the Costs and Benefits of Regulation, we also emphasize the important role that analytic tools have in informing regulatory judgments about precaution. There are offshoots of cost-benefit analysis called value-of-information analysis and decision analysis that were designed precisely for the purpose of analyzing problems with large degrees of scientific uncertainty. These tools are already widely used in engineering and business and are increasingly applied to environmental issues. We urge readers to consult OMB's report for references to this growing analytic literature on precautionary regulation.

Conclusion

In summary, there are two major perils associated with an extreme approach to precaution. One is that technological innovation will be stifled, and we all recognize that innovation has played a major role in economic progress throughout the world. A second peril, more subtle, is that public health and the environment would be harmed as the energies of regulators and the regulated community would be diverted from known or plausible hazards to speculative and ill-founded ones. For these reasons, please do not be surprised if the U.S. government continues to take a precautionary approach to calls for adoption of a universal precautionary principle in regulatory policy.

POSTSCRIPT

Is the Precautionary Principle a Sound Basis for International Policy?

Ronald Bailey, in "Precautionary Tale," *Reason* (April 1999), defines the precautionary principle as "precaution in the face of any actions that may affect people or the environment, no matter what science is able—or unable—to say about that action." "No matter what science says" is not quite the same thing as "lack of full scientific certainty." Indeed, Bailey turns the precautionary principle into a straw man and thereby endangers whatever points he makes that are worth considering. One of those points is that widespread use of the precautionary principle would hamstring the development of the Third World. Roger Scruton, in "The Cult of Precaution," *National Interest* (Summer 2004), calls the Precautionary Principle "a meaningless nostrum" that is used to avoid risk and says it "clearly presents an obstacle to innovation and experiment," which are essential, Bernard D. Goldstein and Russellyn S. Carruth remind us in "Implications of the Precautionary Principle: Is It a Threat to Science?" *International Journal of Occupational Medicine and Environmental Health* (January 2004), for proper assessment of risk. Jonathan Adler, in "The Precautionary Principle's Challenge to Progress," in Ronald Bailey, ed., *Global Warming and Other Eco-Myths* (Prima, 2002), argues that because the precautionary principle does not adequately balance risks and benefits, "the world would be safer without it." Peter M. Wiedemann and Holger Schutz, "The Precautionary Principle and Risk Perception: Experimental Studies in the EMF Area," *Environmental Health Perspectives* (April 2005), report that "precautionary measures may trigger concerns, amplify ... risk perceptions, and lower trust in public health protection." Cass R. Sunstein, *Laws of Fear: Beyond the Precautionary Principle* (Cambridge, 2005), criticizes the precautionary principle in part because, he says, people overreact to tiny risks.

The 1992 Rio Declaration emphasized that the precautionary principle should be "applied by States according to their capabilities" and that it should be applied in a cost-effective way. These provisions would seem to preclude the draconian interpretations that most alarm the critics. Yet, say David Kriebel et al., in "The Precautionary Principle in Environmental Science," *Environmental Health Perspectives* (September 2001), "environmental scientists should be aware of the policy uses of their work and of their social responsibility to do science that protects human health and the environment." Businesses are also conflicted, writes Arnold Brown in "Suitable Precautions," *Across the Board* (January/February 2002), because the precautionary principle tends to slow decision making, but he maintains that "we will all have to learn and practice anticipation."

ISSUE 2

auxiliary prolong (handwritten)

Is Sustainable Development Compatible With Human Welfare?

YES: Jeremy Rifkin, from "The European Dream: Building Sustainable Development in a Globally Connected World," *E Magazine* (March/April 2005)

NO: Ronald Bailey, from "Wilting Greens," *Reason* (December 2002)

ISSUE SUMMARY

YES: Jeremy Rifkin, president of the Foundation on Economic Trends, argues that Europeans pride themselves on their quality of life, and their emphasis on sustainable development promises to maintain that quality of life into the future.

NO: Environmental journalist Ronald Bailey states that sustainable development results in economic stagnation and threatens both the environment and the world's poor.

Over the last 30 years, many people have expressed concerns that humanity cannot continue indefinitely to increase population, industrial development, and consumption. The trends and their impacts on the environment are amply described in numerous books, including historian J. R. McNeill's *Something New Under the Sun: An Environmental History of the Twentieth-Century World* (W. W. Norton, 2000).

"Can we keep it up?" is the basic question behind the issue of sustainability. In the 1960s and 1970s, this was expressed as the "Spaceship Earth" metaphor, which said that we have limited supplies of energy, resources, and room and that we must limit population growth and industrial activity, conserve, and recycle in order to avoid crucial shortages. "Sustainability" entered the global debate in the early 1980s, when the United Nations secretary general asked Gro Harlem Brundtland, a former prime minister and minister of environment in Norway, to organize and chair the World Commission on Environment and Development and produce a "global agenda for change." The resulting report, *Our Common Future* (Oxford University Press, 1987), defined *sustainable development* as "development that meets the needs of the present without compromising the ability of future generations to meet their own needs." It recognized that limits on population size and resource use cannot be known precisely; that problems may arise not suddenly but rather gradually, marked by rising costs; and that limits may be redefined by changes in technology. The report also recognized that limits exist and must be taken into account when governments, corporations, and individuals plan for the future.

The Brundtland report led to the UN Conference on Environment and Development held in Rio de Janeiro in 1992. The Rio conference set sustainability firmly on the global agenda and made it an essential part of efforts to deal with global environmental issues and promote equitable economic development. In brief, sustainability means such things as cutting forests no faster than they can grow back, using groundwater no faster than it is recharged by precipitation, stressing renewable energy sources rather than exhaustible fossil fuels, and farming in such a way that soil fertility does not decline. In addition, economics must be revamped to take into account environmental costs as well as capital, labor, raw materials, and energy costs. Many add that the distribution of the Earth's wealth must be made more equitable as well.

Given continuing growth in population and demand for resources, sustainable development is clearly a difficult proposition. Some think it can be done, but others think that for sustainability to work, either population or resource demand must be reduced. Not surprisingly, many people see sustainable development as in conflict with business and industrial activities, private property rights, and such human freedoms as the freedoms to have many children, to accumulate wealth, and to use the environment as one wishes. Economics professor Jacqueline R. Kasun, in "Doomsday Every Day: Sustainable Economics, Sustainable Tyranny," *The Independent Review* (Summer 1999), goes so far as to argue that sustainable development will require sacrificing human freedom, dignity, and material welfare on a road to tyranny.

In the following selections, Jeremy Rifkin, president of the Foundation on Economic Trends, argues that Europeans pride themselves on a quality of life that in some ways exceeds that of Americans, and their emphasis on inclusivity, diversity, sustainable development, social rights, and individual human rights promises to maintain that quality of life into the future. Ronald Bailey argues that preserving the environment, eradicating poverty, and limiting economic growth are incompatible goals. Indeed, vigorous economic growth provides wealth for all and leads to environmental protection.

YES

Jeremy Rifkin

The European Dream: Building Sustainable Development in a Globally Connected World

A growing number of Americans are beginning to wonder why Europe has leaped ahead of the U.S. to become the most environmentally advanced political space in the world today. To understand why Europe has left America behind in the race to create a sustainable society, we need to look at the very different dreams that characterize the American and European frame of mind.

Ask Americans what they most admire about the U.S.A. and they will likely cite the individual opportunity to get ahead—at least until recently. The American Dream is based on a simple but compelling covenant: Anyone, regardless of the station to which they are born, can leverage a good public education, determination and hard work to become a success in life. We can go from "rags to riches."

Ask Europeans what they most admire about Europe and they will invariably say "the quality of life." Eight out of 10 Europeans say they are happy with their lives and when asked what they believe to be the most important legacy of the 20th century, 58 percent of Europeans picked their quality of life, putting it second only to freedom in a list of 11 legacies.

While the American Dream emphasizes individual success, the European Dream emphasizes collective well-being. The reason for this lies in the divergent histories of the two continents. America's founders came over from Europe 200 years ago in the waning days of the Protestant Reformation and the early days of the European Enlightenment. They took these two streams of European thought, froze them in time, and kept them alive in their purest form until today. Americans are the most devoutly Christian and Protestant people in the industrial world and the fiercest champions of the capitalist marketplace and the nation-state.

Both the Protestant Reformation and the Enlightenment emphasized the central role of the individual in history. John Calvin exhorted the faithful that every person stands alone with his or her God. Adam Smith, in turn, argued that all individuals pursue their own self interest in the marketplace. This individualist strain fit the American context far better than it did the European setting. In a wide-open frontier, every new immigrant did indeed stand alone and had to secure his or her survival with little or no social supports. Americans, even today, are taught by his or her parents that to be free they must learn to be self-sufficient and independent, and that they cannot depend on others.

From *E Magazine*, March/April 2005, pp. 34-39. Copyright © 2005 by Jeremy Rifkin. Reprinted by permission.

Europeans, however, never fully bought the idea of the individual alone in the universe. Europe was already densely populated and without a frontier by the late 18th century. Walled cities and tightly packed human settlement demanded a more communal way of life. While Americans defined freedom in terms of individual autonomy and mobility, Europeans defined freedom by their communal relationships.

In America there was enough cheap and free land and resources so that newcomers could become rich. In Europe, well-defined class boundaries—a remnant of the feudal aristocracy—made it far more difficult for an individual born in a lesser station of life to rise to the top and become wealthy. So while Americans preferred to pursue happiness individually, Europeans pursued happiness collectively by emphasizing the quality of life of the community. Today, Americans devote less than 11 percent of their Gross Domestic Product (GDP) to social benefits, compared to 26 percent in Europe.

Doing It Better

So, what does Europe do better than America? It works hard to create a remarkably high quality of life for all of its people. The European Dream focuses on inclusivity, diversity, sustainable development, social rights and universal human rights. And it works. While Americans are 28 percent wealthier per capita than Europeans, in many ways, Europeans experience a higher quality of life, clear evidence that, in the long run, cooperation rather than competition is sometimes a surer path to happiness.

Europe and the U.S. have nearly opposite approaches to the question of environmental stewardship. At the heart of the difference is the way Americans and Europeans perceive risk. We Americans take pride in being a risk-taking people. We come from immigrant stock, people who risked their lives to journey to the new world and start over, often with only a few coins in their pockets and a dream of a better life. When Europeans and others are asked what they most admire about Americans, our risk-taking, "can-do" attitude generally tops the list. Where others see difficulties and obstacles, Americans see opportunities.

Our optimism is deeply entwined with our faith in science and technology. It has been said that Americans are a nation of tinkerers. When I was growing up, the engineer was held in as high esteem as the cowboy, admired for his efforts to improve the lot of society and contribute to the progress and welfare of civilization.

On the other side of the water, the sensibilities are different. It's not that Europeans aren't inventive. One could make the case that over the course of history Europe has produced most of the great scientific insights and not a few of the major inventions. But with their longer histories, Europeans are far more mindful of the dark side of science and technology.

Saying No to GE Foods

In recent years, the European Union (EU) has turned upside down the standard operating procedure for introducing new technologies and products into the marketplace and society, much to the consternation of the United States. The turn-around started with the controversy over genetically engineered (GE) foods and the introduction of genetically modified organisms (GMOs). The U.S. government gave the green light to the widespread introduction of GE foods in the mid 1990s, and by the end of the decade more than half of America's agricultural land was given over to GE crops. No new laws were enacted to govern the potential harmful environmental and health impacts. Instead, existing statutes were invoked, and no special handling or labeling of the products was required.

In Europe, massive opposition to GMOs erupted across the continent. Farmers, environmentalists and consumer organizations staged protests and political parties and governments voiced concern. A defacto moratorium on the planting of GE crops and sale of GE food products was put into effect. Meanwhile, the major food processors, distributors and retailers pledged not to sell any products containing GE traits.

The EU embarked on a lengthy review process to assess the environmental and health risks of introducing GE food products. In the end, it established tough new protections designed to mitigate the potential harm. The measures included procedures to segregate and track GE grain and food products from the fields to the retail stores to ensure against contamination; labeling of GMOs at every stage of the food process to ensure transparency; and independent testing as well as more rigorous testing requirements by the companies producing GE seeds and other GMOs.

The EU is forging ahead on a wide regulatory front, changing the very conditions and terms by which new scientific and technological pursuits and products are introduced into the marketplace and the environment. Its bold initiatives put the EU far ahead of the rest of the world. Behind all of its newfound regulatory zeal is the looming question of how best to model global risks and create a sustainable and transparent approach to economic development.

Ensuring Safety

In May of 2003, the EU proposed sweeping new regulatory controls on chemicals to mitigate toxic impacts on the environment and human and animal health. The proposed new law would require new companies to register and test for the safety of more than 30,000 chemicals at an estimated cost to the producers of nearly eight billion Euros. Under existing rules, 99 percent of the total volume of chemicals sold in Europe have not passed through any environmental and health testing and review process. In the past, there was no way to even know what kind of chemicals were being used by industry, making it nearly impossible to track potential health risks. The new regulations will change all of that. The "REACH" system—which stands for Registration, Evaluation and Authorization of Chemicals—requires the companies to conduct safety and environmental tests to prove that the products they are producing are safe. If they can't, the products will be banned from the market.

The new procedures represent an about face to the way the chemical industry is regulated in the U.S. In America, new chemicals are generally assessed to be safe and the burden is primarily put on the consumer, the public or the government to prove that they cause harm. The EU has reversed the burden of proof. Former EU Environmental Commissioner Margot Wallstrom makes the point: "No longer do public authorities need to prove they [the products] are dangerous. The onus is now on industry to prove that the products are safe."

BUILDING THE HYDROGEN ECONOMY

At the very top of the list of environmental priorities for the EU is the plan to become a fully integrated renewable-based hydrogen economy by mid-century. The EU has led the world in championing the Kyoto Protocol on Climate Change, and to ensure compliance it has made a commitment to produce 22 percent of its electricity and 12 percent of all of its energy using renewable sources by 2010. Although a number of member states are lagging behind on meeting their renewable energy targets, the very fact that the EU has set benchmarks puts it far ahead of the U.S. in making the shift from fossil fuels to renewable energy sources. The Bush administration has consistently fought Congressional attempts to establish similar benchmarks for ushering in a U.S.-based renewable energy regime.

In June of 2003, EU President Romano Prodi said, "It is our declared goal of achieving a step-by-step shift toward a fully integrated hydrogen economy, based on renewable energy sources, by the middle of the century." He added that creating this economy would be the next critical step in integrating Europe after the introduction of the Euro.

The European hydrogen game plan is being implemented with a sense of history in mind. Great Britain became the world's leading power in the 19th century because it was the first country to harness its vast coal reserves with steam power. The U.S., in turn, became the world's preeminent power in the 20th century because it was the first country to harness its vast oil reserves with the internal-combustion engine. The multiplier effects of both energy revolutions were extraordinary. The EU is determined to lead the world into the third great energy revolution of the modern era.—*J.R.*

Making companies prove that their chemical products are safe before they are sold is a revolutionary change. It's very difficult to conceive of the U.S. entertaining the kind of risk prevention regulatory regime that the EU has rolled out. In a country where corporate lobbyists spend millions of dollars influencing congressional legislation, the chances of ever having a similar regulatory regime to the one being implemented in Europe would be nigh on impossible.

GMOs and chemical products represent just part of the new "risk prevention" agenda taking shape in Brussels. In early 2003, the EU adopted a new rule prohibiting electronics manufacturers from selling products in the EU that contain mercury, lead and other heavy metals. Another new regulation requires the manufacturers of all consumer electronics and household appliances to cover the costs for recycling their products. American companies complain that compliance with the new regulations will cost them hundreds of millions of dollars a year.

All of these strict new rules governing risk prevention would come as a shock to Americans who believe that the U.S. has the most vigilant regulatory oversight regime in the world for governing risks to the environment and public health. Although that was the case 30 years ago, it no longer is today.

The EU is the first governing institution in history to emphasize human responsibilities to the global environment as a centerpiece of its political vision. Europe's new sensitivity to global risks has led it to champion the Kyoto Protocol on climate change, the Biodiversity Treaty, the Chemical Weapons Convention and many others. The U.S. government has refused, to date, to ratify any of the above agreements.

A New Era

In Europe, intellectuals are increasingly debating the question of the great shift from a risk-taking age to a risk-prevention era. That debate is virtually non-existent in the U.S., where risk-taking is seen as a virtue. The new European intellectuals argue that vulnerability is the underbelly of risks. A sense of vulnerability can motivate people to band together in common cause. The EU stands as a testimonial to collective political engagement arising from a sense of risk and shared vulnerability.

What's changed qualitatively in the last half century since the dropping of the atomic bombs on Hiroshima and Nagasaki is that risks are now global in scale, open ended in duration, incalculable in their consequences and not compensational. Their impact is universal, which means that no one can escape their potential effects. Risks have now become truly democratized, making everyone vulnerable. When everyone is vulnerable, then traditional notions of calculating and pooling risks become virtually meaningless. This is what European academics call a risk society.

Americans aren't there yet. While some academics speak to global risks and vulnerabilities and a significant minority of Americans express their concerns about global risks, from climate change to loss of biodiversity, the sense of utter vulnerability just isn't as strong on this side of the Atlantic. Europeans say we have blinders on. In reality, it's more nuanced than that. Call it delusional, but the sense of personal empowerment is so firmly embedded in the American mind, that even when pitted against growing evidence of potentially overwhelming global threats, most Americans shrug such notions off as overly pessimistic and defeatist. "Individuals can move mountains." Most Americans believe that. Fewer Europeans do.

The EU has already institutionalized a litmus test that cuts to the core of the differences between America and Europe. It's called "the precautionary principle" and it has become the centerpiece of EU regulatory policy governing science and technology in a globalizing world.

The Precautionary Principle

In November 2002, the EU adopted a new policy on the use of the precautionary principle to regulate science and new products derived from technology innovations. According to the EU, reviews occur in "cases where scientific evidence is insufficient, inconclusive or uncertain and preliminary scientific evaluation indicates that there are reasonable grounds for concern that the potentially dangerous effects on the environment, human, animal or plant health may be inconsistent with the high level of protection chosen by the EU." The key term is "uncertain." When there is sufficient evidence to suggest a potential negative impact, but not enough to know for sure, the precautionary principle allows regulatory authorities to err on the side of safety. They can suspend the activity altogether, modify it, employ alternative scenarios, monitor the activity or create experimental protocols to better understand its effects.

THE TRANSITION TO ORGANIC AGRICULTURE

Europe is taking the lead in the shift to sustainable farming practices and organic food production. While the organic food sector is soaring in the U.S.—it represents the fastest-growing segment of the food industry—the government has done little to encourage it. Although the U.S. Department of Agriculture fields a small organic food research program, it amounts to only $3 million, less than .004 percent of its $74 billion budget. While American consumers are increasing their purchases of organic food, less than 0.3 percent of total U.S. farmland is currently in organic production.

By contrast, many of the EU member states have made the transition to organic agriculture a critical component of their economic development plans and have even set benchmarks. Germany, which has often been the leader in setting new environmental goals for the continent, has announced its intention to bring 20 percent of its agricultural output into organic production by the year 2020. (Organic agricultural output is now 3.2 percent of all farm output in Germany.)

The Netherlands, Sweden, Great Britain, Finland, Norway, Germany, Switzerland, Denmark, France and Austria also have national programs to promote the transition to organic food production. Denmark and Sweden enjoy the highest consumption of organic vegetables in Europe and both countries project that their domestic markets for organic food will soon reach or exceed 10 percent of domestic consumption.

Sweden has set a goal of having 20 percent of its total cultivated farm area in organic production by 2005. Italy already has 7.2 percent of its farmland under organic production while Denmark is close behind with seven percent.

Great Britain doubled its organic food production in 2002 and now boasts the second-highest sales of organic food in Europe, after Germany. According to a recent survey, nearly 80 percent of British households buy organic food. By comparison, only 33 percent of American consumers buy any organic food.—J.R.

The precautionary principle allows governments to respond with a lower threshold of scientific certainty than in the past. "Scientific certainty" has been tempered by the notion of "reasonable grounds for concern." The precautionary principle gives authorities the flexibility to respond to events in real time, either before or while they are unfolding.

Advocates of the precautionary principle cite the introduction of halocarbons and the tear in the ozone hole in the Earth's upper atmosphere, the outbreak of mad cow disease in cattle, growing antibiotic resistant strains of bacteria caused by the over-administering of antibiotics to farm animals and the widespread deaths caused by asbestos, benzene and polychlorinated biphenyls (PCBs).

The precautionary principle has been finding its way into international treaties and covenants. It was first recognized in 1982 when the United Nations General Assembly incorporated it into the World Charter for Nature. The precautionary principle was subsequently included in the Rio

Declaration on Environment and Development in 1992, the Framework Convention on Climate Change in 1992, the Treaty on EU (Maastricht Treaty) in 1992, the Cartagena Protocol on Biosafety in 2000 and the Stockholm Convention on Persistent Organic Pollutants (POPs) in 2001.

Valuing Nature

Americans, by and large, view nature as a treasure trove of useful resources waiting to be harnessed for productive ends. While Europeans share America's utilitarian perspective, they also have a love for the intrinsic value of nature. One can see it in Europeans' regard for the countryside and their determination to maintain natural landscapes, even if it means providing government assistance in the way of special subsidies, or foregoing commercial development. Nature figures prominently in Europeans' dream of a quality of life. Europeans spend far more time visiting the countryside on weekends and during their vacations than Americans.

The balancing of urban and rural time is less of a priority for most Americans, many of whom are just as likely to spend their weekends at a shopping mall, while their European peers are hiking along country trails. Anyone who spends significant time among Europeans knows that they have a great affinity for rural getaways. Almost everyone I know in Europe—among the professional and business class—has some small second home in the country somewhere—a dacha usually belonging to the family for generations. While working people may not be as fortunate, on any given weekend they can be seen exiting the cities en masse, motoring their way into the nearest rural enclave or country village for a respite from urban pressures.

The strongly held values about rural life and nature is one reason why Europe has been able to support green parties across the continent, with substantial representation in national parliaments as well as in the European Parliament. By contrast, not a single legislator at the federal level in the U.S. is a member of the Green Party.

There is another dimension to the European psyche that makes Europeans supportive of the precautionary principle—their sense of "connectedness."

Because we Americans place such a high premium on autonomy, we are far less likely to see the deep connectedness of things. We tend to see the world in terms of containers, each isolated from the whole and capable of standing alone. We like everything around us to be neatly bundled, autonomous, and self contained. The new view of science that is emerging in the wake of globalization is quite different. Nature is viewed as a myriad of symbiotic relationships, all embedded in a larger whole, of which they are an integral part. In this new vision of nature, nothing is autonomous, everything is connected.

By championing a host of global environmental treaties and accords taking the precautionary approach to regulation, the EU has shown a willingness to act on its commitment to sustainable development and global environmental stewardship. The fact that its commitments in most areas remain weak and are often vacillating is duly noted. But, at least Europe has established a new agenda for conducting science and technology that, if followed, could begin to wean the world from the old ways and toward a second Enlightenment.

Ronald Bailey  **NO**

Wilting Greens

It's clear that we've suffered a number of major defeats," declared Andrew Hewett, executive director of Oxfam Community Aid, at the conclusion of the World Summit on Sustainable Development, held in Johannesburg, South Africa, in September. Greenpeace climate director Steve Sawyer complained, "What we've come up with is absolute zero, absolutely nothing." The head of an alliance of European green groups proclaimed, "We barely kept our heads above water."

It wasn't supposed to be this way. Environmental activists hoped the summit would set the international agenda for sweeping environmental reform over the next 15 years. Indeed, they hoped to do nothing less than revolutionize how the world's economy operates. Such fundamental change was necessary, said the summiteers, because a profligate humanity consumes too much, breeds too much, and pollutes too much, setting the stage for a global ecological catastrophe.

But the greens' disappointment was inevitable because their major goals—preserving the environment, eradicating poverty, and limiting economic growth—are incompatible. Economic growth is a prerequisite for lessening poverty, and it's also the best way to improve the environment. Poor people cannot afford to worry much about improving outdoor air quality, let alone afford to pay for it. Rather than face that reality, environmentalists increasingly invoke "sustainable development." The most common definition of the phrase comes from the 1987 United Nations report *Our Common Future*: development that "meets the needs of the present without compromising the ability of future generations to meet their own needs."

For radical greens, sustainable development means economic stagnation. The Earth Island Institute's Gar Smith told Cybercast News, "I have seen villages in Africa ... that were disrupted and destroyed by the introduction of electricity." Apparently, the natives no longer sang community songs or sewed together in the evenings. "I don't think a lot of electricity is a good thing," Smith added. "It is the fuel that powers a lot of multinational imagery." He doesn't want poor Africans and Asians "corrupted" by ads for Toyota and McDonald's, or by Jackie Chan movies.

Indian environmentalist Sunita Narain decried the "pernicious introduction of the flush toilet" during a recent PBS/BBC television debate hosted by Bill Moyers. Luckily, most other summiteers disagreed with Narain's curious disdain for sanitation. One of the few firm goals set at the confab was that adequate sanitation should be supplied by 2015 to half of the 2.2 billion people now lacking it.

Sustainable development boils down to the old-fashioned "limits to growth" model popularized in the 1970s. Hence Daniel Mittler of Friends of the Earth International moaned that "the summit failed to set the necessary economic and ecological limits to globalization." The *Jo'burg Memo*, issued by the radical green Heinrich Böll Foundation before the summit, summed it up this way: "Poverty alleviation cannot be separated from wealth alleviation."

The greens are right about one thing: The extent of global poverty is stark. Some 1.1 billion people lack safe drinking water, 2.2 billion are without adequate sanitation, 2.5 billion have no access to modern energy services, 11 million children under the age of 5 die each year in developing countries from preventable diseases, and 800 million people are still malnourished, despite a global abundance of food. Poverty eradication is clearly crucial to preventing environmental degradation, too, since there is nothing more environmentally destructive than a hungry human.

Most summit participants from the developing world understood this. They may be egalitarian, but unlike their Western counterparts they do not aim to make everyone equally poor. Instead, they want the good things that people living in industrialized societies enjoy.

That explains why the largest demonstration during the summit, consisting of more than 10,000 poor and landless people, featured virtually no banners or chants about conventional environmentalist issues such as climate change, population control, renewable resources, or biodiversity. Instead, the issues were land reform, job creation, and privatization.

The anti-globalization stance of rich activists widens this rift. Environmentalists claim trade harms the environment and further impoverishes people in the developing world. They were outraged by the dominance of trade issues at the summit.

"The leaders of the world have proved that they work as employees for the transnational corporations," asserted Friends of the Earth Chairman Ricardo Navarro. Indian eco-feminist Vandana Shiva added, "This summit has become a trade summit, it has become a trade show." Yet the U.N.'s own data underscore how trade helps the developing world. As fact sheets issued by the U.N. put it, "During the 1990s the economies of developing countries that were integrated into the world economy grew more than twice as fast as the rich countries. The 'non-globalizers' grew only half as fast and continue to lag further behind."

By invoking a zero sum version of sustainable development, environmentalists not only put themselves at odds with the developing world; they ignore the way in which economic growth helps protect the environment. The real commons from which we all draw is the growing pool of scientific, technological, and institutional concepts, and the capital they create. Past generations have left us far more than they took, and the result has been an explosion in human well-being, longer life spans, less disease, more and cheaper food, and expanding political freedom.

Such progress is accompanied by environmental improvement. Wealthier is healthier for both people and the environment. As societies become richer and more technologically adept, their air and water become cleaner, they set aside more land for nature, their forests expand, they use less land for agriculture, and more people cherish wild species. All indications suggest that the 21st century will be the century of ecological restoration, as humanity uses physical resources ever more efficiently, disturbing the natural world less and less.

In their quest to impose a reactionary vision of sustainable development, the disappointed global greens will turn next to the World Trade Organization, the body that oversees international trade rules. During the summit, the WTO emerged as the greens' bête noire. As Friends of the Earth International's Daniel Mittler carped, "Instead of using the [summit] to respond to global concerns over deregulation and liberalization, governments are pushing the World Trade Organization's agenda." "See you in Cancun!" promised Greenpeace's Steve Sawyer, referring to the location of the next WTO ministerial meeting in September 2003. That confab will build on the WTO's Doha Trade Round, launched last year, which is aimed at reducing the barriers to trade for the world's least developed countries.

The WTO may achieve worthy goals that eluded the Johannesburg summit, such as eliminating economically and ecologically ruinous farm and energy subsidies and opening developed country markets to the products of developing nations. Free marketeers and greens might even form an alliance on those issues.

But environmentalists want to use the WTO to implement their sustainable development agenda: global renewable energy targets, regulation based on the precautionary principle, a "sustainable consumption and production project," a worldwide eco-labeling scheme. According to Greenpeace's Sawyer, nearly everyone at the Johannesburg summit agreed "there is something wrong with unbridled neoliberal capitalism."

Let's hope the greens fail at the WTO just as they did at the U.N. summit. Their sustainable development agenda, supposedly aimed at improving environmental health, instead will harm the natural world, along with the economic prospects of the world's poorest people. The conflicting goals on display at the summit show that at least some of the world's poor are wise to that fact.

POSTSCRIPT

Is Sustainable Development Compatible With Human Welfare?

The first of the Rio Declaration's 22 principles states, "Human beings are at the centre of concerns for sustainable development. They are entitled to a healthy and productive life in harmony with nature." Any solution to the sustainability problem therefore should not infringe human welfare. This makes any solution that involves limiting or reducing human population or blocking improvements in standard of living very difficult to sell. Yet solutions may be possible. David Malin Roodman suggests in *The Natural Wealth of Nations: Harnessing the Market for the Environment* (W. W. Norton, 1998) that taxing polluting activities instead of profit or income would stimulate corporations and individuals to reduce such activities or to discover nonpolluting alternatives. In "Building a Sustainable Society," *State of the World 1999* (W. W. Norton, 1999), he adds recommendations for citizen participation in decision making, education efforts, and global cooperation, without which we are heading for "a world order [that] almost no one wants." (He is referring to a future of environmental crises, not the "new world order" feared by many conservatives, in which national policies are dictated by international [UN] regulators.)

Julie Davidson, in "Sustainable Development: Business as Usual or a New Way of Living?" *Environmental Ethics* (Spring 2000), notes that efforts to achieve sustainability cannot by themselves save the world. But such efforts may give us time to achieve new and more suitable values. It is thus heartening to see that the UN World Summit on Sustainable Development was held in Johannesburg, South Africa, in August 2002. Its aim was to strengthen partnerships between governments, business, nongovernmental organizations, and other stakeholders and to seek to eradicate poverty and make more equal the distribution of the benefits of globalization. See Gary Gardner, "The Challenge for Johannesburg: Creating a More Secure World," *State of the World 2002* (W. W. Norton, 2002), and the United Nations Environmental Programme's Global Environmental Outlook 3 (Earthscan, 2002), prepared as a "global state of the environment report" in preparation for the Johannesburg Summit.

The World Council of Churches brought to the Johannesburg Summit an emphasis on social justice. Martin Robra, in "Justice—The Heart of Sustainability," *Ecumenical Review* (July 2002), writes that the dominant stress on economic growth "has served, first and foremost, the interests of the powerful economic players. It has further marginalized the poor sectors of society, simultaneously undermining their basic security in terms of access to land, water, food, employment, and other basic services and a healthy environment."

Is social justice or equity worth this emphasis? Or is sustainability more a matter of population control, of shielding the natural environment from human impacts, or of economics? A. J. McMichael, C. D. Butler, and Carl Folke, in "New Visions for Addressing Sustainability," *Science* (December 12, 2003), argue that it is wrong to separate—as did the Johannesburg Summit—achieving sustainability from other goals such as reducing fertility and poverty and improving social equity, living conditions, and health. They observe that human population and lifestyle affect ecosystems, ecosystem health affects human health, human health affects population and lifestyle. "A more integrated ... approach to sustainability is urgently needed," they say, calling for more collaboration among researchers and other fields.

Anthony R. Leiserowitz, Robert W. Kates, and Thomas M. Parris, "Do Global Attitudes and Behaviors Support Sustainable Development?" *Environment* (November 2005), find that though the world's people appear to support the component concepts of sustainable development, there is a mismatch between that support and their behavior. In the long term, they say, what is needed is a "shift from materialist to post-materialist values, from anthropocentric to ecological worldview, and a redefinition of the good life." Unfortunately, that shift remains in the future. P. Aarne Vesilind, Lauren Heine, and Jamie Hendry, "The Moral Challenge of Green Technology," *TRAMES: A Journal of the Humanities & Social Sciences* (No.1, 2006), "conclude that the unregulated free market system is incompatible with our search for sustainability. Experience has shown that if green technology threatens profits, green technology loses and profitability wins."

ISSUE 3

Should a Price Be Put on the Goods and Services Provided by the World's Ecosystems?

YES: Jim Morrison, from "How Much Is Clean Water Worth?" *National Wildlife* (February/March 2005)

NO: Marino Gatto and Giulio A. De Leo, from "Pricing Biodiversity and Ecosystem Services: The Never-Ending Story," *BioScience* (April 2000)

ISSUE SUMMARY

YES: Jim Morrison argues that ecosystem services such as cleaning water, controlling floods, and pollinating crops have sufficient economic value to make it profitable to spend millions of dollars to protect natural systems.

NO: Professors of applied ecology Marino Gatto and Giulio A. De Leo contend that the pricing approach to valuing nature's services is misleading because it falsely implies that only economic values matter.

Human activities frequently involve trading a swamp or forest or mountainside for a parking lot or housing development or farm. People generally agree that the parking lot, housing development, or farm is a worthwhile project, for it has obvious benefits—expressible in economic terms—to human beings. But are there costs as well? Construction costs, labor costs, and material costs can easily be calculated, but what about the swamp? The forest? The species living there?

How much is a species worth? One approach to answering this question is to ask people how much they would be willing to pay to keep a species alive. If the question is asked when there are a million species in existence, few people will likely be willing to pay much. But if the species is the last one remaining, they might be willing to pay a great deal. Most people would agree that both answers fail to get at the true value of a species, for nature is not expressible solely in terms of cash values. Yet some way must be found to weigh the effects of human activities on nature against the benefits gained from those activities. If it is not, we will continue to degrade the world's eco-systems and threaten our own continued well-being.

Traditional economics views nature as a "free good." That is, forests generate oxygen and wood, clouds bring rain, and the sun provides warmth, all without charge to the humans who benefit. At the same time, nature has provided ways for people to dispose of wastes—such as dumping raw sewage into rivers or emitting smoke into the air—without paying for the privilege. This "free" waste disposal has turned out to have hidden costs in the form of the health effects of pollution (among other things), but it has been up to individuals and governments to bear the costs associated with those effects. The costs are real, but in general, they have not been borne by the businesses and other organizations that produced them. They have thus come to be known as "external" costs.

Environmental economists have recognized the problem of external costs, and government regulators have devised a number of ways to make those who are responsible accept the bill, such as instituting requirements for pollution control and fining those who exceed permitted emissions. Yet some would say that this approach does not help enough.

The *ecosystem services* approach recognizes that undisturbed ecosystems do many things that benefit us. A forest, for instance, slows the movement of rain and snowmelt into streams and rivers; if the forest is removed, floods may follow (a connection that recently forced China to deemphasize forest exploitation). Swamps filter the water that seeps through them. Food chains cycle nutrients necessary for the production of wood and fish and other harvests. Bees pollinate crops and make food production possible. These services are valuable—even essential—to us, and anything that interferes with them must be seen as imposing costs just as significant as the illnesses associated with pollution.

How can those costs be assessed? In 1997 Robert Costanza and his colleagues published an influential paper entitled "The Value of the World's Ecosystem Services and Natural Capital" in the May 15 issue of the journal *Nature*. In it, the authors listed a variety of ecosystem services and attempted to estimate what it would cost to replace those services if they were somehow lost (such as by building a sewage treatment plant to replace a swamp). The total bill for the entire biosphere came to $33 trillion (the middle of a $16–54 trillion range), compared to a global gross national product of $25 trillion. Costanza et al. stated that this was surely an underestimate. Janet N. Abramovitz, "Putting a Value on Natures 'Free' Services," *WorldWatch* (January/February 1998), argues that nature's services are responsible for the vast bulk of the value in the world's economy and that attaching economic value to those services may encourage their protection.

In the following selections, journalist Jim Morrison argues that ecosystem services such as cleaning water, controlling floods, and pollinating crops have sufficient economic value to make it profitable to spend millions of dollars to protect natural systems. Marino Gatto and Giulio A. De Leo argue that the pricing approach to valuing nature's services is misleading because it ignores equally important "nonmarket" values.

How Much Is Clean Water Worth?

The water that quenches thirsts in Queens and bubbles into bathtubs in Brooklyn begins about 125 miles north in a forest in the Catskill Mountains. It flows down distant hills through pastures and farmlands and eventually into giant aqueducts serving 9 million people with 1.3 billion gallons daily. Because it flows directly from the ground through reservoirs to the tap, this water—long regarded as the champagne of city drinking supplies—comes from what's often called the largest "unfiltered" system in the nation.

But that's not strictly true. Water percolating through the Catskills is filtered naturally—for free. Beneath the forest, fine roots and microorganisms break down contaminants. In streams, plants absorb nutrients from fertilizer and manure. And in meadows, wetlands filter nutrients while breaking down heavy metals.

New York City discovered how valuable these services were 15 years ago when a combination of unbridled development and failing septic systems in the Catskills began degrading the quality of the water that served Queens, Brooklyn and the other boroughs. By 1992, the U.S. Environmental Protection Agency (EPA) warned that unless water quality improved, it would require the city to build a filtration plant, estimated to cost between $6 and $8 billion and between $350 and $400 million a year to operate.

Instead, the city rolled the dice with nature in a historic experiment. Rather than building a filtration plant, officials decided to restore the health of the Catskills watershed, so it would do the job naturally.

What's this ecosystem worth to the city of New York? So far, $1.3 billion. That's what the city has committed to build sewage treatment plants upstate and to protect the watershed through a variety of incentive programs and land purchases. It's a lot of money. But it's a fraction of the cost of the filtration plant—a plant, city officials note, that wouldn't work as tirelessly or efficiently as nature.

"It was a stunning thing for the New York City council to think maybe we should invest in natural capital," says Stanford University researcher Gretchen Daily. Daily is one of a growing number of academics—some from

From *National Wildlife*, February/March 2005, pp. 24, 26–28. Copyright © 2005 by Jim Morrison. Reprinted by permission of the author.

economics, some from ecology—who are putting dollar figures on the services that ecosystems provide. She and other "ecological economists" look not only at nature's products—food, shelter, raw materials—but at benefits such as clean water, clean air, flood control and storm mitigation, irreplaceable services that have been taken for granted throughout history. "Much of Mother Nature's labor has enormous and obvious value, which has failed to win respect in the marketplace until recently," Daily writes in the book *The New Economy of Nature: The Quest to Make Conservation Profitable.*

Ecological economist Geoffrey Heal, a professor of public policy and business responsibility at Columbia University, became interested in the field as an economist who was concerned about the environment. "The idea of ecosystem services is an interesting framework for thinking why the environment matters," says Heal, author of *Nature and the Marketplace: Capturing the Value of Ecosystem Services.* "The traditional argument for environmental conservation had been essentially aesthetic or ethical. It was beautiful or a moral responsibility. But there are powerful economic reasons for keeping things intact as well."

Daily notes that beyond providing clean water, the Catskills ecosystem has value for its beauty, as wildlife habitat and for recreation, particularly trout fishing. Such values are not inconsequential. While no one has assessed the total worth of the watershed, even a partial look reveals that habitat and wildlife are powerful economic engines.

Restored habitat for trout and other game fish, for example, attracts fishermen, and angling is big business in this state. According to a report by the U.S. Fish and Wildlife Service (FWS), more than 1.5 million people fished in New York during 2001, yielding an economic benefit to the state of more than $2 billion and generating the equivalent of 17,468 full-time jobs and more than $164 million in state, federal, sales and motor fuel taxes. Though not as easily measured, individual Catskills species also have value. Beavers, for instance, create wetlands that are vital to filtering water and to biodiversity.

Ecological economists maintain that ecosystems are capital assets that, if managed well, provide a stream of benefits just as any investment does. The FWS report, for example, notes that 66 million Americans spent more than $38 billion in 2001 observing, feeding or photographing wildlife. Those expenditures resulted in more than a million jobs with total wages and salaries of $27.8 billion. The analysis found that birders alone spent an estimated $32 billion on wildlife watching that year, generating $85 billion of economic benefits. In Yellowstone National Park, the reintroduction of gray wolves that began in 1995 has already increased revenues in surrounding communities by $10 million a year, with total benefits projected to reach $23 million annually as more visitors come to catch a glimpse of these charismatic predators.

When it comes to water quality, EPA projects that the United States will have to spend $140 billion over the next 20 years to maintain minimum required standards for drinking water quality. No wonder, then, that 140 U.S. cities have studied using an approach similar to New York's. Under that agreement, finalized in 1997, the city promised to pay farmers, landowners and businesses that abided by restrictions designed to protect the watershed. (The city owns less than 8 percent of the land in the 2,000-square-mile watershed; the vast majority is in private hands.) "In the case of the Catskills, it was a matter of coming up with a way to reward the stewards of the natural asset for something they had been providing for free," Daily says. "As soon as they got paid even a little bit, they were much happier and inclined to go about their stewardship." There's no guarantee this experiment will work, of course; it may be another decade before the city finds out.

Elsewhere, other governing bodies are also recognizing the value of ecosystem services. The U.S. Army Corps of Engineers, for example, bought 8,500 acres of wetlands along Massachusetts' Charles River for flood control. The land cost $10 million, a tenth of the $100 million the Corps estimated it would take to build the dam and levee originally proposed. To fight floods in Napa, California, county officials spent $250 million to reconnect the Napa River to its historical floodplains, allowing the river to meander as it once did. The cost was a fraction of the estimated $1.6 billion that would have been needed to repair flood damage over the next century without the project. Within a year, notes Daily, flood insurance rates in the county dropped 20 percent and real estate prices rose 20 percent, thanks to the flood protection now promised by nature.

Even insects supply vital ecosystem services. More than 218,000 of the world's 250,000 flowering plants, including 70 percent of all species of food plants, rely on pollinators for reproduction—and more than 100,000 of these pollinators are invertebrates, including bees, moths, butterflies, beetles and flies. Another 1,000 or more vertebrate species, including birds, mammals and reptiles, also pollinate plants. According to University of Arizona entomologist Stephen Buchmann, author of *The Forgotten Pollinators,* one of every three bites of food we eat comes courtesy of a pollinator.

A Cornell University study estimated the value of pollination by honeybees in the United States alone at $14.6 billion in 2000. Yet honeybee populations are dropping everywhere, as much as 25 percent since 1990, according to one study. Now many farms and orchards are paying to have the bees shipped in.

Today's interest in assigning dollar values to pollination and other ecosystem services was spawned by publication of a controversial 1997 report in *Nature* that estimated the total global contribution of ecosystems to be $33 trillion or more each year—roughly double the combined gross national product of all countries in the world. The study became a lightning rod. Detractors scoffed at the idea that one could put a dollar value on something people weren't willing to purchase. One report by researchers at the University of Maryland, Bowden College and Duke University called the estimate "absurd," noting that if taken literally, the figure suggests that a family earning $30,000 annually would pay $40,000 annually for ecosystem protection.

Other researchers, including Daily and Heal, charged that the $33 trillion figure greatly underestimates nature's value. "If you believe, as I do, that ecosystem services are necessary for human survival, they're invaluable really," Heal says. "We would pay anything we could pay."

Daily doesn't believe the absolute value of an ecosystem can ever be measured. Heal agrees, yet both scientists say that pricing ecosystem services is an important tool for making decisions about nature—and for making the case for conservation. "Valuation is just one step in the broader politics of decision making," she says. "We need to be creative and innovative in changing social institutions so we are aligning economic forces with conservation."

Indeed, as dollar values for nature's services become available, environmentalists increasingly use them to bolster arguments for conservation. One high-profile example is the contentious dispute over whether to tear down four dams on the lower Snake River in southeastern Washington to restore salmon habitat, and thus the region's lucrative salmon fishery. Ed Whitelaw, a professor of economics at the University of Oregon, notes that estimates of the economic impact of breaching the dams range from $300 million in net costs to $1.3 billion in net benefits, largely due to the wide range of projections about recreational spending.

A 2002 report by the respected, nonprofit think-tank RAND Corporation concluded the dams could be breached without hurting economic growth and employment. Energy lost as a result of the breaches could be replaced with more efficient sources, including natural gas, resulting in 15,000 new jobs. Further, the report noted that recreation, retail, restaurants and real estate would experience a marked growth. Recreational activities alone would increase by an estimated $230 million over 20 years.

There's no question that returning the salmon runs would have a major impact on the region. When favorable ocean conditions increased the runs in 2001, Idaho's Department of Fish and Game estimated the salmon season that year alone generated more than $90 million of revenue in the state, most of it in rural communities that badly needed the funds.

"Some people think it sounds crass to put a price tag on something that's invaluable, careening down the slippery slope of the market economy," says Daily. "In fact, the idea is to do something elegant but tricky: to finesse the economic system, the system that drives so much of our individual and collective behavior, so that without even thinking it makes natural sense to invest in and protect our natural assets, our ecosystem capital."

What Daily and other ecological economists want is to insinuate consideration of ecosystem services into daily decision making, whether it takes the form of financial incentive or penalty. "At a practical level, decisions are made at the margin, not at the 'should we sterilize the Earth' level," she says. "It's in all the little decisions—whether to farm here or leave a few trees, whether to build the shopping mall there or leave the wetland, whether to buy an SUV or a Prius—that ecosystem service values need to be incorporated."

Heal agrees. "Although ecosystem services have been with us for millennia," he says, "the scale of human activity is now sufficiently great that we can no longer take their continuation for granted."

**Marino Gatto and
Giulio A. De Leo**

 **NO**

Pricing Biodiversity and Ecosystem Services: The Never-Ending Story

In 1844, the French engineer Jules Juvénal Dupuit introduced cost–benefit analysis to evaluate investment projects.... The application of cost–benefit analysis to ecological issues fell out of favor three decades ago, and it was gradually replaced by multicriteria analysis in the decision-making process for projects that have an impact on the environment. Although multicriteria analysis is currently used for environmental impact assessments [EIA] in many nations, [recently] the concept of cost–benefit analysis has again become fashionable, along with the various pricing techniques associated with it, such as contingent valuation methods, hedonic prices, and costs of replacement of ecological services.... Economists have generated a wealth of virtuosic variations on the theme of assessing the societal value of biodiversity, but most of these techniques are invariably based on price—that is, on a single scale of values, that of goods currently traded on world markets.

Perhaps the most famous recent study on the issue of pricing biodiversity and ecological services is that by Costanza et al., who argued that if the importance of nature's free benefits could be adequately quantified in economic terms, then policy decisions would better reflect the value of ecosystem services and natural capital. Drawing on earlier studies aimed at estimating the value of a wide variety of ecosystem goods and services, Costanza et al. estimated the current economic value of the entire biosphere at \$16–54 trillion per year, with an average value of approximately \$33 trillion per year. By contrast, the gross national product of the United States totals approximately \$18 trillion per year. The paper, as its authors intended, stimulated much discussion, media attention, and debate. A special issue of *Ecological Economics* (April 1998) was devoted to commentaries on the paper, which, with few exceptions, were laudatory. Some economists have questioned the actual numbers, but many scientists have praised the attempt to value biodiversity and ecosystem functions.

Although Costanza et al. acknowledged that their estimates were crude and imperfect, they also pointed the way to improved assessments. In particular, they noted the need to develop comprehensive ecological economic models that could adequately incorporate the complex interdependencies between ecosystems and economic systems, as well as the complex individual dynamics of both types of systems. Despite the authors' caveats and the fact that many economists have been circumspect in applying their own tools to decisions regarding natural systems, the monetary approach is perceived by scientists, policymakers, and the general public as extremely appealing; a number of biologists are also of the opinion that attaching economic values to ecological services is of paramount importance for preserving the biosphere and for effective decision-making in all cases where the environment is concerned.

In this article, we espouse a contrary view, stressing that, for most of the values that humans attach to biodiversity and ecosystem services, the pricing approach is inadequate—if not misleading and obsolete—because it implies erroneously that complex decisions with important environmental impacts can be based on a single scale of values. We contend that the use of cost–benefit analysis as the exclusive tool for decision-making about environmental policy represents a setback relative to the existing legislation of the United States, Canada, the European Union, and Australia on environmental impact assessment, which explicitly incorporates multiple criteria (technical, economic, environmental, and social) in the process of evaluating different alternatives. We show that there are sound methodologies, mainly developed in business and administration schools by regional economists and by urban planners, that can assist decision-makers in evaluating projects and drafting policies while accounting for the nonmarket values of environmental services.

The Limitations of Cost–Benefit Analysis and Contingent Valuation Methods

Historically, the first important implementation of cost–benefit analysis at the political level came in 1936, with passage of the US Flood Control Act. This legislation stated that a public project can be given a green light if the benefits, to whomsoever they accrue, are in excess of estimated costs. This concept implies that all benefits and costs are to be considered, not just actual cash flows from and to government coffers. However, public agencies (e.g., the US Army Corps of Engineers) quickly ran into a problem: They were not able to give a monetary value to many environmental effects, even those that were predictable in quantitative terms. For instance, engineers could calculate the reduction of downstream water flow resulting from construction of a dam, and biologists could predict the river species most likely to become extinct as a consequence of this flow reduction. However, public agencies were not able to calculate the cost of each lost species. Therefore, many ingenious techniques for the monetary valuation of environmental goods and services have been devised since the 1940s. These techniques fall into four basic categories.

- **Conventional market approaches.** These approaches, such as the replacement cost technique, use market prices for the environmental service that is affected. For example, degradation of vegetation in developing countries leads to a decrease in available fuelwood. Consequently, animal dung has to be used as a fuel instead of a fertilizer, and farmers must therefore replace dung with chemical fertilizers. By computing the cost of these chemical fertilizers, a monetary value for the degradation of vegetation can then be calculated.

- **Household production functions.** These approaches, such as the travel cost method, use expenditures on commodities that are substitutes or complements for the environmental service that is affected. The travel cost method was first proposed in 1947 by the economist Harold Hotelling, who, in a letter to the director of the US National Park Service, suggested that the actual traveling costs incurred by visitors could be used to develop a measure of the recreation value of the sites visited.
- **Hedonic pricing.** This form of pricing occurs when a price is imputed for an environmental good by examining the effect that its presence has on a relevant market-priced good. For instance, the cost of air and noise pollution is reflected in the price of plots of land that are characterized by different levels of pollution, because people are willing to pay more to build their houses in places with good air quality and little noise....
- **Experimental methods.** These methods include contingent valuation methods, which were devised by the resource economist Siegfried V. Ciriacy-Wantrup. Contingent valuation methods require that individuals express their preferences for some environmental resources by answering questions about hypothetical choices. In particular, respondents to a contingent valuation methods questionnaire will be asked how much they would be willing to pay to ensure a welfare gain from a change in the provision of a nonmarket environmental commodity, or how much they would be willing to accept in compensation to endure a welfare loss from a reduced provision of the commodity.

Among these pricing techniques, the contingent valuation methods approach is the only one that is capable of providing an estimate of existence values, in which biologists have a special interest. Existence value was first defined by Krutilla as the value that individuals may attach to the mere knowledge that rare and diverse species, unique natural environments, or other "goods" exist, even if these individuals do not contemplate ever making active use of or benefiting in a more direct way from them. The name "contingent valuation" comes from the fact that the procedure is contingent on a constructed or simulated market, in which people are asked to manifest, through questionnaires and interviews, their demand function for a certain environmental good (i.e., the price they would pay for one extra unit of the good versus the availability of the good)....

The limits of cost–benefit analysis were discussed in the 1960s, after more than two decades of experimentation. In particular, many authors pointed out that cost–benefit analysis encouraged policymakers to focus on things that can be measured and quantified, especially in cash terms, and to disregard problems that are too large to be assessed easily. Therefore, the associated price might not reflect the "true" value of social equity, environmental services, natural capital, or human health. In particular, economists themselves recognize that the increasingly popular contingent valuation methods are undermined by several conceptual problems, such as free-riding, overbidding, and preference reversal.

When it comes to monetary valuation of the goods and services provided by natural ecosystems and landscapes specifically, a number of additional problems undermine the effectiveness of pricing techniques and cost–benefit analysis. These problems include the very definition of "existence" value, the dependence of pricing techniques on the composition of the reference group, and the significance of the simulated market used in contingent valuation.

The definition of "existence" value   A classic example of contingent valuation methods is to ask for the amount of money individuals are willing to pay to ensure the continued existence of a species such as the blue whale. However, the existence value of whales does not take into account potential indirect services and benefits provided by these mammals. It is just the value of the existence of whales for humans, that is, the satisfaction that the existence of blue whales provides to people who want them to continue to exist. Therefore, there is a real risk that species with very low or no aesthetic appeal or whose biological role has not been properly advertised will be given a low value, even if they play a fundamental ecological function. Without adequate information, most people do not understand the extent, importance, and gravity of most environmental problems. As a consequence, people may react emotionally and either underestimate or overestimate risks and effects.

Therefore, it is not surprising that five of the seven guidelines issued by the National Oceanic and Atmospheric Administration [NOAA] about how to conduct contingent valuation discuss how to properly inform and question respondents to produce reliable estimates (e.g., in-person interviews are preferred to telephone surveys to elicit values). Of course, acquisition of reliable and complete information is always possible in theory, but in practice strict adherence to NOAA guidelines makes contingent valuation methods expensive and time consuming.

Difficulties with the reference group for pricing   Pricing techniques such as contingent valuation methods provide information about individual willingness to pay or willingness to accept, which must be summed up in the final balance of cost–benefit analysis. Therefore, the outcome of cost–benefit analysis depends strongly on the group of people that is taken as a reference for valuation—particularly on their income. Van der Straaten noted that the Exxon *Valdez* oil spill in 1989 provides a good example of this dependence. The population of the United States was used as a reference group to calculate the damage to the existence value of the affected species and ecosystems using contingent valuation methods. Exxon was ultimately ordered to pay $5 billion to compensate the people of Alaska for their losses. This huge figure was a consequence of the high income of the US population. If the same accident had occurred in Siberia, where salaries are lower, the outcome would certainly have been different.

This example shows that contingent valuation methods simply provide information about the preferences of a particular group of people but do not necessarily reflect the ecological importance of ecosystem goods and services. Moreover, the outcome of cost–benefit analysis depends on which individual willingness to pay or willingness to accept are included in the cost–benefit analysis. If the quality of the Mississippi River is at issue, should the analysis be restricted to US citizens living close to the river, or should the willingness to pay of Californians and New Yorkers be included too? According to Krutilla's definition of existence value, for many environmental goods and ecological services that may ultimately affect ecosystem integrity at the global level, the preferences of the entire human population should potentially be considered in the analysis. Because practical reasons obviously preclude doing so, contingent valuation methods will inevitably only provide information about the preferences of specific groups of people. For many of the ecological services that may be considered the heritage of humanity, contingent valuation methods analyses performed locally in a particular economic situation should be extrapolated only with great caution to other areas. The process of placing a monetary value on biodiversity and ecosystem functioning through nonuser willingness to pay is performed in the same way as for user willingness to pay, but the identification of people who do not use an environmental good directly and still have a legitimate interest in its preservation is problematic.

Significance of the simulated market Contingent valuation methods are contingent on a market that is constructed or simulated, not real. It is difficult to believe in the efficiency of what Adam Smith called the "invisible hand" of the market for a process that is the artificial production of economic advisors and does not possess the dynamic feedback that characterizes real competitive markets. Is it even possible to simulate a market where units of biodiversity are bought and sold? As Friend stated, "these contingency evaluation methods (CVM) tend to create an illusion of choice based on psychology (willingness) and ideology (the need to pay) which is supposed, somewhat mysteriously, to reflect an equilibrium between the consumer demand for and producer supply of environmental goods and services."

Many additional criticisms of pricing ecological services are more familiar to biologists. For many ecological services, there is simply no possibility of technological substitution. Moreover, the precise contribution of many species is not known, and it may not be known until the species is close to extinction.... In addition, specific ecosystem services, as evaluated by Costanza et al., should not be separated from one another and valued individually because the importance of any piece of biodiversity cannot be determined without considering the value of biodiversity in the aggregate. And finally, the use of marginal value theory may be invalidated by the erratic and catastrophic behavior of many ecological systems, resulting in potentially detrimental effects on the health of humans, the productivity of renewable resources, and the vitality and stability of societies themselves.

Despite the efforts of many economists, we believe that some goods and services, especially those related to ecosystems, cannot reasonably be given a monetary value, although they are of great value to humans. Economists coined the term "intangibles" to define these goods. Cost–benefit analysis cannot easily deal with intangibles. As Nijkamp wrote, more than 20 years ago, "the only reasonable way to take account of intangibles in the traditional cost–benefit analysis seems to be the use of a balance with a debit and a credit side in which all intangible project effects (both positive and negative) are represented in their own (qualitative or quantitative) dimensions" as secondary information. In other words, the result of cost–benefit analysis is primarily a single number, the net monetary benefit that comprises all the effects that can be sensibly converted into monetary returns and costs.

Commensurability of Different Objectives and Multicriteria Analysis

Cost–benefit analysis includes intangibles in the decision-making process only as ancillary information, with the main focus being on those effects that can be converted to monetary value. This approach is not a balanced solution to the problem of making political decisions that are acceptable to a wide number of social groups with a range of legitimate interests....

However, even if the attempt to put a price on everything is abandoned, it is not necessary to give up the attempt to reconcile economic issues with social and environmental ones. Social scientists long ago developed multicriteria techniques to reach a decision in the face of multiple different and structurally incommensurable goals. The most important concept in multicriteria analysis was actually conceived by an Italian economist, Vilfredo Pareto, at the end of the nineteenth century. It is best explained by a simple example. Suppose that a natural area hosting several rare species is a target for the development of a mining activity. Alternative mining projects can have different effects in terms of profits from mining (measured in dollars) and in terms of sustained biodiversity (measured in suitable units, for instance, through the Shannon index). Profit from mining can be corrected using welfare economics to include those environmental and social effects that can be priced (e.g., the benefit of providing jobs to otherwise unemployed people, the cost of treating lung disease of miners, and the cost of the loss of the tourists who used to visit the natural area)....

The methods of multicriteria analysis are intended to assist the decision-maker in choosing among ... alternatives ... (a task that is particularly difficult when there are several incommensurable objectives, not just two). Nevertheless, the initial step of determining [these] alternatives is of enormous importance, for three reasons. First, [doing so] makes perfect sense even if there is no way of pricing a certain environmental good because each objective can be expressed in its own proper units without reduction to a common scale. Second, the determination of all the feasible alternatives ... requires the joint effort of a multidisciplinary team that includes, for example, economists, engineers, and biologists and that must predict the effects of alternative decisions on all of the different environmental and social components to which humans are sensitive and which, therefore, deserve consideration. Third, the determination of [feasible alternatives] allows the objective elimination of inadequate alternatives because [they are] independent of the subjective perception of welfare ... [and] in essence describe the tradeoff between the various incommensurable objectives when every effort is made to achieve the best results in all respects; the attention of the authority that must make the final decision is thus directed toward genuine potential solutions because nonoptimal decisions have already been discarded.

It should be noted that a cost–benefit analysis does not elicit tradeoffs between incommensurable goods because it also gives a green light to projects ..., provided that the benefits that can be converted into a monetary scale exceed the costs.... Cost–benefit analysis, however, is not useful for eliciting the tradeoffs between two incommensurable goods, neither of which is monetary. For instance, there might be a conflict between the goals of preserving wildlife within a populated area and minimizing the risk that wild animals are vectors of dangerous diseases. A multicriteria analysis can describe this tradeoff, whereas a cost–benefit analysis cannot.

Another philosophical point concerning the issue of commensurability is the question of implicit pricing. Economists often argue that to make a decision is to put an implicit price on such intangibles as human life or aesthetics and, therefore, to reduce their value to a common scale (as pointed out also by Costanza et al.)....

Environmental Impact Assessment and Multiattribute Decision-Making

Because of the flaws of cost–benefit analysis, many countries have taken a different approach to decision-making through the use of environmental impact assessment legislation (e.g., the United States in 1970, with the signing of the National Environmental Policy Act, NEPA; France in 1976, with the act 76/629; the European Union in 1985, with the directive 85/337). Environmental impact assessment procedures, if properly carried out, represent a wiser approach than setting an a priori value of biodiversity and ecosystem services because these procedures explicitly recognize that each situation, and every regulatory decision, responds to different ethical, economic, political, historical, and other conditions and that the final decision must be reached by giving appropriate consideration to several different objectives. As Canter noted, all projects, plans, and policies that are expected to have a significant environmental impact would ideally be subject to environmental impact assessment.

The breadth of goals embraced by environmental impact assessment is much wider than that of cost–benefit analysis. Environmental impact assessment provides a conceptual framework and formal procedures for comparing different alternatives to a proposed project (including the possibilities of not development a site, employing different management rules, or using mitigation measures); for fostering interdisciplinary team formation to investigate all possible environmental, social, and economic consequences of a proposed activity; for enhancing administrative review procedures and coordination among the agencies involved in the process; for producing the necessary documentation to enhance transparency in the decision-making process and the possibility of reviewing all the objective and subjective steps that resulted in a given conclusion; for encouraging broad public participation and the input of different interest groups; and for including monitoring and feedback procedures. Classical multiattribute analysis can be used to rank different alternatives.... Ranking usually requires the use of value functions to transform environmental and other indicators (e.g., biological oxygen demand or animal density) to levels of satisfaction on a normalized scale, and the weighting of factors to combine value functions and to rank the alternatives. These weights explicitly reflect the relative importance of the different environmental, social, and economic compartments and indicators.

A wide range of software packages for decision support can assist experts in organizing the collected information; in documenting the various phases of EIA; in guiding the assignment of importance weights; in scaling, rating, and ranking alternatives; and in conducting sensitivity analysis for the overall decision-making process. This last step, of testing the robustness and consistency of multiattribute analysis results, is especially important because it shows how sensitive the final ranking is to small or large changes in the set of weights and value functions, which often reflect different and subjective perspectives. It is important to stress that, although the majority of environmental impact assessments have been conducted on specific projects, such as road construction or the location of chemical plants, there is no conceptual barrier to extending the procedure to evaluation of plans, programs, policies, and regulations. In fact, according to NEPA, the procedure is mandatory for any federal action with an important impact on the environment. The extension of environmental impact assessment to a level higher than a single project is termed "strategic environmental assessment" and has received considerable attention.

Conclusions

An impressive literature is available on environmental impact assessment and multiattribute analysis that documents the experience gained through 30 years of study and application. Nevertheless, these studies seem to be confined to the area of urban planning and are almost completely ignored by present-day economists as well as by many ecologists. Somewhere between the assignment of a zero value to biodiversity (the old-fashioned but still used practice, in which environmental impacts are viewed as externalities to be discarded from the balance sheet) and the assignment of an infinite value (as advocated by some radical environmentalists), lie more sensible methods to assign value to biodiversity than the price tag techniques suggested by the new wave of environmental economists. Rather than collapsing every measure of social and environmental value onto a monetary axis, environmental impact assessment and multiattribute analysis allow for explicit consideration of intangible nonmonetary values along with classical economic assessment, which, of course, remains important. It is, in fact, possible to assess ecosystem values and the ecological impact of human activity without using prices. Concepts such as Odum's eMergy [the available energy of one kind previously required to be used up directly and indirectly to make the product or service] and Rees' ecological footprint [the area of land and water required to support a defined economy or population at a specified standard of living], although perceived by some as naive, may aid both ecologists and economists in addressing this important need.

To summarize our viewpoint, economists should recognize that cost–benefit analysis is only part of the decision-making process and that it lies at the same level as other considerations. Ecologists should accept that monetary valuation of biodiversity and ecosystem services is possible (and even helpful) for part of its value, typically its use value. We contend that the realistic substitute for markets, when they fail, is a transparent decision-making process, not old-style cost–benefit analysis. The idea that, if one could get the price right, the best and most effective decisions at both the individual and public levels would automatically follow is, for many scientists, a sort of Panglossian obsession. In reality, there is no simple solution to complex problems. We fear that putting an a priori monetary value on biodiversity and ecosystem services will prevent humans from valuing the environment other than as a commodity to be exploited, thus reinvigoraing the old economic paradigm that assumes a perfect substitution between natural and human-made capital. As Rees wrote, "for all its theoretical attractiveness, ascribing money values to nature's services is only a partial solution to the present dilemma and, if relied on exclusively, may actually be counterproductive."

POSTSCRIPT

Should a Price Be Put on the Goods and Services Provided by the World's Ecosystems?

In "Can We Put a Price on Nature's Services?" *Report From the Institute for Philosophy and Public Policy* (Summer 1997), Mark Sagoff objects that trying to attach a price to ecosystem services is futile because it legitimizes the accepted cost-benefit approach and thereby undermines efforts to protect the environment from exploitation. The March 1998 issue of *Environment* contains environmental economics professor David Pearce's detailed critique of the 1997 Costanza et al. study. Pearce objects chiefly to the methodology, not the overall goal of attaching economic value to ecosystem services. Costanza et al. reply to Pearce's objections in the same issue. Pearce and Edward B. Barbier have published *Blueprint for a Sustainable Economy* (Earthscan, 2000), in which they discuss how governments worldwide are now applying economics to environmental policy.

Despite the controversy over the worth of assigning economic values to various aspects of nature, researchers continue the effort. Gretchen C. Daily et al., in "The Value of Nature and the Nature of Value," *Science* (July 21, 2000), discuss valuation as an essential step in all decision making and argue that efforts "to capture the value of ecosystem assets ... can lead to profoundly favorable effects." Daily and Katherine Ellison continue the theme in *The New Economy of Nature: The Quest to Make Conservation Profitable* (Island Press, 2002). In "What Price Biodiversity?" *Ecos* (January 2000), Steve Davidson describes an ambitious program funded by the Commonwealth Scientific and Industrial Research Organization (CSIRO) and the Myer Foundation that is aimed at developing principles and methods for objectively valuing "ecosystem services—the conditions and processes by which natural ecosystems sustain and fulfil human life—and which we too often take for granted." These include such services as flood and erosion control, purification of air and water, pest control, nutrient cycling, climate regulation, pollination, and waste disposal."

Stephen Farber, et al., "Linking Ecology and Economics for Ecosystem Management," *Bioscience* (February 2006), find "the valuation of ecosystem services ... necessary for the accurate assessment of the trade-offs involved in different management options."

On the Internet . . .

ECOLEX: A Gateway to Environmental Law

This site, sponsored by the United Nations and the World Conservation Union, is a comprehensive resource for environmental treaties, national legislation, and court decisions.

http://www.ecolex.org/ecolex/

EarthTrends

The World Resources Institute offers data on biodiversity, fisheries, agriculture, population, and a great deal more.

http://earthtrends.wri.org/

Environmental Defense

Environment Defense (once The Environmental Defense Fund) is dedicated to "protecting the environmental rights of all people, including future generations." Guided by science, Environmental Defense evaluates environmental problems and works "to create solutions that win lasting economic and social support because they are nonpartisan, cost-efficient, and fair."

http://www.environmentaldefense.org/home.cfm

Office of Environmental Justice

The U.S. Environmental Protection Agency (EPA) pursues environmental justice under the Office of Enforcement and Compliance Assurance as part of its "firm commitment to the issue of environmental justice and its integration into all programs, policies, and activities, consistent with existing environmental laws and their implementing regulations."

http://www.epa.gov/compliance/environmentaljustice/index.html

The Heritage Foundation

The Heritage Foundation is a think-tank whose mission is to formulate and promote conservative public policies on many issues, including environmental ones. It bases its work on the principles of free enterprise, limited government, individual freedom, and traditional American values.

http://www.heritage.org/

The Military Toxics Project

The mission of the Military Toxics Project is to unite activists, organizations, and communities in the struggle to clean up military pollution, safeguard the transportation of hazardous materials, and develop and implement preventative solutions to the toxic and radioactive pollution caused by military activities. MTP's mission is based on mutual respect and justice for all peoples, free from any form of discrimination or bias.

http://www.miltoxproj.org/

PART 2

Principles versus Politics

*I*n *many environmental issues, it is easy to tell what basic principles apply and therefore determine what is the right thing to do. Ecology is clear on the value of species to ecosystem health. Sociology and politics have agreed that racism is an evil to be avoided. Medicine makes no bones about the ill effects of pollution. But are the environmental problems so bad that we must act immediately? How much of the "right thing" should we do? How should we do it? Such questions arise in connection with every environmental issue, not just the four in this part of the book, but these four will serve to introduce the theme of principles versus politics.*

- Is Biodiversity Overprotected?

- Should Environmental Policy Attempt to Cure Environmental Racism?

- Will Pollution Rights Trading Effectively Control Environmental Problems?

- Should the Military Be Exempt from Environmental Regulations?

ISSUE 4

Is Biodiversity Overprotected?

YES: David N. Laband, from "Regulating Biodiversity: Tragedy in the Political Commons," *Ideas on Liberty* (September 2001)

NO: Howard Youth, from "Silenced Springs: Disappearing Birds," *Futurist* (July/August 2003)

ISSUE SUMMARY

YES: Professor of economics David N. Laband argues that the public demands excessive amounts of biodiversity largely because decision makers and voters do not have to bear the costs of producing it.

NO: Wildlife conservation researcher and writer Howard Youth argues that the actions needed to protect biodiversity not only have economic benefits but also are the same actions needed to ensure a sustainable future for humanity.

Extinction is normal. Indeed, 99.9 percent of all the species that have ever lived are extinct, according to some estimates. But the process is normally spread out over time, with the formation of new species by mutation and selection balancing the loss of old ones to disease, new predators, climate change, habitat loss, and other factors. Today, human activities are an important cause of species loss mostly because humans destroy or alter habitat but also because of hunting, the introduction of competitors, and the introduction of diseases. According to Martin Jenkins, "Prospects for Biodiversity," *Science* (November 14, 2003), some 350 (3.5 percent) of the world's bird species may vanish by 2050. Other categories of living things may suffer greater losses, leading to a "biologically impoverished" world. He states that the consequences for human life are "unforeseeable but probably catastrophic."

Awareness of the problem has been growing. In 1973 the United States adopted the Endangered Species Act to protect species that were so reduced in numbers or restricted in habitat that a single untoward event could wipe them out. The act barred construction projects that would further threaten endangered species. In one famous case, construction on the Tellico Dam on the Little Tennessee River in Loudon County, Tennessee, was halted because it threatened the snail darter, a small fish. Another case involved the spotted owl, which was threatened by logging in the Northwest. Those in favor of the dam or the timber industry felt that the endangered species was trivial compared to the human benefits at stake. Those in favor of the act argued that the loss of a single species might not matter to the world, but where one species went, others would follow. Protecting one species also protects others. However, the number of U.S. threatened and endangered species has not diminished. In fact, that number has increased more than sevenfold, from 174 in 1976 to 1311 in May 2006 (see http://ecos.fws.gov/tess_public/ TESSBoxscore for a current tally).

Internationally, species protection is covered by the Convention on International Trade in Endangered Species of Wild Fauna and Flora (CITES). This agreement has banned trade in such natural products as elephant ivory to prevent the continued slaughter of elephants. Less successfully, it has also tried to protect rhinoceroses (killed for their horns) and about 5,000 other species of animals and 25,000 species of plants, including some whole groups, such as primates, cetaceans (whales, dolphins, and porpoises), sea turtles, parrots, corals, cacti, and orchids. The 2006 IUCN Red List (http:// www.iucn.org/themes/ssc/redlist2006/redlist2006.htm) says that worldwide the number of species threatened with extinction is 16,119.

Is it enough to stop construction projects and ban trade? Some argue that efforts should be made to undo some of the damage that has already been done. For example, there is a movement to tear down dams that block the path of migratory fish, such as salmon and shad, so that they may once more breed and multiply (see http://www.americanrivers.org). For another example, urbanization and agricultural development have greatly altered the Everglades in Florida: rivers have been straightened, water has been diverted, and land has been drained for farms. This activity has resulted in low water tables, increased fire danger, smaller bird populations, and the decline of the Florida panther, among other negative impacts. Currently, the Army Corps of Engineers is planning to undo some of the changes to the Everglades' water flow in order to restore the natural habitat as much as possible. The Comprehensive Everglades Restoration Plan (CERP) was approved by Congress in 2000, will cost almost $8 billion, and will take more than 30 years. See Phyllis McIntosh, "Reviving the Everglades," *National Parks* (January 2002).

Is too much being done to protect the species with which we share the earth? In the following selection, David N. Laband argues that it is, largely because the people who set environmental policy do not need to pay for protection efforts themselves. Instead, the costs of protecting biodiversity are unfairly laid upon landowners. In the second selection, Howard Youth reviews the declining state of the world's birds, describes what is causing the decline and what can be done, and argues that actions that protect birds can have economic benefits. These actions are also needed to ensure a sustainable future for humanity.

YES

David N. Laband

Regulating Biodiversity: Tragedy in the Political Commons

Last summer, lightning struck and killed an enormous pine tree on one side of my backyard. At about the same time, voracious pine bark beetles girdled and killed an equally impressive pine tree on the other side. Now bereft of needles, these two arboreal giants pose a potential threat to my house: if they were to fall at just the right angle, the damage could be substantial. In the interest of safety, my wife wants to have the trees removed; for the sake of promoting biodiversity on my two-acre lot, I do not.

Our personal dilemma mirrors a much larger struggle that quietly threatens to destroy the rights of private timberland owners across the United States—the desire of urban dwellers to have their cake and eat it too. They demand houses made of wood, wood furniture, paper and paper products, and so on, while also demanding environmental amenities such as aesthetically pleasing landscape views, biodiversity, and animal habitat. At a personal level this can't be done. If the trees are removed, my wife has peace of mind, but the many animals that depend on dead pine trees for their existence, either directly or indirectly, will vanish. If the trees stay, we will be promoting the ecological diversity of our property, but my wife will worry about our house with every gust of wind. We can't have it both ways. Similarly, at a macro level, there is a tradeoff between production/consumption of timber and production/consumption of related environmental amenities.

The Role of Intensively Managed Forests

The problem of how to grow and harvest increasing amounts of timber while simultaneously producing a steadily increasing array and level of environmental amenities associated with forested land has resulted in an industry-wide discussion of how to simultaneously achieve both objectives. There is a growing appreciation within the forestry community for the prospect that intensively managed forests may yield increasing amounts of wood while minimizing the total acreage from which wood is harvested. This maximizes the amount of acreage available to meet other demands—such as agricultural production, animal habitat, and other environmental amenities associated with natural forests.

However, intensively managed forests have come under heavy fire from self-proclaimed environmentalists. In these so-called plantation forests, man, not nature, regenerates the trees, which accordingly grow in even-aged stands. Their well-being is affected by the application of herbicides and pesticides, as well as by occasional thinning and fire management. In contrast to naturally (re)generated timberland, plantation timberland has been described as an "ecological desert," with the stated or implied conclusion that the nature and extent of biological diversity associated with natural forests is both greater and therefore more desirable than that associated with plantation forests.[1]

The Threat to Private Landowners and Social Welfare

Such pejorative rhetoric is both misleading and counterproductive. The unfortunate but nonetheless compelling truth is that we can't have our cake and eat it too. We must make responsible choices about what to produce and how to produce it. A serious threat to private landowners develops when citizens living in urban areas demand that private owners of timberland (definitionally located in rural areas) produce environmental amenities such as aesthetically pleasing views, biodiversity, animal habitat, and the like, *provided the urbanites don't have to pay for it.*

Further, they seek to enforce their demands by using the political process to pass regulations that require landowners disproportionately to bear the cost of producing these environmental amenities. For example, Oregon law requires private timberland owners to replant within two years areas from which they cut trees. Other regulations forbid clearcutting of timberland. Federal regulations pertaining to endangered species are incredibly restrictive and intrusive with respect to an individual's property rights. The pursuit of environmental amenities that we are told are vital to some vaguely defined public interest through policies that impose virtually all the costs on relatively small numbers of private landowners generates what might be termed a "tragedy of the political commons."

Garrett Hardin introduced us to the tragedy of the commons.[2] Hardin developed a stylized example of a communal pasture open to all comers. There are no private property rights to the pasture, or rules, customs, or norms for shared use. In this setting, each shepherd, seeking to maximize the value of his holdings, keeps adding sheep to his flock as long as doing so adds an increment of gain. Further, the shepherds graze their sheep on the commons as long as the pasture provides any sustenance. Ignorant of the effects of their individual actions on the others, the shepherds collectively (and innocently) destroy the pasture. As Hardin concludes: "Therein is the tragedy. Each man is locked into a system that compels him to increase his herd without limit—in a world that is limited. Ruin is the destination toward which all men rush, each pursuing his own best interest in a society that believes in freedom of the commons."

Man's exploitation of the political commons is analogous to his exploitation of natural-resource commons. Our majority-rule voting process, which permits a majority of citizens to impose differential costs on the minority, encourages overprotection of endangered species, and overproduction of biodiversity, animal habitat, and landscape views. This occurs because each individual who bears a negligible portion of the costs of providing environmental amenities has a private incentive to keep demanding additional environmental protections as long as there is *any* perceived marginal benefit. As with the overgrazed pasture, the result of overprotecting Bambi is, as has become apparent all over the eastern United States, disastrous. Moreover, and not surprisingly, we are starting to hear real concern voiced about the recent proliferation of other animal species such as black bears, mountain lions, and coyotes. We are creating social tragedies that result from the political commons.

The tragedy is compounded by the incentives generated for private landowners by the heavy hand of command-and-control policies. When government abrogates property rights without compensation, landowners have strong incentives to mitigate their expected losses. They can do so by changing their land use from timber production to housing or commercial development. There is no incentive to promote habitat for endangered species; doing so means only that use of one's land will be seriously compromised by the highly restrictive provisions of the Endangered Species Act. Instead, a landowner who finds a member of an endangered species on his property has a well-understood incentive to "shoot, shovel, and shut up." Such behaviors are not likely to further environmental objectives.

Other People's Costs

It is relatively easy to demonstrate that because private timberland owners bear the cost of producing biodiversity, nonland-owners demand excessive amounts of it. The first point to be made in this regard is that urbanites do not in fact place a high value on biodiversity. One need look no further than the readily observable behavior of urbanites for proof of this claim. Urbanites have the ability and prerogative to produce biodiversity on their own residential property. That is, they could let their residential lots grow wild with natural flora and fauna. This would, without question, promote ecological diversity. In practice, virtually no residential property owners, living anywhere in the United States, do this. Instead, they invest (implicitly through their time and explicitly by purchase) hundreds, if not thousands, of dollars annually in the care and maintenance of their lawns and grounds in a decidedly unnatural state. Like owners of intensively managed timberland, owners of residential property chemically treat and harvest the growth on their property. In so doing, they create a landscape with relatively little floral or faunal diversity. What this behavior reveals, of course, is that urban dwellers place a higher value on having their own aesthetically pleasing ecological deserts than on personally promoting local biodiversity, even when the latter would save them hundreds, perhaps thousands, of dollars each year. The clear implication is that urbanites simply do not attach much importance to biodiversity.

This leads directly to a second point: notwithstanding that biodiversity is of little importance to them personally, urbanites may favor local, state, and federal statutes that ostensibly enhance biodiversity, provided such statutes impose the cost burden on rural landowners. The feel-good benefit of such regulation may be small, but with no personal costs to worry about, urbanites can be convinced to vote for them. However, if there were even a moderate cost to urban dwellers, we can be reasonably certain that restrictive regulations would not be passed. This explains why, for example, Oregon's replanting regulations are not imposed on owners of residential properties who cut down trees.

Earth's limited resources cannot provide all things to all people simultaneously. For that matter, the earth cannot provide all things just to self-proclaimed environmentalists. Consequently, responsible choices about the use of resources must be made. It is irresponsible to enact environmental policies that impose costs disproportionately on private timberland owners. Such policies lead to overproduction of environmental protection because urban voters who place little value on environmental amenities support regulations that impose little or no cost on themselves personally. Further, these policies create incentives for private timberland owners to minimize, not maximize, their production of environmental amenities. This problem of incompatible incentives makes it less likely that public policy will actually attain its stated objectives.

Notes

1. National Audubon Society, `www.audubon.org/campaign/fh/chipmills.htm`, no date.
2. Garrett Hardin, "The Tragedy of the Commons," *Science* (162), 1968, pp. 1243–48; see `www.dieoff.org/page95.htm`.

Howard Youth  **NO**

Silenced Spring: Disappearing Birds

Almost 1,200 bird species—about 12% of those remaining in the world—may face extinction within the next century. Most struggle against a deadly mixture of threats, including habitat loss, human disasters, and disease. Although some bird extinctions now seem imminent, many can still be avoided with deep commitment to conservation as an integral part of a sustainable development strategy. Such a commitment would be in humanity's best interest.

What ornithologists have already tallied is alarming. Human-related factors threaten 99% of the species in greatest danger. Bird extinctions are increasing, already topping 50 times the natural rate of loss. At least 128 species vanished over the last 500 years—103 of which became extinct since 1800.

If we focus solely on the prospects of extinction, we partly miss the point. From an ecological perspective, extinction is but the last stage in a spiraling degeneration that sends a thriving species slipping toward oblivion. Species stop functioning as critical components of their ecosystems well before they completely disappear.

Although birds are probably the best-studied animal class, a great deal remains to be learned about them—from their life histories to their vulnerability to environmental change. In the tropics, where both avian diversity and habitat loss are greatest, experts just do not know the scope of bird declines because many areas remain poorly surveyed, if at all. Species may vanish even before scientists can classify them or study their behavior, let alone determine their ecological importance.

Several new bird species are described every year. One of this century's earliest was an owl discovered in Sri Lanka in 2001, the first new bird species found there in 132 years. These scarce and newly described birds sit at a crossroads, as does humanity. One path leads toward continued biodiversity and sustainability. The other heads toward extinction.

Habitat Loss: The Greatest Threat

Many problems faced by birds and other wildlife stem from how we handle our real estate. The human population explosion from 1.6 billion to 6 billion during the last century fueled widespread habitat loss that chiseled once-extensive wilderness into habitat islands. Today, loss or damage to species' living spaces poses by far the greatest threat to birds and biodiversity in general.

Originally published in the November 1999 issue of *The Futurist*. Used with permission from the World Future Society, 7910 Woodmont Avenue, Suite 450, Bethesda, MD 20814. Telephone: 301/656-8274; Fax: 301/951-0394; http://www.wfs.org. Copyright © 1999.

Forests. Timber operations, farms, pastures, and settlements have already claimed almost half the world's forests. Between the 1960s and 1990s, about 4.5 million square kilometers (1,100 acres), or 20% of the world's tropical forest cover, were cut or burned. Habitat loss jeopardizes 1,008 (85%) of the world's most-threatened bird species.

Foresters herald regrowth of temperate forests as an environmental success story, and in recent decades, substantial reforestation did take place in the eastern United States, China, and Europe. Forest management profoundly affects diversity and natural balances, however, and satellite images of tree cover do not tell us how much regrown habitat is indeed quality habitat. In the southeastern United States, foresters replace the clearcut area with rows of same-age, same-species pine saplings. For many native animals and plants, simplified plantation monocultures are no substitute for more-complex natural forests.

Grasslands. Once cloaking more than a third of the earth's surface, grasslands sustain bird populations found nowhere else, but they also host almost one-sixth of the human population. Few large, undisturbed grassland areas remain. In North America, where less than 4% remains, bird populations continue to shrivel. According to the U.S. Geological Survey, between 1966 and 1998, 15 of 28 grassland bird species steadily declined. The last strongholds for many grassland species in Europe face severe pressure from increased irrigation and modernization programs subsidized by the European Union's common agricultural policy.

Wetlands. Draining, filling, and conversion to farmlands or cities destroyed an estimated half of the world's wetlands during the twentieth century. Estimates within individual countries are often much higher. Spain, for instance, has lost an estimated 60%–70% of its wetland area since the 1940s. Even wilderness areas such as the Everglades and Spain's Donana National Park have not been spared.

Outside protected areas, changes have been far more dramatic. Over the last 70 years, Armenia's Lake Sevan suffered dramatic lowering due to water diversion, and Lake Gill was drained entirely. With their vital wetlands destroyed, at least 31 locally breeding bird species abandoned the lakes.

Mountains. Mountains often hold their habitats longer against human endeavors. In many countries, including Jamaica and Mexico, much of the remaining habitat is found only in prohibitively steep terrain. Once targeted, though, mountain habitats and wildlife are extremely vulnerable. Altitude and moisture levels dictate vegetation and wildlife occurrence there, creating narrow ribbons of habitat. Humans and migrant birds alike particularly favor temperate and rainsoaked middle elevations. In the Andes, Himalayas, and Central American highlands, among other areas, middle-elevation forests are highly degraded, creating severe erosion problems, fouling watersheds vital to human populations, and providing less area for wintering and resident birds.

Exotic Animals and Plants

Even in otherwise undisturbed wildlife habitats, a new order is taking hold as exotic (nonnative) species are introduced. Today, exotics threaten birds and their ecosystems in myriad ways, constituting the second most intense threat to birds worldwide, after habitat loss and degradation. Once introduced, some exotic predators became all the more lethal on islands, where endemic species evolved with few or no defenses against such hunters. To date, 93% of bird extinctions have occurred on islands, where extremely vulnerable endemic species succumbed to habitat loss, hunting, and, in most cases, exotic species that unsettle unique island ecological balances.

The yellow crazy ant, for example, a frenetic, fast-multiplying insect, is marching across Australia's Christmas Island following its introduction there during the twentieth century. As they spread across the island, crazy ants will likely kill young native birds, including those of two critically endangered species, the Christmas Island hawk owl and Abbott's booby, a seabird that nests nowhere else but in the island's forest canopy. Both species are expected to decline 80% due to ant invasion.

Exotic birds compete with native birds both genetically and directly. People around the world have dumped familiar domesticated mallard ducks into ponds and other wetlands, where they vigorously interbreed with closely related species. Such hybridization affects South Africa's yellow-billed ducks, endangered Hawaiian ducks, American black ducks, and mottled ducks.

Introduced plants change birds' habitats until they are eventually uninhabitable. Exotic plant species have gone wild in many parts of the world at the expense of birds and other wildlife. Cheatgrass brought over from Eurasia has spread far and wide since its introduction to North America in the late 1800s. As it overtakes sagebrush and bunchgrass habitats, cheatgrass fuels the decline of birds that nest among sagebrush shrubs and depend on them for food. Cheatgrass now covers an area of North America larger than the area of Germany. Perhaps 5% of 283 million hectares (700 million acres) of public land is "seriously infested" in the United States, where at least 400 exotic plant species have gone out of control.

Dealing with exotic introductions often requires active management, including hunting, poisoning, herbicide spraying, and in some cases introducing natural predators of the out-of-control exotic—activities that can also potentially disturb or harm native birds and other wildlife. In the United States alone, estimates of the annual cost of damage caused by exotics and the measures to control them reach an estimated $137 billion.

Bullets, Cages, Chemicals, and Climate

Other threats come from human activities, such as unregulated hunting and trapping. On the island of Malta, hunters take aim at island-hopping birds during migration. Officially protected birds, such as swallows, bee-eaters, harriers, and herons, fall to Maltese shooters in staggering numbers. Most of this hunting is just target practice and hurts already declining European nesting bird populations.

The wild bird pet trade has also been devastating. Parrots have long been loved by people the world over for their colorful plumages, potential affection toward their owners, and, in many species, adept "talking" abilities. Almost a third of the world's 330 parrot species are threatened with extinction due to habitat loss and collecting pressures, part of a burgeoning illegal wildlife trade valued at billions of dollars a year.

Other human activities have had indirect but nonetheless harmful effects on birds and their habitats, such as oil spills, factory effluents, and lead poisoning, as well as tall buildings, towers, and power lines interfering with birds' migratory paths or daily movements.

Various solutions to the problems posed by human attention to resident and migrant birds have met with some success. Over the last decade, for example, protection measures helped reduce the international trade in wild parrots. But protection laws in many parrot-rich countries often go unheeded, and parrot poaching and smuggling remain widespread.

Since Canadian and U.S. efforts to stem industrial contaminants such as PCBs [polychlorinated biphenyls] began in the late 1970s, herring gull and double-crested cormorant populations have grown, and the bald eagle returned to the Great Lakes region. After U.S. law banned the pesticide DDT in 1972, the country's peregrine falcon, bald eagle, osprey, and brown pelican populations rebounded. Similar rebounds occurred in Britain in such raptors as sparrowhawks after a ban was initiated there. In 2001, 120 countries signed a pesticide treaty that included phasing out DDT except for limited use in controlling malaria.

Global warming is another threat to birds. Global warming is hastened by many of the same activities that destroy habitat: forest clearing, rampant forest fires, road building, and urban expansion. Scientists estimate that the earth's climate warmed 0.3–0.6 [degrees]C over the past century, and that temperature change will continue and possibly intensify Already, ecological changes seem to be under way in ecosystems around the world. Temperate fauna and flora seem to be changing their schedules. Over the past few decades, scientists have documented earlier flower blooming, butterfly emergence, and frog calling—and earlier bird migration and egg-laying dates in Europe and North America.

Global climate change will also likely increase the frequency and severity of weather anomalies that pound bird populations. El Niño events, for example, could finish off rare, localized, and declining species such as the Galapagos penguin, which has evolved and thrived on an equatorial archipelago flushed by cool, fish-rich currents. In addition, intensified and more-frequent droughts and fires could accompany El Niño and other cycles, both in the tropics and as far-north as Canada's boreal forests.

Biological Hot Spots

Decades of fieldwork, computer modeling, and satellite imagery analysis have pinpointed "hot spots"—areas that harbor disproportionately high diversity and high numbers of imperiled bird species. BirdLife International, working with organizations, agencies, and biologists around the world, identified 7,000 important bird areas in 140 countries—critical bird breeding and migration spots—and 218 endemic bird areas, which are places with the highest numbers of restricted-range and endemic species. While not conferring formal protection, these designations offer a framework from which to set international, national, and local protection priorities.

Linking these bird hot spots and other key habitats and striking a balance between developed and undeveloped areas will be key in saving birds in our ever more crowded world. Over the past 20 years, the emergence of the multidisciplinary field of conservation biology has changed the focus of biodiversity protection efforts from the park to the landscape level, incorporating not just protected areas but also adjacent lands and water resources and the people who inhabit and use them. This landscape focus increasingly brings conservation goals alongside business plans.

The approach is not only progressive but also pragmatic, since most of the world's remaining wild areas remain in private hands or are managed by no one at all. All told, between 6.4% and 8.8% of the earth's land area falls under some category of formal habitat protection. These areas are sprinkled across the globe, many are quite small, and management varies from one place to the next. In general, the largest and most biologically diverse parks are the least well staffed and protected, as they are in some of the world's poorest regions. An example is Peru's Manu National Park, where up to 1,000 species—about 10% of the world's bird species—have been recorded. Local support is critical for these areas—and for the buffer zones and green corridors needed to protect them adequately.

Most of the world remains open to alteration, and people who are hungry and lack alternatives cannot embrace or focus on efforts to protect natural resources unless they clearly benefit in the bargain. Boosting economic prospects and educational opportunities will allow local people to focus on saving birds and other natural resources for the future.

Conservation Programs That Work

The growing awareness that biodiversity protections can be combined with moneymaking ventures seems to be bringing enterprise and environmentalism together. Shade-grown coffee is increasingly popular, for instance. This crop is grown the traditional way, beneath a tropical forest canopy that also shelters resident and migratory birds. In addition, cultivating various fruits, cork, cacao, and other crops supports many bird species. Farm operations that minimize use of harmful pesticides provide more diverse food sources and safer habitats for birds.

Some successful incentive programs pay farmers to set aside land for wildlife, water, and soil conservation purposes. From 2002 to 2007, for example, about 15.9 million hectares will be enrolled in the U.S. Department of Agriculture's Conservation Reserve Program. Hundreds of thousands of farmers enroll land for 10 to 15 years—taking it out of production, planting grasses and trees, restoring wetlands, or grazing or harvesting hay in a way compatible with wildlife and erosion control. Although some of the grasses used in this program are invasive exotics, since its inception in 1985, the program has helped many declining grassland birds regain ground.

In the Netherlands, a program set up by Dutch biologists offers dairy farmers payments to protect and encourage nesting birds as a farm product. An experiment conducted between 1993 and 1996 found that it was cheaper to pay farmers to monitor and manage breeding wild birds as if they were a crop rather than compensate them for restricting farming practices for the sake of bird protection. The project resulted in increased breeding success of meadow-nesting lapwings, godwits, ruffs, and redshanks, while not interrupting the dairy business. By 2002, about 36,000 hectares of Dutch farmland were enrolled in this program.

Ecotourism, which first arose in Costa Rica and Kenya in the early 1980s, is loosely defined as nature-oriented travel that does not harm the environment and that benefits both the traveler and the local host community. Most nations now court ecotourists. Although nature-oriented tourism is not always light on the environment, this industry shows signs of improving and is often an economically viable alternative to resource extraction.

Success in Florida

The increasingly crowded state of Florida provides a compelling example of how local, state, federal, and private concerns prioritized conservation while struggling with relentless development and population growth. Florida is one of the most biologically diverse and environmentally challenged U.S. states. Fortunately, since the 1980s, careful study and planning have been hallmarks of growing conservation efforts there.

One 2000 study by three University of Florida biologists plotted out an interconnected web of wildlife habitat called the Florida Ecological Network, which embraces the state's most-diverse remaining habitats and wildlife. More than half of the network is already under protected status. With the most-critical areas mapped out and many of them targeted, planners should be better able to steer and concentrate development into the many areas outside the park and corridor network and incorporate protected lands into landscapes that combine compatible forms of agriculture.

Another state study plotted private lands needed to ensure a secure future for the most threatened wildlife, including Florida's 117 rare and endangered listed animals. The researchers deduced that a specifically targeted 33% of the state's land area would need protection to lower significantly the chances of rare species extinctions. They included the 20% of the state that already falls under protection. Florida has identified at least 6% more land for future acquisition or protection through easements.

As prime wild real estate becomes more expensive and hard to find, conservationists have stepped up efforts to secure targeted Florida lands. In 2001, The Nature Conservancy announced that it had helped protect its millionth Florida acre. This organization secures funding to buy acreage that is later turned over to government protection or kept as private preserves.

Meanwhile, Florida's government runs a land-buying program called Florida Forever, an aggressive 10-year effort that targets properties most in need of conservation. Under this program, the state spends about $105 million each year to acquire critical conservation lands, protect watersheds, restore polluted or degraded areas, and provide public recreation. Some properties are held in conservation easements, under which property owners receive state payments or tax incentives in return for managing property as wildlife habitat.

A good part of Florida's economy derives from tourism, and more than 40 million people flood into the state each year on vacation. Meanwhile, almost 20% of the state's population is over age 65, and many are retired and are frequent visitors to state tourist attractions. Combining its huge tourism infrastructure and highway system with a newly honed focus on wild places, the state identified nature watching as vital tourism with The Great Florida Birding Trail. Slated for completion in 2005, this sign-marked driving route of some 3,000 kilometers (1,864 miles) winds its way past most of the state's bird hot spots, including county parks, ranches, state forests, private preserves, an alligator farm or two, and federal lands.

Bird Watching

Birding trails follow decades of growing interest in birding, a hobby that turns most of its participants into supporters of conservation efforts that protect birds and other wildlife. One survey noted that more than 66 million Americans aged 16 or older observed, fed, or photographed wildlife (particularly birds) during the year, spending an estimated $40 billion on equipment and travel expenses. Another report estimated that at least 70.4 million U.S. residents 16 or older go outdoors to watch birds sometime during the year, and that these numbers more than doubled between 1983 and 2001. Surveys conducted in Britain yielded similar results.

Economic impact aside, the burgeoning ranks of birders also provide a powerful infusion of eyes and ears that assist scientists in monitoring bird and other wildlife populations around the world. For example, more than 50,000 volunteers a year now participate in the 103-year-old National Audubon Society Christmas Bird Count, the largest and probably longest-running bird census. These knowledgeable birders identify and tally birds wintering at more than 1,800 local census sites throughout North America and in an increasing number of other regions as well. More than a century's worth of wintering bird data gives ornithologists a telling picture of bird abundance and distribution. As bird surveyors note, many bird species are in decline, and prospects remain bleak for many of the world's most-threatened bird species. Governmental and private efforts to save some, however, are bearing fruit, setting good examples for future endeavors elsewhere:

- The Seychelles magpie-robin is rebounding after being reintroduced to predator-free islands and after reductions in pesticide use in its habitat.
- The whooping crane has been a hallmark of conservation efforts between Canada and the United States—up to about 200 birds after a

low of 14 adults in 1938. A nonmigratory population was reintroduced to Florida, providing an extra hedge against extinction.

- In 1999, the peregrine falcon was lifted from the U.S. endangered species list following the ban on DDT in the 1970s and decades of protection, captive breeding, and reintroduction programs.
- Protection combined with apparent adaptability to changed landscapes enabled red kites to return to former haunts in western Europe.
- Four threatened parrot species on three Caribbean islands are inching back from the brink thanks to government and other protections, public-education campaigns, and some captive-breeding efforts.
- On the island of Mauritius, habitat protection and exotic plant and animal eradication efforts benefit now-growing populations of the endemic Mauritius cuckoo-shrike and Mauritius kestrel, a species that also benefited from captive breeding and release programs until the early 1990s.
- The bright blue Lear's macaw, a rare parrot of northeastern Brazil, appears to be steadily rising in number, from about 170 in the late 1990s to about 250. A local landowner, Brazilian conservation organizations, the World Parrot Trust, and funding from the Disney Conservation Initiative help conservationists plant essential foods for the birds, monitor the population, and protect nest sites.

The actions needed to ensure a secure future for birds are the same ones needed to achieve a sustainable human future: preserving and restoring ecosystems, cleaning up polluted areas, reducing use of harmful pesticides, reversing global climate change, restoring ecological balances, and controlling exotic species. Wildlife conservation must be compatible with rural, suburban, and urban planning efforts that improve prospects for the world's poor while making our cities and industries safer for all living beings.

POSTSCRIPT

Is Biodiversity Overprotected?

The question of whether biodiversity is overprotected cannot really be answered without first defining how much protection is too much. Timothy H. Tear, et al., address this problem in "How Much Is Enough? The Recurrent Problem of Setting Measurable Objectives in Conservation," *Bioscience* (October 2005), and conclude that the current state of our knowledge is inadequate; a great deal of research is needed.

There is debate over whether the best way to protect biodiversity is to protect individual endangered species or to protect habitat, which must also shield all the species that share that habitat. In 1998 a bill that would have substituted habitat conservation plans for the protection of individual species was proposed in the U.S. Senate but never came to a vote. It was opposed by environmentalists and House representatives, who argued that habitat conservation does not adequately protect endangered species, and by conservatives who wanted the bill to compensate property owners who lost property value or income opportunities because of habitat protection restrictions.

Mark L. Shaffer, J. Michael Scott, and Frank Casey, in "Noah's Options: Initial Cost Estimates of a National System of Habitat Conservation Areas in the United States," *Bioscience* (May 2002), state, "Solving the habitat portion of the endangered species and biodiversity conservation problems is neither trivial nor overwhelming. A national system of habitat conservation areas in the United States could be secured for an initial annual investment between $5 billion and $8 billion, sustained over 30 years, or roughly one-fourth to one-third the cost of maintaining our national highway system over the same period." The high cost of adequate conservation efforts has prompted Michael Jenkins, Sara J. Scherr, and Mira Inbar, "Markets for Biodiversity Services," *Environment* (July/August 2004) to consider market-based funding. They conclude that "Conservation of biodiversity and of the services biodiversity provides to humans and to the ecological health of the planet requires financing on a scale many times larger than is feasible from public and philanthropic sources. It is essential to find new mechanisms by which resource owners and managers can realize the economic values created by good stewardship of biodiversity." Stuart L. Pimm and Clinton Jenkins, "Sustaining the Variety of Life," *Scientific American* (September 2005), note that many conservation techniques are inexpensive and have economic benefits.

The sixth International Conference on Biological Diversity was held in The Hague in April 2002. Sponsored by the United Nations Environment Programme, it emphasized efforts to protect genetic resources and addressed the hazards posed to biodiversity by exotic (nonnative) species, a topic that is itself controversial. Peter Warshall, in "Green Nazis?" *Whole Earth* (March

2001), says that conservatives are attacking attempts to control or eradicate nonnative species as motivated by a kind of environmental racism.

Although the Endangered Species Act was due for reauthorization in 1993, the necessary legislation has not yet been enacted. Useful articles in favor of reauthorization include John Volkman's "Making Room in the Ark," *Environment* (May 1992) and T. H. Watkins's "What's Wrong With the Endangered Species Act?" *Audubon* (January/February 1996). Bonnie B. Burgess discusses the history and future of the Endangered Species Act in *Fate of the Wild: The Endangered Species Act and the Future of Biodiversity* (University of Georgia Press, 2001). Burgess feels that the act is itself endangered because of attacks from conservatives and obstacles erected by the government. According to Margaret Kritz, "Species Act in Their Sights," *National Journal* (December 20, 2003), the Bush administration, conservative Republicans, and business groups want to change the Endangered Species Act to give greater weight to the economic impact of species protections and less to protecting species. In September 2005, the U.S. House of Representatives passed the Threatened and Endangered Species Recovery Act despite charges that it would weaken protections. See "House Passes Bill Reforming Endangered Species Act," *Human Events* (October 31, 2005).

ISSUE 5

Should Environmental Policy Attempt to Cure Environmental Racism?

YES: **Julian Agyeman**, from "Where Justice and Sustainability Meet," *Environment* (July/August 2005)

NO: **David Friedman**, from "The 'Environmental Racism' Hoax," *The American Enterprise* (November/December 1998)

ISSUE SUMMARY

YES: Professor Julian Agyeman argues that although there is much debate over whether sustainable development means addressing environmental issues or environmental justice, equity, human rights, and poverty reduction, the two can and must be integrated.

NO: Writer and social analyst David Friedman denies the existence of environmental racism. He argues that the environmental justice movement is a government-sanctioned political ploy that will hurt urban minorities by driving away industrial jobs.

Archeologists delight in our forebears' habit of dumping their trash behind the house or barn. Today, however, most people try to arrange for their junk to be disposed of as far away from home as possible. Landfills, junkyards, recycling centers, and other operations with large negative environmental impacts tend to be sited in low-income and minority areas. Boston's Great Molasses Flood (see http://www.mv.com/ipusers/arcade/molasses.htm and Stephen Puleo, *Dark Tide: The Great Boston Molasses Flood of 1919* [Beacon Press, 2003]) happened when a two-million-gallon molasses storage tank burst; the tank had been built in a neighborhood crowded with immigrant laborers who lacked the political influence to say "Not in My Back Yard." Was such a location mere coincidence? Or was it deliberate?

Does the paucity of poor people and minorities in the environmental movement indicate that these people do not really care? (See Robert Emmett Jones, "Blacks Just Don't Care: Unmasking Popular Stereotypes About Concern for the Environment Among African-Americans," *International Journal of Public Administration* [vol. 25, nos. 2 & 3, 2002]).

The environmental movement has, in fact, been charged with having been created to serve the interests of white middle- and upper-income people. Native

Americans, blacks, Hispanics, and poor whites were not well represented among early environmental activists. It has been suggested that the reason for this is that these people were more concerned with more basic needs, such as jobs, food, health, and safety. However, the situation has been changing. In 1982, for example, in Warren County, North Carolina, poor black and Native American communities held demonstrations in protest of a poorly planned PCB (polychlorinated biphenyl) disposal site. This incident kicked off the environmental justice movement, which has since grown to include numerous local, regional, national, and international groups. The movement's target is systematic discrimination in the setting of environmental goals and in the siting of polluting industries and waste disposal facilities—also known as environmental racism. The global reach of the problem is discussed by Jan Marie Fritz in "Searching for Environmental Justice: National Stories, Global Possibilities," *Social Justice* (Fall 1999).

In 1990 the Environmental Protection Agency (EPA) published "Environmental Equity: Reducing Risks for All Communities," a report that acknowledged the need to pay attention to many of the concerns raised by environmental justice activists. At the 1992 United Nations Earth Summit in Rio de Janeiro, a set of "Principles of Environmental Justice" was widely discussed. In 1993 the EPA opened an Office of Environmental Equity (now the Office of Environmental Justice) with plans for cleaning up sites in several poor communities. In February 1994 President Bill Clinton made environmental justice a national priority with an executive order. Since then, many complaints of environmental discrimination have been filed with the EPA under Title VI of the federal Civil Rights Act of 1964; and in March 1998 the EPA issued guidelines for investigating those complaints. However, in April 2001 the U.S. Supreme Court ruled that individuals cannot sue states by charging that federally funded policies unintentionally violate the Civil Rights Act of 1964. The decision is expected to limit environmental justice lawsuits (see Franz Neil, "Supreme Court Ruling May Hurt Environmental Justice Claims," *Chemical Week* [May 2, 2001]). At present, "the sole relief available for victims of environmental civil rights violations is through a private action against a state if the community can prove intentional discrimination. To date, no such action has been successful." See the Environmental Justice Coalition's home page at `http:// groups.msn.com/environmentaljusticecoalition/homepage.msnw`.

Critics of the environmental justice movement contend that inequities in the siting of sources of pollution are the natural consequence of market forces that make poor neighborhoods (whether occupied by whites or minorities) the economically logical choice for locating such facilities. Critics also charge that such facilities depress property values and drive more prosperous people away while attracting a poorer population. In the following selections, Julian Agyeman, assistant professor of Urban and Environmental Policy and Planning at Tufts University, describes the history of the environmental justice movement and argues that improving justice, equity, and human rights, and reducing poverty, are essential components of sustainable development. David Friedman, on the other hand, asserts that the environmental justice movement is a politically inspired movement that is unsupported by scientific facts. He calls environmental racism a hoax and argues that attacking it will harm the urban poor by denying them the industrial jobs they need.

YES

Julian Agyeman

Where Justice and Sustainability Meet

Environmentalism and its twenty-first century variant, environmental sustainability, are going through hard times. Berated from within and without, from left and right, in developed and developing countries alike, the movement is struggling to come to terms with its increasingly vociferous critics.

What is at stake here is no less than the power to frame the broad and emerging sustainable development agenda,[2] and a key question emerges from this debate: Should organizations and institutions focus more on "green" or environmental strategies, or should they focus on environmental justice, equity, human rights, and poverty reduction—in short, on human security?[3]

Traditionally, the relationship between mainstream environmental sustainability groups and environmental justice groups in the United States and elsewhere has been uneasy. What might at first glance seem like an obvious case for partnership or coalition in the greater pursuit of more just and sustainable communities is fraught with ideological and internecine concerns. At the global level, too, the international and millennium development goals of achieving "sustainable livelihoods" are clouded by this same uneasy relationship. In this case, it is being played out between institutions and organizations representing the "green" environmental agenda of developed countries and the "brown" poverty reduction and human rights, or more recently, human security agenda typical of developing countries.

Taken as a whole, the goals of developing sustainable communities or sustainable livelihoods are important ends upon which most would agree. The dispute is in the means of achieving them. Environmental sustainability advocates want a focus on more and better environmental stewardship and protection as well as policy-reform strategies such as debt-for-nature swaps. Frequently, the environmental justice, poverty reduction, human rights, and human security advocates are their adversaries, focusing instead on inequities, racism, and classism as well as paradigm-shifting strategies such as per-capita resource redistribution. . . .

Perhaps, however, instead of looking at environmental sustainability and environmental justice as polarities and running the risk of essentializing their

From *Environment*, July/August 2005, pp. 11-18, 20. Reprinted with permission of the Helen Dwight Reid Educational Foundation. Published by Heldref Publications, 1319 Eighteenth St., NW, Washington, DC 20036-1802. Copyright © (2005).

differences, it is helpful to examine organizations that are exploring common ground or even a common agenda between justice and sustainability. The practices of one such organization, Alternatives for Community and Environment (ACE), offer some clues as to how this might happen. Based in Boston's diverse Roxbury section, ACE employs the discourse of "just sustainability"—with its overlapping threads of environmental justice and environmental sustainability—to leverage influence and thereby bring positive, demonstrable benefits to this low-income and minority community.

However, to give a full picture of the bridge that needs to be built to link environmental justice and environmental sustainability, it is helpful to look at their histories in greater detail.

Background

Environmental justice and sustainability are two concepts that have evolved over the past two decades to provide new, exciting, and challenging directions for public policy and planning in the United States and throughout the world. Both concepts are highly contested and have been characterized as either being too broad and difficult to define or lacking precision in their policy recommendations. Nevertheless, both have tremendous potential to effect long-lasting change on a variety of levels, from local to global.

The Environmental Justice Movement

Environmental justice can best be understood as a grassroots community reaction to external threats to a given community in its broadest sense.[4] These threats have been shown in the United States and a growing number of other countries, such as England, Scotland, and South Africa, and regions, such as Central and Eastern Europe, to affect people of color and low-income neighborhoods disproportionately.[5] In the United States, the landmark 1987 United Church of Christ study *Toxic Wastes and Race in the United States* was the first large-scale research to show this.[6] The report also coined the term that became a rallying cry: environmental racism. The report's findings were confirmed by later research, including a meta-analysis of 64 U.S. studies, which "provide an overwhelming body of empirical evidence that people of color and lower incomes face disproportionate environmental impacts."[7]

Environmental justice organizations emerged from grassroots activism in and around the U.S. civil rights movement. They expanded the dominant traditional environmental discourse, which was based around environmental stewardship, to include social justice and equity considerations. In doing this, these organizations redefined what are considered environmental issues so that the dominant wilderness and natural resource focus now includes urban disinvestment, racism, homes, jobs, neighborhoods, and communities. This redefinition, combined with the conclusion of a 1992 *National Law Journal* article that there is unequal protection and enforcement of environmental law by the U.S. Environmental Protection Agency (EPA),[8] has ensured a full-fledged environmental justice movement made up of tenant associations, religious groups, civil rights groups, farm workers, professional nonprofits, university centers and academics, and labor unions, among others. This movement occurs across the country, driven by grassroots activism of African-American, Latino, Asian and Pacific American, Native American, and poor white communities concerned about the intragenerational inequities embedded deep within our risk society, such as the siting of waste facilities and transfer, storage, and disposal facilities; lead contamination; pesticides; water and air pollution; workplace safety; and public transportation. More recently, the environmental justice movement has begun targeting such issues as sprawl and smart growth, sustainability, and climate change.[9]

At the federal level, EPA oversees the Office of Environmental Justice, leads the Interagency Working Group on Environmental Justice (IWG) (which comprises 11 federal agencies and several White House offices), and is advised by the National Environmental Justice Advisory Council (NEJAC). In 1994, President Bill Clinton's Executive Order 12898 on environmental justice reinforced the 1964 Civil Rights Act Title VI (which prohibits discriminatory practices in programs receiving federal funds) and directed all federal agencies to begin developing policies to reduce environmental inequity.

Sustainable Development

Sustainability and sustainable development, often prefaced by the word "environmental" or "environmentally," rest on the "futurity principle" and intergenerational equity. As defined by the World Commission on Environment and Development, "[s]ustainable development is development that meets the needs of the present without compromising the ability of future generations to meet their own needs."[10] Such concepts emerged in large part from top-down international processes and committees, governmental structures, think tanks, and international nongovernmental organization (NGO) networks. Two major conferences boosted sustainability and sustainable development to their current primary status in policy-circle rhetoric, if not in practice: the 1992 United Nations Conference on Environment and Development (UNCED) in Rio de Janeiro, Brazil, followed by the 2002 World Summit on Sustainable Development (WSSD) in Johannesburg, South Africa.

The major policy outcome of UNCED was Agenda 21, a global agenda for sustainable development in the twenty-first century that was adopted by more than 178 governments, except, significantly, the United States. It was quickly realized that the tasks outlined at UNCED would be enormously difficult to achieve on a global scale. However, Agenda 21 addresses social, economic, and environmental problems that also are manifested locally, and this fact coupled with the "subsidiarity principle"—which holds that local governments are closest to the people—indicated a need to find local solutions. In response, many local governments and their communities are now adopting more transparent decisionmaking processes, involving more citizens, sharing control, and inventorying and adopting the principles of sustainability at the local level through Local Agenda 21 (LA21).[11] . . . Such is the success of LA21 that, by February 2002, there were 6,416 local governments in 113 countries worldwide that had either made a formal commitment to the principles or were actively undertaking the process.[12] Another 600 local governments worldwide are involved in the Cities for Climate Protection program.[13]

Not surprisingly, sustainable development and sustainable community advocates have largely come from the environmental movement and are generally professionally qualified, often in a related discipline. They are usually from a different social location than are people in the environmental justice movement. Wary of interest-group pluralism, advocates of sustainable development promote the use of innovative, deliberative, and democratic forums known as deliberative inclusionary processes. . . .[14] The overall aim is to involve a broad cross section of lay citizens in developing shared values, consensus, and a vision of "the common good."

While not without problems—for example, who should convene these processes, define their objectives, and frame the problems they address?[15]— such processes are being increasingly used on a local level in sustainable community initiatives.[16]. . . In addition, these processes are used in sustainable livelihoods programs in developing countries.[17]

Common Ground?

Despite the historical, geographic, and ideological differences in origin between the environmental justice and sustainability paradigms, there exists an area of theoretical, conceptual, and practical common ground between them called "just sustainability". . . .

Regardless of one's perspective—looking at the issue from a global, statewide, or more local focus or approaching it from a moral or practical standpoint—inequity and injustice resulting from racism and classism are bad for the environment and bad for sustainability. Research has shown that globally, countries with more equal income distribution, greater civil liberties and political rights, and higher literacy levels tend to have higher environmental quality (measured in lower concentrations of air and water pollutants, and access to clean water and sanitation) than those with less equal income distributions, fewer rights and civil liberties, and lower levels of literacy.[18] Good examples are the Scandinavian countries Sweden, Denmark, Norway, and Finland. A survey of the 50 U.S. states found that those with greater inequalities in power distribution (measured by voter participation, tax fairness, Medicaid access, and educational attainment levels) such as Tennessee, Mississippi, and Alabama had among the least stringent environmental policies, greater levels of environmental stress, and higher rates of infant mortality and premature deaths.[19] At a more local level, a 1997 study of counties in California showed that highly segregated counties— segregated in terms of income, class, and race—had higher levels of hazardous air pollutants.[20] However, as consultants Michael Shellenberger and Ted Nordhaus indicate, there has been a disconnect:

> Why . . . is a human-made phenomenon like global warming—which may kill hundreds of millions of human beings over the next century—considered

"environmental"? Why are poverty and war not considered environmental problems while global warming is? What are the implications of framing global warming as an environmental problem—and handing off the responsibility for dealing with it to "environmentalists"?[21]

Unfortunately, the environmental sustainability movement, typified in the United States by the National Audubon Society, World Wildlife Fund, and The Nature Conservancy, generally lacks core principles or systems—and therefore strategies—for dealing with these issues. . . . Outside the United States, other countries such as the United Kingdom have run into the same problems. To be effective, organizations promoting sustainable communities and sustainable livelihoods must be more cognizant of the relationships between human inequality and environmentally unsustainable practices—and need to respond by refocusing their work around just sustainability.

Alternatives for Community and Environment

One organization that is doing this is Alternatives for Community and Environment (ACE). Operating out of Dudley Square in Boston's Roxbury district, ACE has established a growing local and national presence over the last 10 years. From its inception in 1994, ACE has been dedicated to working primarily within the local community to promote empowerment in decisionmaking for environmental, social, and economic issues.

The Roxbury Neighborhood

The Roxbury community is 5 percent white, 63 percent black, 24 percent Hispanic, 1 percent Asian or Pacific Islander, less than 1 percent Native American, 3 percent other, and 4 percent multiracial.[22] By contrast, the city of Boston is 50 percent white, 24 percent black, 14 percent Hispanic, 8 percent Asian or Pacific Islander, less than 1 percent Native American, 1 percent other, and 3 percent multiracial.[23] However, in 2003 the city officially became a "majority-minority" city, with people of color making up 50.5 percent of the population.

In addition to this demographic difference, compared to Boston's overall population, higher percentages of Roxbury residents speak languages other than English. In particular, Spanish is much more prevalent in Roxbury than in the city as a whole, and to a lesser degree, this is true for French and French Creole, Portuguese and Portuguese Creole, and African languages. In 1999, 73.2 percent of Roxbury residents were classified as low to moderate income, as compared to 56.2 percent for the city as a whole. These statistics reveal one of metro Boston's poorest neighborhoods and an inner urban population that meets Massachusetts's criteria for an "environmental justice population"—those segments of the population that the Massachusetts Executive Office of Environmental Affairs has determined to be most at risk of being unaware of or unable to participate in environmental decisionmaking or to gain access to state environmental resources.[24]. . .

Empowering a Community of Decisionmakers

ACE's mission, as defined by its 2002–2007 strategic plan, is:

> ACE builds the power of communities of color and lower income com-
> munities in New England to eradicate environmental racism and classism
> and achieve environmental justice. We believe that everyone has the
> right to a healthy environment and to be decision makers in issues affect-
> ing communities.[25]

Although ACE initially began locally, it has now broadened its base to
include the New England region. To do this, it has sprouted many programs
focusing broadly on youth, transportation networks, and environmental health.
. . . It typically involves cross-boundary coalitions, which are currently
managed by members of ACE's staff of 14. The work done by the organization is
multifaceted and includes popular education, coalition building, movement
building, and what board and staff members call "family building"—the
nurturing of talent within its staff and local community.

ACE has been extremely successful in bringing about local and regional
change: It has won funding for "Air Beat," real-time pollution monitoring
equipment in Dudley Square; pressured officials to enact stricter regulations
on dumpster storage lots, junkyards, and recycling facilities to stop health
code violations; convinced the Massachusetts Bay Transportation Authority to
purchase 350 compressed natural gas buses; and urged a policy change for
free bus-to-bus transfers. At least three of its strategies have led to this success:
the power of and mutual respect within its multicultural workforce, its ability
to be heard by the community while effectively communicating policy issues
to policymakers, and its practice in intermeshing justice and sustainability
issues. All three of these strategies have broader implications for uniting the
environmental sustainability and environmental justice movements. . . .

[A]lso, a crucial part of the group's ethos is the belief that justice and
sustainability are inseparable. In response to the question, "Is ACE an
environmental justice organization or is it a sustainable development
organization?" Bob Terrell, director of the Washington Street Corridor
Coalition, a community-based transit advocacy project in Roxbury, and ACE's
chair, responds:

> I think ACE is both, because . . . we come out of the context of the African-
> American community and other communities of color. Most of us doing
> this kind of work see sustainability and environmental justice as two sides
> of one coin, but you really can't have one without the other. Now we
> understand why other people think differently. To some people it's justice,
> get it out of my neighborhood, that's it, or sustainability people who have
> wonderful ideas about solar energy. (29)

He then continues with a very specific line of reasoning about the links
between justice and sustainability, using affordable housing as an example.

He sees ACE's future as a lead organization in this justice-sustainability linkage:

> We've already said, we're all in favor of affordable housing. ACE participated in the Roxbury Master Plan. We talked to other people in the neighborhood about where the housing, should be built, how affordable it should be. Now we've got to ask how energy efficient is that housing, because 30 percent of your costs of housing are wrapped up in energy. Most of the affordable housing people are not thinking along those lines.[30]

Conclusions

ACE is in the vanguard of a growing movement looking to integrate justice and sustainability. It does this in a practical grassroots way by building power in Roxbury, metro Boston, and the wider New England region. It is not alone in this. It works in long-term coalition with ideologically related partners such as The Dudley Street Neighborhood Initiative, Coalition to Protect Chinatown, and City Life-Vida Urbana. In addition, there are other, similar just sustainability organizations around the United States . . . and the world that are effectively multicultural alliances pragmatic enough to avoid the trap of essentialism by steering a "middle way," exploring the common ground between an environmental justice philosophy and that of traditional environmental sustainability organizations, seamlessly and flexibly overlapping the discourses of both. The main problem is that there are not enough of them and they are generally locally based.[31]

What can be learned from ACE? There are three key interrelated messages: the power of a multicultural workforce, especially in local activism; the ability to use a community-empowering, communitarian justice-and-rights discourse and to overlap this with a more personalized, policy-based, academic, individualized "down town power broker" talk; and the inseparability of justice and sustainability. . . .

From the 1990 "Letter to the Big Ten"—a letter from environmental justice leaders accusing the major environmental sustainability groups of racist hiring practices and a disregard for wider equity and justice issues—to Shellenberger and Nordhaus's withering critique in their essay "The Death of Environmentalism" and the environmental justice response, "The Soul of Environmentalism," the call for just sustainability has grown.[34] ACE and other organizations recognize these "prefigurative politics" and are shaping not only their mission statements and programs but their entire organizations around this in terms of staffing, strategy, external relations, and image.

And it works. As Bill Shutkin, says:

> The Boston Foundation stepped up and then the Public Welfare Foundation and everybody stepped up because, what were we doing? We were bringing neighborhood environmental justice organizations together with mainstream environmental and sustainability organizations.[35]

Notes

... 2. Robert M. Entman, professor of communication and political science at North Carolina State University, argues that "to frame is to select some aspects of a perceived reality and make them more salient in a communicating text, in such a way as to promote a particular problem definition, causal interpretation, moral evaluation, and/or treatment recommendation for the item described." (R. M. Entman, "Framing: Toward Clarification of a Fractured Paradigm," *Journal of Communication* 43, no. 4 (1993): 52.)

3. The authors of a 2003 Harvard University paper on human security argue that "human security offers much to this vibrant field of sustainable development. Most notably, human security—like human development—highlights the social dimension of sustainable development's 'three pillars' (environment, economy, society). Moreover, the high importance and urgency given to the elimination of destitution and deprivation over the short-term that is core to human security reminds proponents of sustainable development that intra-generational equity must not be sacrificed to the altar of inter-generational equity . . . thus, the field of security should be broadened to a more comprehensive notion of 'sustainable security.' Sustainable security is less anthropocentric because it values the environment in itself and not merely as a set of risks. This more expanded field facilitates critical integrations of state, human and environmental security, and parallels the three linked pillars of society, economy, and nature central to the field of sustainable development." (S. Khagram, W. C. Clark, and D. F. Raad, "From the Environment and Human Security to Sustainable Security and Development," *Journal of Human Development* 4, no. 2 (2003): 290.)

4. According to the U.S. Environmental Protection Agency (EPA), "Environmental Justice is the fair treatment and meaningful involvement of all people regardless of race, color, national origin, or income with respect to the development, implementation, and enforcement of environmental laws, regulations, and policies. Fair treatment means that no group of people, including a racial, ethnic, or a socioeconomic group, should bear a disproportionate share of the negative environmental consequences resulting from industrial, municipal, and commercial operations or the execution of federal, state, local, and tribal programs and policies. Meaningful involvement means that: (1) potentially affected community residents have an appropriate opportunity to participate in decisions about a proposed activity that will affect their environment and/or health; (2) the public's contribution can influence the regulatory agency's decision; (3) the concerns of all participants involved will be considered in the decision making process; and (4) the decision makers seek out and facilitate the involvement of those potentially affected." (EPA, *EPA—Environmental Justice,* accessible via archive at http://permanent.access.gpo.gov/websites/epagov/www.epa.gov/environmentaljustice/ last updated 18 November 2004).)

5. In England, studies on the disproportionate effect of external threats include Friends of the Earth, *Pollution Injustice* (London, 2000); J. Agyeman, "Constructing Environmental (In)justice: Transatlantic Tales," *Environmental Politics* 11, no. 3 (2002): 31–53; and J. Agyeman and B. Evans, "'Just Sustainability': The Emerging Discourse of Environmental Justice in Britain?" *The Geographical Journal* 170, no. 2 (2004): 155–64. In Scotland, studies include K. Dunion and E. Scandrett, "The Campaign for Environmental Justice in Scotland as a Response to Poverty in a Northern Nation," in J. Agyeman, R. D. Bullard, and B. Evans, eds., *Just Sustainabilities: Development in an Unequal World* (London: Earthscan/MIT Press, 2003). In South Africa, studies include D. McDonald, ed., *Environmental Justice in South Africa* (Athens, OH, and Cape Town,

South Africa: Ohio University Press and Cape Town University Press, 2002); and D. Roberts, "Sustainability and Equity: Reflections of a Local Government Practitioner in South Africa," in Agyeman, Bullard, and Evans, this note above. In Eastern and Central Europe, studies include A. Costi "Environmental Protection, Economic Growth and Environmental Justice: Are They Compatible in Central and Eastern Europe?" in Agyeman, Bullard, and Evans, this note above.

6. United Church of Christ Commission for Racial Justice, *Toxic Wastes and Race in the United States: A National Report on the Racial and Socioeconomic Characteristics of Communities with Hazardous Waste Sites* (New York: United Church of Christ, 1987).

7. H. L. White, "Race, Class and Environmental Hazards," in D. E. Camacho, ed., *Environmental Injustices, Political Struggles: Race, Class, and the Environment* (Durham, NC: Duke University Press, 1998). White's study was an adaptation of B. Goldman, *Not Just Prosperity: Achieving Sustainability with Environmental Justice* (Washington, DC: National Wildlife Federation, 1993), 8. Some of the original national-level studies include studies on pesticides, hazardous waste, lead, incinerators, and multiple pollution hazards (indication of racial or income disparity or both in parentheses): On pesticides, see F. W. Kutz, A. R. Yobs, and S. C. Strassman, "Racial Stratification of Organochlorine Insecticide Residues in Human Adipose Tissue," *Journal of Occupational Medicine* 19, no. 9 (1977): 619–22 (race). On hazardous waste, see J. M. Gould, "Quality of Life in American Neighborhoods: Levels of Affluence, Toxic Waste, and Cancer Mortality in Residential Zip Code Areas" (Boulder, CO: Westview Press, 1986) (income); United Church of Christ Commission for Racial Justice, note 6 above (both); P. Costner and J. Thornton, *Playing with Fire: Hazardous Waste Incineration* (Washington, DC: Greenpeace, 1990) (both); L. J. Fitton, "A Study of the Correlation between the Siting of Hazardous Waste Facilities and Racial and Socioeconomic Characteristics" (undergraduate thesis for Government 500, Cornell University-in-Washington Program, Fall 1992) (both); C. J. McDermott, "Environmental Equity: A Waste Manager's Perspective," *Land Use Forum* 2, no. 1 (1993): 12–17 (race); J. T. Hamilton, "Politics and Social Costs: Estimating the Impact of Collective Action on Hazardous Waste Facilities," *Rand Journal of Economics* 24, no. 1 (1993): 101–25 (both); and R. Zimmermann, "Social Equity and Environmental Risk," Risk Analysis 13, no. 6 (1993): 649–66 (race). On lead, see O. Carter-Pokras, J. Pirkle, G. Chavez, and E. Gunter, "Blood Lead Levels of 4-11-Year-Old Mexican American, Puerto Rican, and Cuban Children," *Public Health Reports* 105, no. 4 (1990): 388–93. On incinerators, see M. R. Greenberg, "Proving Environmental Inequity in Siting Locally Unwanted Land Uses," *Risk Issues In Health & Safety* 4, no. 2 (1993): 235–52 (race). On multiple pollution hazards, see B. Goldman, *The Truth about Where You Live: An Atlas for Action on Toxins and Mortality* (New York: Random House, 1992) (race); and P. Mohai and B. Bryant, "Environmental Racism: Reviewing the Evidence," in B. Bryant and P. Mohai, *Race and the Incidence of Environmental Hazards: A Time for Discourse* (Boulder, CO: Westview Press, 1992), 163–76 (both). Regional or city studies included studies on air pollution, pesticides, toxic fish, lead, toxic releases, waste incineration, and multiple pollution hazards. On air pollution, see Council on Environmental Quality, *The Second Annual Report of the Council on Environmental Quality* (Washington, DC: U.S. Government Printing Office, 1971) (Chicago, IL, income); A. M. Freeman, "The Distribution of Environmental Quality," in A. V. Kneese and B. T. Bower, eds., *Environmental Quality Analysis* (Baltimore, MD: Johns Hopkins University Press for Resources for the Future, 1972) (Kansas City, MO, St. Louis, MO, and Washington, DC, both); J. M. Zupan, *The Distribution of Air Quality in the New York Region* (Baltimore, MD: Johns Hop-

kins University Press for Resources for the Future, Inc., 1973) (New York, race); W. J. Kruvant, "People, Energy, and Pollution," in D. K. Newman and D. Day, eds., *The American Energy Consumer* (Cambridge, MA: Ballinger, 1975) (Washington, DC, both); W. R. Burch, "The Peregrine Falcon and the Urban Poor: Some Sociological Interrelations," in P. J. Richerson and J. McEvoy III, eds., *Human Ecology, An Environmental Approach* (Belmont, CA: Duxbury Press, 1976) (New Haven, CT, income); P. Asch and J. J. Seneca, "Some Evidence on the Distribution of Air Quality," *Land Economics* 54, no. 3 (1978): 278–97 (urban areas, both); M. Gelobter, "The Distribution of Outdoor Air Pollution by Income and Race: 1970–1984" (master's thesis, Energy and Resource Group, University of California, 1987) (urban areas, both); E. Mann, *L.A.'s Lethal Air: New Strategies for Policy, Organizing, and Action* (Los Angeles: Labor/Community Strategy Center, 1991) (Los Angeles, race); and D. R. Wernette and L. A. Nieves, "Breathing Polluted Air" *EPA Journal* 18, no. 3/4 (1992): 16–17 (urban areas, race). On pesticides, see J. Davies, W. Edmundson, A. Raffonelli, J. Cassady, and C. Morgade, "The Role of Social Class in Human Pesticide Pollution," *American Journal of Epidemiology* 96 (1972): 334–41 (Dade County, FL, income); and J. E. Burns, "Pesticides in People: Organochlorine Pesticides and Polychlorinated Biphenyl Residues in Biopsied Human Tissue—Texas 1969–72," *Pesticide Monitoring Journal* 7, no. 3/4 (1974): 1222–26 (southern states, race). On toxic fish, see H. W. Puffer, S. P Azen, M. J. Duda, and D. R. Young, *Consumption Rates of Potentially Hazardous Marine Fish Caught in the Metropolitan Los Angeles Area*, EPA-600/3-82-070 (Corvallis, OR: EPA, Environmental Research Laboratory, 1982) (Los Angeles, race); M. McAllum, *Recreational and Subsistence Catch and Consumption of Selected Seafood from Three Urban Industrial Bays of Puget Sound: Port Gardner, Elliot Bay, and Sinclair Inlet* (Olympia, WA: Washington State Department of Health, 1985) (Puget Sound, WA, race); National Oceanic and Atmospheric Administration (NOAA), *Potential Toxicant Exposure Among Consumers of Recreationally Caught Fish from Urban Embayments of Puget Sound* (Washington, DC: NOAA, 1985) (Puget Sound, WA, race); and P. C. West, J. M. Fly, F. Larkin, and R. Marans, "Minority Anglers and Toxic Fish Consumption: Evidence from a State-Wide Survey of Michigan," in Bryant and Mohai, this note above, 100–13. On hazardous waste, see U.S. General Accounting Office, *Siting of Hazardous Waste Landfills and Their Correlation with Racial and Economic Status of Surrounding Communities* (Washington, DC: Government Printing Office, 1983) (southeastern states, race); M. Greenberg and R. F. Anderson, *Hazardous Waste Sites: The Credibility Gap* (New Brunswick, NJ: Rutgers University Center for Urban Policy Research, 1984) (New Jersey, both); P. Mohai and B. Bryant, "Environmental Injustice: Weighing Race and Class as Factors in the Distribution of Environmental Hazards," *University of Colorado Law Review* 63, no. 4 (1992): 921–32 (Detroit, MI, both); D. G. Unger, A. Wandersman, and W. Hallman "Living Near a Hazardous Waste Facility: Coping with Individual and Family Distress," *American Journal of Orthopsychiatry* 55 (1992): 62 (Pinewood, SC, both); K. Ketkar, "Hazardous Waste Sites and Property Values in the State of New Jersey," *Applied Economics* 24 (1992): 647–53 (New Jersey, race); and V. Been, "What's Fairness Got To Do with It? Environmental Justice and the Siting of Locally Undesirable Land Uses," *Cornell Law Review* 78 (1993): 1001–85 (southeastern states, both). On lead, see Agency for Toxic Substances Disease Registry (ATSDR), *The Nature and Extent of Lead Poisoning in Children in the United States: A Report to Congress* (Atlanta, GA: Centers for Disease Control, 1988) (urban areas, both). On toxic releases, see M. Belliveau, M. Kent, and B. Rosenblum, *Richmond at Risk: Community Demographics and Toxic Hazards from Industrial Polluters* (San Francisco: Citizens for a Better Environment, 1989) (Richmond, CA, both); K. L. Brown, "Environmental Discrimination: Myth or Reality?" unpublished manuscript written 29 March 1991, on file with *North-*

western University Law Review, Chicago, IL, 1993 (St. Louis, MO, race); J. Kay, "Minorities Bear Brunt of Pollution," *San Francisco Examiner*, 7–10 April 1991 (Los Angeles, race); W. M. Bowen, M. J. Salling, E. J. Cyran, and H. A. Moody, "The Spatial Association between Race, Income and Industrial Toxic Emissions in Cuyahoga County, Ohio," presentation for the Annual Meetings of the Association of American Geographers, Atlanta, GA, May 1993 (Cuyahoga, OH, income); and L. M. Burke, "Race and Environmental Equity: A Geographic Analysis in Los Angeles," *Geographic Information Systems* 3, no. 9 (1993); 44–50 (Los Angeles, both). On waste incineration, see E. Holtzman, *Smokescreen: How the Department of Sanitation's Solid Waste Plan and Environmental Impact Statement Cover Up the Poisonous Health Effects of Burning Garbage* (New York: Office of the Comptroller, 1992) (New York, race). On multiple pollution hazards, see B. J. Berry, et al., *The Social Burdens of Environmental Pollution: A Comparative Metropolitan Data Source* (Cambridge, MA: Ballinger Publishing Co., 1977) (urban areas, both); and D. Pfaff, "Pollution and the Poor," *The Detroit News*, 27 November 1989 (Detroit, MI, income). Others include R. D. Bullard, *Dumping in Dixie: Race, Class, and Environmental Quality* (Boulder, CO: Westview Press, 1990); R. D. Bullard, "Ecological Inequalities and the New South: Black Communities Under Siege," *The Journal of Ethnic Studies* 17, no. 4 (1990): 101–15; R. D. Bullard, "Overcoming Racism in Environmental Decisionmaking," *Environment*, May 1994, 10–20, 39–44; F. O. Adeola, "Environmental Hazards, Health, and Racial Inequity in Hazardous Waste Distribution," *Environmental and Behavior* 26, no. 1 (1994): 99–126; A. Szasz and M. Meuser, "Environmental Inequalities: Literature Review and Proposals for New Directions in Research and Theory," *Current Sociology* 45, no. 3 (1997): 99–120; and P. Mohai, "Dispelling Old Myths: African American Concern for the Environment," *Environment*, June 2003, 10–26.

8. M. Lavelle and M. Coyle, eds., "Unequal Protection: The Racial Divide in Environmental Law," *National Law Journal* 15 (1992): S1–S7.

9. For sprawl and smart growth, see, for example, R. D. Bullard, G. S. Johnson, and A. O. Torres, eds., *Sprawl City: Race, Politics, and Planning in Atlanta* (Washington, DC: Island Press, 2000). For sustainability, see, for example, J. Agyeman, R. D. Bullard, and B. Evans, note 5 above. For climate change, see, for example, International Climate Justice Network, "Bali Principles of Climate Justice," press release (Johannesburg, South Africa, 29 August 2002), accessible via http://www.corpwatch.org/article.php?id=3748; and Congressional Black Caucus Foundation, Inc. (CBCF), *African Americans and Climate Change: An Unequal Burden* (Washington, DC: CBCF, 2004).

10. World Commission on Environment and Development, *Our Common Future* (Oxford, UK: Oxford University Press, 1987), 43.

11. At the Local Government Summit of the World Summit on Sustainable Development held in Johannesburg in August/September 2002, Local Agenda 21 (LA21) was renamed "Local Action 21." Local Action 21 was launched as a motto for the second decade of LA21, a mandate to local authorities worldwide to move from agenda to action and ensure an accelerated implementation of sustainable development, and to strengthen the LA21 movement of local governments to create sustainable communities and cities while protecting global common goods.

12. International Council for Local Environmental Initiatives (ICLEI), *Second Local Agenda 21 Survey*, (New York: United Nations Commission on Sustainable Development, 2002).

13. ICLEI, *CCP Five-Milestone Framework*, 2002, http://www.iclei.org/ccp/five_milestones.htm (accessed 22 December 2004).

14. O. Renn, T. Webler, and P. Wiedermann, eds., *Fairness and Competence in Citizen Participation* (Dordrecht, Netherlands: Kluwer, 1995); J. Dryzek, "Green Reason: Communicative Ethics and the Biosphere," *Environmental Ethics* 12 (1990): 195–210; T. Holmes and I. Scoones, "Participatory Environmental Policy Processes: Experiences from North and South," IDS Working Paper 113 (Brighton, UK: Institute of Development Studies, 2000); and D. Smith, *Deliberative Democracy and the Environment* (London: Routledge, 2003).

15. Holmes and Scoones, ibid.

16. L. C. Hempel, "Conceptual and Analytical Challenges in Building Sustainable Communities," in D. A. Mazmanian and M. E. Kraft, eds., *Toward Sustainable Communities: Transition and Transformations in Environmental Policy* (Cambridge, MA: MIT Press, 1999), 43–74; and M. Roseland, *Toward Sustainable Communities: Resources for Citizens and Their Governments* (Gabriola Island, BC: New Society Publishers, 1998).

17. R. Chambers and G. Conway, "Sustainable Rural Livelihoods: Practical Concepts for the 21st Century," IDS Discussion Paper 296 (Brighton, UK: Institute for Development Studies, 1991); and Khagram, Clark, and Raad, note 3 above.

18. M. Torras and J. K. Boyce, "Income, Inequality, and Pollution: A Reassessment of the Environmental Kuznets' Curve," *Ecological Economics* 25, no. 2 (1998): 147–60.

19. J. K. Boyce, A. R. Klemer, P. H. Templet, and C. E. Willis, "Power Distribution, the Environment, and Public Health: A State Level Analysis," *Ecological Economics* 29, no. 1 (1999): 127–40.

20. R. Morello-Frosch, "Environmental Justice and California's 'Riskscape': The Distribution of Air Toxics and Associated Cancer and Non Cancer Risks Among Diverse Communities" (PhD dissertation, University of California, Berkeley, Department of Health Sciences, 30 November 1997).

21. Shellenberger and Nordhaus, note 1 above.

22. U.S. Bureau of the Census, *Roxbury Data Profile, 2000,* http://www.cityofboston. gov/DND/PDFs/Profiles/Roxbury_PD_Profile.pdf (accessed 12 June 2004).

23. Ibid.

24. Commonwealth of Massachusetts, *Environmental Justice Policy,* State House, Boston, MA, 2002.

25. ACE Strategic Plan 2002, http://ace-ej.org/mission.html,1. ...

29. Bob Terrell, director, Washington Street Corridor Coalition, in interview with the author, Boston, Massachusetts, 5 June 2004.

30. Ibid.

31. There are some national organizations which, like ACE, have a Just Sustainability Index of 3 (see the box on page 19). They include Center for Health, Environment, and Justice; Center for a New American Dream; Environmental Defense; and Redefining Progress. ...

34. Shellenberger and Nordhaus, note 1 above; and M. Gelobter, et al., *The Soul of Environmentalism,* http://www.soulofenvironmentalism.org/ (accessed 27 May 2005).

35. Shutkin, note 32 above.

NO

David Friedman

The "Environmental Racism" Hoax

When the U.S. Environmental Protection Agency (EPA) unveiled its heavily criticized environmental justice "guidance" earlier this year, it crowned years of maneuvering to redress an "outrage" that doesn't exist. The agency claims that state and local policies deliberately cluster hazardous economic activities in politically powerless "communities of color." The reality is that the EPA, by exploiting every possible legal ambiguity, skillfully limiting debate, and ignoring even its own science, has enshrined some of the worst excesses of racialist rhetoric and environmental advocacy into federal law.

"Environmental justice" entered the activist playbook after a failed 1982 effort to block a hazardous-waste landfill in a predominantly black North Carolina county. One of the protesters was the District of Columbia's congressional representative, who returned to Washington and prodded the General Accounting Office (GAO) to investigate whether noxious environmental risks were disproportionately sited in minority communities.

A year later, the GAO said that they were. Superfund and similar toxic dumps, it appeared, were disproportionately located in non-white neighborhoods. The well-heeled, overwhelmingly white environmentalist lobby christened this alleged phenomenon "environmental racism," and ethnic advocates like Ben Chavis and Robert Bullard built a grievance over the next decade.

Few of the relevant studies were peer-reviewed; all made critical errors. Properly analyzed, the data revealed that waste sites are just as likely to be located in white neighborhoods, or in areas where minorities moved only after permits were granted. Despite sensational charges of racial "genocide" in industrial districts and ghastly "cancer alleys," health data don't show minorities being poisoned by toxic sites. "Though activists have a hard time accepting it," notes Brookings fellow Christopher H. Foreman, Jr., a self-described black liberal Democrat, "racism simply doesn't appear to be a significant factor in our national environmental decision-making."

❧

This reality, and the fact that the most ethnically diverse urban regions were desperately trying to *attract* employers, not sue them, constrained the environmental racism movement for a while. In 1992, a Democrat-controlled Congress ignored environmental justice legislation introduced by then-Senator Al Gore. Toxic racism made headlines, but not policy.

All of that changed with the Clinton-Gore victory. Vice President Gore got his former staffer Carol Browner appointed head of the EPA and brought Chavis, Bullard, and other activists into the transition government. The administration touted environmental justice as one of the symbols of its new approach.

Even so, it faced enormous political and legal hurdles. Legislative options, never promising in the first place, evaporated with the 1994 Republican takeover in Congress. Supreme Court decisions did not favor the movement.

So the Clinton administration decided to bypass the legislative and judicial branches entirely. In 1994, it issued an executive order—ironically cast as part of Gore's "reinventing government" initiative to streamline bureaucracy—which directed that every federal agency "make achieving environmental justice part of its mission."

At the same time, executive branch lawyers generated a spate of legal memoranda that ingeniously used a poorly defined section of the Civil Rights Act of 1964 as authority for environmental justice programs. Badly split, confusing Supreme Court decisions seemed to construe the 1964 Act's "nondiscrimination" clause (prohibiting federal funds for states that discriminate racially) in such a way as to allow federal intervention wherever a state policy ended up having "disparate effects" on different ethnic groups.

Even better for the activists, the Civil Rights Act was said to authorize private civil rights lawsuits against state and local officials on the basis of disparate impacts. This was a valuable tool for environmental and race activists, who are experienced at using litigation to achieve their ends.

Its legal game plan in place, the EPA then convened an advocate-laden National Environmental Justice Advisory Council (NEJAC), and seeded activist groups (to the tune of $3 million in 1995 alone) to promote its policies. Its efforts paid off. From 1993, the agency backlogged over 50 complaints, and environmental justice rhetoric seeped into state and federal land-use decisions.

❧◈❧

Congress, industry, and state and local officials were largely unaware of these developments because, as subsequent news reports and congressional hearings established, they were deliberately excluded from much of the agency's planning process. Contrary perspectives, including EPA-commissioned studies highly critical of the research cited by the agency to justify its environmental justice initiative in the first place, were ignored or suppressed.

The EPA began to address a wider audience in September 1997. It issued an "interim final guidance" (bureaucratese for regulation-like rules that agencies can claim are not "final" so as to avoid legal challenge) which mandated that environmental justice be incorporated into all projects that file federal environmental impact statements. The guidance directed that applicants pay particular attention to potential "disparate impacts" in areas where minorities live in "meaningfully greater" numbers than surrounding regions.

The new rules provoked surprisingly little comment. Many just "saw the guidance as creating yet another section to add to an impact statement," explains Jennifer Hernandez, a San Francisco environmental attorney. In response, companies wanting to build new plants had to start "negotiating with community advocates and federal agencies, offering new computers, job training, school or library improvements, and the like" to grease their projects through.

In December 1997, the Third Circuit Court of Appeals handed the EPA a breathtaking legal victory. It overturned a lower court decision against a group of activists who sued the state of Pennsylvania for granting industrial permits in a town called Chester, and in doing so the appeals court affirmed the EPA's extension of Civil Rights Act enforcement mechanisms to environmental issues.

(When Pennsylvania later appealed, and the Supreme Court agreed to hear the case, the activists suddenly argued the matter was moot, in order to avoid the Supreme Court's handing down an adverse precedent. This August, the Court agreed, but sent the case back to the Third Circuit with orders to dismiss the ruling. While activists may have dodged a decisive legal bullet, they also wiped from the books the only legal precedent squarely in their favor.)

Two months after the Third Circuit's decision, the EPA issued a second "interim guidance" detailing, for the first time, the formal procedures to be used in environmental justice complaints. To the horror of urban development, business, labor, state, local, and even academic observers, the guidance allows the federal agency to intervene at any time up to six months (subject to extension) after any land-use or environmental permit is issued, modified, or renewed anywhere in the United States. All that's required is a simple allegation that the permit in question was "an act of intentional discrimination or has the effect of discriminating on the basis of race, creed, or national origin."

The EPA will investigate such claims by considering "multiple, cumulative, and synergistic risks." In other words, an individual or company might not itself be in violation, but if, combined with previous (also legal) land-use decisions, the "cumulative impact" on a minority community is "disparate," this could suddenly constitute a federal civil rights offense. The guidance leaves important concepts like "community" and "disparate impact" undefined, leaving them to "case by case" determination. "Mitigations" to appease critics will likewise be negotiated with the EPA case by case.

This "guidance" subjects virtually any state or local land-use decision—made by duly elected or appointed officials scrupulously following validly enacted laws and regulations—to limitless ad hoc federal review, any time there is the barest allegation of racial grievance. Marrying the most capricious elements of wetlands, endangered species, and similar environmental regulations with the interest-group extortion that so profoundly mars urban ethnic politics, the guidance transforms the EPA into the nation's supreme land-use regulator.

Reaction to the Clinton administration's gambit was swift. A coalition of groups usually receptive to federal interventions, including the U.S. Conference of Mayors, the National Association of Counties, and the National Association of Black County Officials, demanded that the EPA withdraw the guidance. The House amended an appropriations bill to cut off environmental justice enforcement until the guidance was revised. This August, EPA officials were grilled in congressional hearings led by Democratic stalwarts like Michigan's John Dingell.

Of greatest concern is the likelihood the guidance will dramatically increase already-crippling regulatory uncertainties in urban areas where ethnic populations predominate. Rather than risk endless delay and EPA-brokered activist shakedowns, businesses will tacitly "redline" minority communities and shift operations to white, politically conservative, less-developed locations.

Stunningly, this possibility doesn't bother the EPA and its environmentalist allies. "I've heard senior agency officials just dismiss the possibility that their policies might adversely affect urban development," says lawyer Hernandez. Dingell, a champion of Michigan's industrial revival, was stunned when Ann Goode, the EPA's civil rights director, said her agency never considered the guidance's adverse economic and social effects. "As director of the Office of Civil Rights," she lectured House lawmakers, "local economic development is not something I can help with."

Perhaps it should be. Since 1980, the economies of America's major urban regions, including Cleveland, Chicago, Milwaukee, Detroit, Pittsburgh, New Orleans, San Francisco, Newark, Los Angeles, New York City, Baltimore, and Philadelphia, grew at only one-third the rate of the overall American economy. As the economies of the nation's older cities slumped, 11 million new jobs were created in whiter areas.

Pushing away good industrial jobs hurts the pocketbook of urban minorities, and, ironically, harms their health in the process. In a 1991 *Health Physics* article, University of Pittsburgh physicist Bernard L. Cohen extensively analyzed mortality data and found that while hazardous waste and air pollution exposure takes from three to 40 days off a lifespan, poverty reduces a person's life expectancy by an average of 10 *years*. Separating minorities from industrial plants is thus not only bad economics, but bad health and welfare policy as well.

✎◉✐

Such realities matter little to environmental justice advocates, who are really more interested in radical politics than improving lives. "Most Americans would be horrified if they saw NEJAC [the EPA's environmental justice advisory council] in action," says Brookings's Foreman, who recalls a council meeting derailed by two Native Americans seeking freedom for an Indian activist incarcerated for killing two FBI officers. "Because the movement's main thrust is toward ... 'empowerment'..., scientific findings that blunt or conflict with that goal are ignored or ridiculed."

Yet it's far from clear that the Clinton administration's environmental justice genie can be put back in the bottle. Though the Supreme Court's dismissal of the Chester case eliminated much of the EPA's legal argument for the new rules, it's likely that more lawsuits and bureaucratic rulemaking will keep the program alive. The success of the environmental justice movement over the last six years shows just how much a handful of ideological, motivated bureaucrats and their activist allies can achieve in contemporary America unfettered by fact, consequence, or accountability, if they've got a President on their side.

POSTSCRIPT

Should Environmental Policy Attempt to Cure Environmental Racism?

The problems that led to the environmental justice movement have been documented in many reports. For example, in "Who Gets Polluted? The Movement for Environmental Justice," *Dissent* (Spring 1994), Ruth Rosen presents a history of the environmental justice movement, stressing how the movement has woven together strands of the civil rights and environmental struggles. Rosen argues that racial discrimination plays a significant role in the unusually intense exposure to industrial pollutants experienced by disadvantaged minorities, and she expresses the hope that "greening the ghetto will be the first step in greening our entire society." In addition, Bullard's *Dumping in Dixie: Race, Class and Environmental Quality* (Westview Press, 1990, 1994, 2000) has become a standard text in the environmental justice field. Also see his *Unequal Protection: Environmental Justice and Communities of Color* (Sierra Club Books, 1994); Michael Heiman's "Waste Management and Risk Assessment: Environmental Discrimination Through Regulation," *Urban Geography* (vol. 17, no. 5, 1996); and Luke W. Cole and Sheila R. Foster's *From the Ground Up: Environmental Racism and the Rise of the Environmental Justice Movement* (New York University Press, 2000). David W. Allen, in "Social Class, Race, and Toxic Releases in American Counties, 1995," *Social Science Journal* (vol. 38, no. 1, 2001), finds that the data support the existence of environmental racism but that the effect is strongest in the southern portion of the United States (the Sun Belt). Robert J. Brulle and David N. Pellow, "Environmental Justice: Human Health and Environmental Inequalities," *Annual Review of Public Health* (No. 1, 2006), review unequal exposures to environmental pollution and call for more research. Bradley C. Parks and J. Timmons Roberts, "Globalization, Vulnerability to Climate Change, and Perceived Injustice," *Society & Natural Resources* (April 2006), link environmental injustice to vulnerability to global climate change. Jeffrey D. Sachs, "Can Extreme Poverty Be Eliminated?" *Scientific American* (September 2005), argues that though market economics and globalization are improving the lot of most poor people, the extreme poverty that still afflicts a sixth of the world's population can be eliminated by 2025 with an appropriate aid program.

Those who criticize the environmental justice movement tend to focus on other studies. In "Green Redlining: How Rules Against 'Environmental Racism' Hurt Poor Minorities Most of All," *Reason* (October 1998), Henry Payne labels the Environmental Protection Agency's efforts to impose environmental equity "redlining" and, like Friedman, argues that the practice reduces job opportunities and economic benefits for minorities.

There is great contrast in the sides to this debate. In such cases, the reader must not ignore the social values and political commitments of the debaters. The reader must also be careful to consider the data relied on by the debaters and to watch for unsupported claims and simplistic explanations for events whose causes are likely to be more complicated.

Where is government policy going? Jim Motavalli, in "Toxic Targets: Polluters That Dump on Communities of Color Are Finally Being Brought to Justice," *E: The Environmental Magazine* (July–August 1998), states that although minorities and the poor have been forced to bear a disproportionate share of the burden of industrial pollution, changes in environmental policy and law are finally offering remedies. And in an August 9, 2001, memorandum regarding the Environmental Protection Agency's stance on environmental justice, EPA administrator Christine Todd Whitman wrote, "The Environmental Protection Agency has a firm commitment to the issue of environmental justice and its integration into all programs, policies, and activities, consistent with existing environmental laws and their implementing regulations.... [E]nvironmental justice is the goal to be achieved for all communities and persons across this Nation. Environmental justice is achieved when everyone, regardless of race, culture, or income, enjoys the same degree of protection from environmental and health hazards and equal access to the decision-making process to have a healthy environment in which to live, learn, and work." However, Robert D. Bullard, "Environmental Justice for All," *Crisis (The New)* (January/February 2003), argues that environmental racism remains a genuine problem and present policies of the U.S. government threaten to undo what progress has been achieved. Indeed, recent federal budgets have slashed EPA funding. See Chris Mooney, "Earth Last," *The American Prospect* (May 2004).

ISSUE 6

Can Pollution Rights Trading Effectively Control Environmental Problems?

YES: Charles W. Schmidt, from "The Market for Pollution," *Environmental Health Perspectives* (August 2001)

NO: Brian Tokar, from "Trading Away the Earth: Pollution Credits and the Perils of 'Free Market Environmentalism,'" *Dollars & Sense* (March/April 1996)

ISSUE SUMMARY

YES: Freelance science writer Charles W. Schmidt argues that economic incentives such as emissions rights trading offer the most useful approaches to reducing pollution.

NO: Author, college teacher, and environmental activist Brian Tokar maintains that pollution credits and other market-oriented environmental protection policies do nothing to reduce pollution while transferring the power to protect the environment from the public to large corporate polluters.

Following World War II the United States and other developed nations experienced an explosive period of industrialization accompanied by an enormous increase in the use of fossil fuel energy sources and a rapid growth in the manufacture and use of new synthetic chemicals. In response to growing public concern about the pollution and other forms of environmental deterioration resulting from this largely unregulated activity, the U.S. Congress passed the National Environmental Policy Act of 1969. This legislation included a commitment on the part of the government to take an active and aggressive role in protecting the environment. The next year the Environmental Protection Agency (EPA) was established to coordinate and oversee this effort. During the next two decades an unprecedented series of legislative acts and administrative rules were promulgated, placing numerous restrictions on industrial and commercial activities that might result in the pollution, degradation, or contamination of land, air, water, food, and the workplace.

Such forms of regulatory control have always been opposed by the affected industrial corporations and developers as well as by advocates of a free-market policy. More moderate critics of the government's regulatory program recognize that adequate environmental protection will not result from completely voluntary policies. They suggest that a new set of strategies is needed. Arguing that "top down, federal, command and control legislation" is not an appropriate or effective means of preventing ecological degradation, they propose a wide range of alternative tactics, many of which are designed to operate through the economic marketplace. The first significant congressional response to these proposals was the incorporation of tradable pollution emission rights into the 1990 Clean Air Act amendments as a means for achieving the set goals for reducing acid rain–causing sulfur dioxide emissions. More recently, the 1997 international negotiations on controlling global warming in Kyoto, Japan, resulted in a protocol that includes emissions trading as one of the key elements in the plan to limit the atmospheric buildup of greenhouse gases.

Despite past difficulties in obtaining compliance with or enforcing strict statutory pollution limits, the idea of using such market-based strategies as the trading of pollution control credits or the imposition of pollution taxes has won limited acceptance from some major mainstream environmental organizations. Many environmentalists, however, continue to oppose the idea of allowing anyone to pay to pollute, either on moral grounds or because they doubt that these tactics will actually achieve the goal of controlling pollution. Diminishment of the acid rain problem is often cited as an example of how well emission rights trading can work, but in "Dispelling the Myths of the Acid Rain Story," *Environment* (July–August 1998), Don Munton argues that other control measures, such as switching to low-sulfur fuels, deserve much more of the credit for reducing sulfur dioxide emissions.

In "A Low-Cost Way to Control Climate Change," *Issues in Science and Technology* (Spring 1998), Byron Swift argues that the "cap-and-trade" feature of the U.S. Acid Rain Program has been so successful that a similar system for implementing the Kyoto Protocol's emissions trading mandate as a cost-effective means of controlling greenhouse gases should work. In March 2001 the U.S. Senate Committee on Agriculture, Nutrition, and Forestry held a "Hearing on Biomass and Environmental Trading: Opportunities for Agriculture and Forestry," in which witnesses urged Congress to encourage trading for both its economic and its environmental benefits. Richard L. Sandor, chairman and chief executive officer of Environmental Financial Products LLC, said that "200 million tons of CO_2 could be sequestered through soils and forestry in the United States per year. At the most conservative prices of $20–$30 per ton, this could potentially generate $4–$6 billion in additional agricultural income."

In the following selections, Charles W. Schmidt describes the use of economic incentives to motivate corporations to reduce pollution, and he argues that emissions trading schemes represent "the most significant developments" in this area. Brian Tokar has a much more negative assessment of sulfur dioxide pollution credit trading. He argues that such "free-market environmentalism" tactics fail to reduce pollution while turning environmental protection into a commodity that corporate powers can manipulate for private profit.

YES

Charles W. Schmidt

The Market for Pollution

Throughout much of its short history, environmental protection in the United States has been guided by a traditional paradigm based on strict regulatory guidelines for reducing emissions and punishments for noncompliance. Experts credit this traditional approach with improvements in air and water quality evident since the U.S. Environmental Protection Agency (EPA) was created more than 30 years ago. Tough environmental standards imposed under programs such as the Clean Water Act and the Clean Air Act filled a regulatory void and forced industries to cut their emissions or face heavy fines. Many of the greatest gains were seen with respect to point sources such as smokestacks and effluent pipes that could be easily monitored. But beyond the avoidance of penalties, industries regulated under those so-called command-and-control programs had little motivation to develop advanced pollution control technologies, which produced little economic gain.

Today, many stakeholders believe a more modern framework based on economic incentives that allow companies to profit from achieving environmental goals will build on the achievements of the past and allow for even greater improvements in environmental protection. Types of incentives vary widely, but they all share one thing in common: they attach a monetary value to the act of reducing pollution. In a January 2001 document titled *The United States Experience with Economic Incentives for Protecting the Environment*, the EPA described several types of incentives, including fees and taxes levied on pollutant releases, tax rebates for environmental technologies, and the trading of air emissions permits on the open market.

Attention is increasingly turning to the use of economic incentives in the wake of President George W. Bush's pledge to make them a foundation of his environmental policy. During the 2000 presidential campaign, Bush said that under his watch government would "set high environmental standards and provide market-based incentives to develop new technologies ... so that Americans could meet and exceed those standards."

Business organizations have responded warmly to the administration's support for incentives. For example, the Business Roundtable, a Washington, D.C.—based nonprofit organization of "CEOs committed to improving public policy," released a statement on 17 May 2001 that "applauds President Bush for incorporating the use of new technologies, as well as incentives that spur

From *Environmental Health Perspectives*, vol. 109, no. 8, August 2001. Reproduced with permission from Environmental Health Perspectives.

technological innovation, as the cornerstone of the administration's national energy policy."

Among the environmental community, the idea that market instruments could be used to control pollution was initially greeted with skepticism and even hostility. But over time, support has risen to a level that Joseph Goffman, a senior attorney with the public interest group Environmental Defense in Washington D.C., describes as "lukewarm to enthusiastic in many cases."

According to Goffman, economic incentives motivate companies to reduce pollution quickly and to exceed environmental standards whenever possible. This is in contrast to command-and-control approaches, which he says stifle innovation while encouraging polluters to do little more than meet minimum requirements. Under a traditional system, the EPA not only sets environmental standards, it often describes how companies should achieve them—a scenario sometimes described as "technology forcing."

Goffman suggests the downside to this approach is that the EPA usually only sets standards that can be met with current technology. This means companies have to wait for the agency to finish a technology review before either the EPA or the states revise a given standard. "With incentive programs," he says, "you don't have this kind of chicken-and-egg mentality. The agency sets a target and leaves the means of compliance up to industry. Companies want to profit from pollution control, so they invest more resources in technology development." Furthermore, Goffman adds, market forces naturally gravitate toward the least-cost option for reducing pollution, while traditional regulatory strategies lock companies into technologies that become progressively less effective, and thus less attractive, over time.

Most experts suggest it's too soon to gauge where and how incentive programs will grow under the Bush administration. This is because a host of key positions at the EPA and other agencies remain unfilled, and policy directions have yet to be fully clarified. However, Bush's commitment to market forces is undiminished, as indicated by comments from White House spokesperson Marcy Viana, who, referring to the president's position on global warming during an interview on 4 June 2001, said, "[He is] committed to reducing greenhouse gas emissions by drawing on the power of the market and the power of technology."

Emissions Trading Schemes

The most significant developments in incentive programs have occurred in the area of emissions trading, through which air pollutants are viewed as tradable commodities, each with its own regional, national, and even international markets. In an emissions trading program, companies that emit less than their assigned limits, or caps, of a pollutant can sell residual allowances on the open market or bank them for future transactions. This gives other, higher-polluting facilities a choice: either buy allowances and continue releasing the same pollutant or clean their own emissions—whichever is cheaper. The only stipulation is that regional environmental quality continue to meet mandated standards.

These so-called cap-and-trade schemes aren't new. The best-known example is the Acid Rain Program established under the Clean Air Act amendments of 1990, which allows electric utilities to trade allowance credits in sulfur dioxide (SO_2). Many experts point to this initiative, which achieved dramatic reductions in SO_2 at lower costs than expected, as an emissions trading success story. The EPA estimates that since the program was formalized in 1995, annual emissions of SO_2 have fallen by 4 million tons, while rainfall acidity in the Northeast has dropped by 25%. Dallas Burtraw, a senior fellow at Resources for the Future in Washington, D.C., says the program works well because it's simple, it sets firm environmental targets, it keeps transaction costs to a minimum, and it's transparent—meaning that information on available allowances and credit trades is freely available to the public.

The success of the Acid Rain Program has fueled the development of similar initiatives within the private sector. Undeterred by President Bush's rejection of the Kyoto Protocol, a diverse group of 34 major companies called the Chicago Climate Exchange (CCX) recently announced an emissions trading scheme for carbon dioxide and other greenhouse gases. Boasting high-profile members such as BP, Ford Motor Company, DuPont, and International Paper, this effort aims to reduce greenhouse gas emissions to 5% below 1999 levels by 2005. The CCX's role will be similar to that of an organized commodity exchange—it will establish the requisite technical infrastructure, common standards, and a computerized platform through which participants can trade in emissions reductions.

Richard Sandor, project leader at the CCX, points to the following hypothetical trade as an example of how the system will work: Two companies, a manufacturer with advanced pollution control technology and a power plant with older controls, agree to cut their combined emissions of greenhouse gases by three tons each for a total of six tons. Taking advantage of its superior technology, the manufacturer can cut its own emissions by five tons at minimal cost while the power plant can only reduce its own emissions cost-effectively by one ton. But by purchasing the rights to the additional two tons from the manufacturer, the power plant pays for another company to reduce greenhouse gases on its behalf. In this win—win situation, the manufacturer takes in revenues for reducing pollution while the power plant avoids higher costs by passing off its emissions reductions agreement to another source.

According to Sandor, the CCX will facilitate trades among seven midwestern states that together comprise the fourth-largest trading bloc in the world. The CCX also plans to include Brazil as a member, indicating the organization hopes to achieve an international presence. Says Sandor, "We've had a fantastic response from industry. We expect to be in the design phase for 12 months and to begin trading by 2002."

The states have also gotten into the game. In Southern California, a cap-and-trade program known as the Regional Clean Air Incentives Market, or RECLAIM, is being used to control SO_2 and nitrogen oxide (NO_x) air emissions from 360 industrial facilities, including power plants, in Los Angeles and the San Bernardino Valley. A coalition known as the Ozone Transport Commission, comprising the environmental agencies from 13 northeastern and midwestern states and the federal EPA, has developed a cap-and-trade program for NO_x. And elsewhere, in Chicago, a cap-and-trade program for volatile organic compounds was established by the Illinois EPA in early 2000.

The states have, for the most part, had a measure of success with these programs. The Ozone Transport Commission announced on 10 May 2001 that NO_x emissions for 1999 and 2000 were less than half those reported in 1990, before the cap-and-trade system was implemented. California's RECLAIM system has been in operation since 1993 but is just now beginning to demonstrate results. The reason for the delay, says Sam Atwood, spokesperson for the Diamond Bar–based South Coast Air Quality Management District, which coordinates RECLAIM, is that state-mandated "allocations" (a state term that defines the emissions that can be traded under the cap) for SO_2 and NO_x have only recently been set at levels below actual emissions released by industry. For several years after the program was initiated, facilities regulated under RECLAIM were allowed to emit SO_2 and NO_x at unusually high levels to cushion the economic shock of a recession that took place during the early 1990s. "By dropping the allocation levels below real emissions, we're just starting to cross over to the point where the incentive begins to kick in," says Atwood. "This is when we expect to see voluntary improvements in technology."

The Question of Mobile Sources

In a recent and somewhat controversial trend, emissions trading schemes have begun incorporating mobile sources, such as cars and trucks. Under this approach, stationary sources such as factories can obtain emission credits from regulators by paying to have old, highly polluting vehicles taken off the road. For example, RECLAIM recently issued a rule allowing stationary sources to receive mobile source credits by replacing diesel-fueled heavy-duty vehicles with cleaner-running alternatives.

Burtraw suggests this practice provides a major opportunity for cost savings. "It can be a lot less expensive to reduce emissions from mobile sources than stationary sources," he explains. But he concedes that adding mobile sources to the mix doesn't come without its own unique set of challenges. "People are all too willing to bring in an old lemon that barely runs so they can collect $500 from a utility company," he says. In a case like this, the emissions reduction is negligible because the car isn't driveable anyway.

Goffman says programs that include mobile sources need to incorporate safeguards to prevent this kind of abuse. The challenges exist, he says, but solutions are available if the systems are well designed at the outset. The South Coast Air Quality Management District, for example, only agrees to pay credits for cars that could continue running for three years or more.

Trading Issues

Despite a generally positive response from the stakeholder community, emissions trading still raises a number of important concerns. Perhaps the greatest worry is that it might lead to "hot spots," or areas of high pollutant exposure.

A company that cuts its emissions in half might help reduce average air pollution concentrations in a particular region, but this means little to those who live close to an older facility that buys credits rather than upgrading its pollution control technology.

John Walke, director of clean air projects with the National Resources Defense Council in Washington, D.C., suggests that environmental justice problems could arise if the dirtier facilities are located close to poor communities. "There are a lot of fundamental issues that need to be addressed with these systems," he says. "One is the extent to which pollution sources may be heavily localized in a particular area. It's important to consider how much pollution the neighboring communities are already saddled with."

And what about facilities located upwind of residential communities? Should they be allowed to purchase air pollution credits if downwind populations don't experience the benefit of cleaner emissions? Experts suggest the answer is no, and that hot spots can be avoided with effective planning. Suellen Keiner, director of the Center for the Economy and the Environment at the National Academy of Public Administration, a public interest group based in Washington, D.C., says potential solutions include discouraging trades across long distances and on-site review of credit uses to protect against hot spots.

Another incentive category that tends to trouble environmentalists is "open market" emissions trading, which is a scheme developed by the EPA in 1995. Unlike cap-and-trade programs, neither the overall sectors nor the individual trading sources regulated under an open market trading system are subject to a cap. Rather, any source that finds that its actual rate of emissions is below permitted levels for even a short time is eligible for credit that it can save for later or sell to another source. A chief concern is that under these schemes industry sets the standard for emissions allowances—not the regulatory agency. This is critical, given widespread agreement among stakeholders that health-protective standards should be set by the government on behalf of the public, while the means of compliance is left to the regulated community.

Burtraw says monitoring emissions under an open market system is particularly challenging. "Unlike cap-and-trade programs, which are often targeted toward large stationary sources that can be monitored at the stack, open trading is geared toward smaller sources, for example dry cleaners," he explains. "It's difficult and expensive to monitor actual emissions from these sources, so they tend to be estimated based on economic activity and the use of a given technology. On paper, open market trading seems promising, but in practice monitoring is often poor, and emissions inventories are weak."

Responding to New Jersey's announcement of an open market trading system for NO_x, approved by the EPA in July 2001, Environmental Defense called on the agency to withhold additional pending approvals in states including Michigan, New Hampshire, and Illinois. Also critical of open market trading is the Washington, D.C.–based organization Public Employees for Environmental Responsibility. This group, which says it represents anonymous EPA employees who fear the repercussions of speaking out publicly, issued a white paper in June 2000 called *Trading Thin Air* in which they claim that state and federal agencies don't have the ability to monitor these programs. According to the paper, open market trading could "cripple enforcement of the Clean Air Act against stationary sources of pollution."

Despite the uproar, many experts believe open market systems will improve over time. "I do have a healthy dose of skepticism about open market trading," says Burtraw. "It isn't based on sound policy and shouldn't be used on a wide scale. But I also see it as a way to include in trading programs a variety of smaller sources of emissions for which there do not exist emission inventories. At best, open market trading should be viewed as a transitional stepping stone to some better-developed institution that will emerge in the future."

Outlook for the Future

When applied to the nation as a whole, the EPA suggests in its April 2001 report that "the potential savings from widespread use of economic incentives ... could be almost one-fourth of the approximately $200 billion per year currently spent on environmental pollution control in the United States." In applying these tools, the EPA recommends that regulators consider their use in the context of political acceptability, potential for stimulating technological improvements, and enforceability. A number of important questions need to be considered. How many sources are there for each pollutant? Does a unit of pollution from each source have the same health and ecologic impact regardless of where it's released? Who's being affected by the pollution, and will the program reduce these impacts?

A key point raised by Burtraw is that incentives are a tool—not a solution. "You can compare incentives to a hammer," he says. "You can use a hammer to build a house, or you can use it to pull out the nails. This is the big issue we're facing now—if we use the incentives to back away from emissions reductions, then we're using the hammer to pull out the nails. But if we use incentives to aggressively pursue emissions reduction in the most cost-effective way, then we're building a stronger house for the future."

Brian Tokar  **NO**

Trading Away the Earth: Pollution Credits and the Perils of "Free Market Environmentalism"

The Republican takeover of Congress has unleashed an unprecedented assault on all forms of environmental regulation. From the Endangered Species Act to the Clean Water Act and the Superfund for toxic waste cleanup, laws that may need to be strengthened and expanded to meet the environmental challenges of the next century are instead being targeted for complete evisceration.

For some activists, this is a time to renew the grassroots focus of environmental activism, even to adopt a more aggressively anti-corporate approach that exposes the political and ideological agendas underlying the current backlash. But for many, the current impasse suggests that the movement must adapt to the dominant ideological currents of the time. Some environmentalists have thus shifted their focus toward voluntary programs, economic incentives and the mechanisms of the "free market" as means to advance the cause of environmental protection. Among the most controversial, and widespread, of these proposals are tradeable credits for the right to emit pollutants. These became enshrined in national legislation in 1990 with President George Bush's amendments to the 1970 Clean Air Act.

Even in 1990, "free market environmentalism" was not a new phenomenon. In the closing years of the 1980s, an odd alliance had developed among corporate public relations departments, conservative think tanks such as the American Enterprise Institute, Bill Clinton's Democratic Leadership Council (DLC), and mainstream environmental groups such as the Environmental Defense Fund. The market-oriented environmental policies promoted by this eclectic coalition have received little public attention, but have nonetheless significantly influenced debates over national policy.

Glossy catalogs of "environmental products," television commercials featuring environmental themes, and high profile initiatives to give corporate officials a "greener" image are the hallmarks of corporate environmentalism in the 1990s. But the new market environmentalism goes much further than these showcase efforts. It represents a wholesale effort to recast environmental protec-

From Brian Tokar, "Trading Away the Earth: Pollution Credits and the Perils of 'Free Market Environmentalism,'" *Dollars & Sense* (March/April 1996). Copyright © 1996 by Economic Affairs Bureau, Inc. Reprinted by permission. *Dollars & Sense* is a progressive economics magazine published six times a year.

tion based on a model of commercial transactions within the marketplace. "A new environmentalism has emerged," writes economist Robert Stavins, who has been associated with both the Environmental Defense Fund and the DLC's Progressive Policy Institute, "that embraces ... market-oriented environmental protection policies."

Today, aided by the anti-regulatory climate in Congress, market schemes such as trading pollution credits are granting corporations new ways to circumvent environmental concerns, even as the same firms try to pose as champions of the environment. While tradeable credits are sometimes presented as a solution to environmental problems, in reality they do nothing to reduce pollution—at best they help businesses reduce the costs of complying with limits on toxic emissions. Ultimately, such schemes abdicate control over critical environmental decisions to the very same corporations that are responsible for the greatest environmental abuses.

How It Works, and Doesn't

A close look at the scheme for nationwide emissions trading reveals a particular cleverness; for true believers in the invisible hand of the market, it may seem positively ingenious. Here is how it works: The 1990 Clean Air Act amendments were designed to halt the spread of acid rain, which has threatened lakes, rivers and forests across the country. The amendments required a reduction in the total sulfur dioxide emissions from fossil fuel burning power plants, from 19 to just under 9 million tons per year by the year 2000. These facilities were targeted as the largest contributors to acid rain, and participation by other industries remains optional. To achieve this relatively modest goal for pollution reduction, utilities were granted transferable allowances to emit sulfur dioxide in proportion to their current emissions. For the first time, the ability of companies to buy and sell the "right" to pollute was enshrined in U.S. law.

Any facility that continued to pollute more than its allocated amount (roughly half of its 1990 rate) would then have to buy allowances from someone who is polluting less. The 110 most polluting facilities (mostly coal burners) were given five years to comply, while all the others would have until the year 2000. Emissions allowances were expected to begin selling for around $500 per ton of sulfur dioxide, and have a theoretical ceiling of $2000 per ton, which is the legal penalty for violating the new rules. Companies that could reduce emissions for less than their credits are worth would be able to sell them at a profit, while those that lag behind would have to keep buying credits at a steadily rising price. For example, before pollution trading every company had to comply with environmental regulations, even if it cost one firm twice as much as another to do so. Under the new system, a firm could instead choose to exceed the mandated levels, purchasing credits from the second firm instead of implementing costly controls. This exchange would save money, but in principle yield the same overall level of pollution as if both companies had complied equally. Thus, it is argued, market forces will assure that the most cost-effective means of reducing acid rain will be implemented first, saving the economy billions of dollars in "excess" pollution control costs.

Defenders of the Bush plan claimed that the ability to profit from pollution credits would encourage companies to invest more in new environmental technologies than before. Innovation in environmental technology, they argued, was being stifled by regulations mandating specific pollution control methods. With the added flexibility of tradeable credits, companies could postpone costly controls—through the purchase of some other company's credits—until new technologies became available. Proponents argued that, as pollution standards are tightened over time, the credits would become more valuable and their owners could reap large profits while fighting pollution.

Yet the program also included many pages of rules for extensions and substitutions. The plan eliminated requirements for backup systems on smokestack scrubbers, and then eased the rules for estimating how much pollution is emitted when monitoring systems fail. With reduced emissions now a marketable commodity, the range of possible abuses may grow considerably, as utilities will have a direct financial incentive to manipulate reporting of their emissions to improve their position in the pollution credits market.

Once the EPA actually began auctioning pollution credits in 1993, it became clear that virtually nothing was going according to their projections. The first pollution credits sold for between $122 and $310, significantly less than the agency's estimated minimum price, and by 1995, bids at the EPA's annual auction of sulfur dioxide allowances averaged around $130 per ton of emissions. As an artificial mechanism superimposed on existing regulatory structures, emissions allowances have failed to reflect the true cost of pollution controls. So, as the value of the credits has fallen, it has become increasingly attractive to buy credits rather than invest in pollution controls. And, in problem areas air quality can continue to decline, as companies in some parts of the country simply buy their way out of pollution reductions.

At least one company has tried to cash in on the confusion by assembling packages of "multi-year streams of pollution rights" specifically designed to defer or supplant purchases of new pollution control technologies. "What a scrubber really is, is a decision to buy a 30-year stream of allowances," John B. Henry of Clean Air Capital Markets told the *New York Times*, with impeccable financial logic. "If the price of allowances declines in future years," paraphrased the *Times*, "the scrubber would look like a bad buy."

Where pollution credits have been traded between companies, the results have often run counter to the program's stated intentions. One of the first highly publicized deals was a sale of credits by the Long Island Lighting Company to an unidentified company located in the Midwest, where much of the pollution that causes acid rain originates. This raised concerns that places suffering from the effects of acid rain were shifting "pollution rights" to the very region it was coming from. One of the first companies to bid for additional credits, the Illinois Power Company, canceled construction of a $350 million scrubber system in the city of Decatur, Illinois. "Our compliance plan is based almost totally on purchase of credits," an Illinois Power spokesperson told the *Wall Street Journal*. The comparison with more traditional forms of commodity trading came full circle in 1991, when the government announced that the entire system for trading and auctioning emissions allowances would be administered by the Chicago Board of Trade, long famous for its ever-frantic markets in everything from grain futures and pork bellies to foreign currencies.

Some companies have chosen not to engage in trading pollution credits, proceeding with pollution control projects, such as the installation of new scrubbers, that were planned before the credits became available. Others have switched to low-sulfur coal and increased their use of natural gas. If the 1990 Clean Air Act amendments are to be credited for any overall improvement in the air quality, it is clearly the result of these efforts and not the market in tradeable allowances.

Yet while some firms opt not to purchase the credits, others, most notably North Carolina-based Duke Power, are aggressively buying allowances. At the 1995 EPA auction, Duke Power alone bought 35% of the short-term "spot" allowances for sulfur dioxide emissions, and 60% of the long-term allowances redeemable in the years 2001 and 2002. Seven companies, including five utilities and two brokerage firms, bought 97% of the short term allowances that were auctioned in 1995, and 92% of the longer-term allowances, which are redeemable in 2001 and 2002. This gives these companies significant leverage over the future shape of the allowances market.

The remaining credits were purchased by a wide variety of people and organizations, including some who sincerely wished to take pollution allowances out of circulation. Students at several law schools raised hundreds of dollars, and a group at the Glens Falls Middle School on Long Island raised $3,171 to purchase 21 allowances, equivalent to 21 tons of sulfur dioxide emissions over the course of a year. Unfortunately, this represented less than a tenth of one percent of the allowances auctioned off in 1995.

Some of these trends were predicted at the outset. "With a tradeable permit system, technological improvement will normally result in lower control costs and falling permit prices, rather than declining emissions levels," wrote Robert Stavins and Brad Whitehead (a Cleveland-based management consultant with ties to the Rockefeller Foundation) in a 1992 policy paper published by the Progressive Policy Institute. Despite their belief that market-based environmental policies "lead automatically to the cost-effective allocation of the pollution control burden among firms," they are quite willing to concede that a tradeable permit system will not in itself reduce pollution. As the actual pollution levels still need to be set by some form of regulatory mandate, the market in tradeable allowances merely gives some companies greater leverage over how pollution standards are to be implemented.

Without admitting the underlying irrationality of a futures market in pollution, Stavins and Whitehead do acknowledge (albeit in a footnote to an Appendix) that the system can quite easily be compromised by large companies' "strategic behavior." Control of 10% of the market, they suggest, might be enough to allow firms to engage in "price-setting behavior," a goal apparently sought by companies such as Duke Power. To the rest of us, it should be clear that if pollution credits are like any other commodity that can be bought, sold and traded, then the largest "players" will have substantial control over the entire "game." Emissions trading becomes yet another way to assure that large corporate interests will remain free to threaten public health and ecological survival in their unchallenged pursuit of profit.

Trading the Future

Mainstream groups like the Environmental Defense Fund (EDF) continue to throw their full support behind the trading of emissions allowances, including the establishment of a futures market in Chicago. EDF senior economist Daniel Dudek described the trading of acid rain emissions as a "scale model" for a much more ambitious plan to trade emissions of carbon dioxide and other gases responsible for global warming. This plan was unveiled shortly after the passage of the 1990 Clean Air Act amendments, and was endorsed by then-Senator Al Gore as a way to "rationalize investments" in alternatives to carbon dioxide-producing activities.

International emissions trading gained further support via a U.N. Conference on Trade and Development study issued in 1992. The report was coauthored by Kidder and Peabody executive and Chicago Board of Trade director Richard Sandor, who told the *Wall Street Journal*, "Air and water are simply no longer the 'free goods' that economists once assumed. They must be redefined as property rights so that they can be efficiently allocated."

Radical ecologists have long decried the inherent tendency of capitalism to turn everything into a commodity; here we have a rare instance in which the system fully reveals its intentions. There is little doubt that an international market in "pollution rights" would widen existing inequalities among nations. Even within the United States, a single large investor in pollution credits would be able to control the future development of many different industries. Expanded to an international scale, the potential for unaccountable manipulation of industrial policy by a few corporations would easily compound the disruptions already caused by often reckless international traders in stocks, bonds and currencies.

However, as long as public regulation of industry remains under attack, tradeable credits and other such schemes will continue to be promoted as market-savvy alternatives. Along with an acceptance of pollution as "a by-product of modern civilization that can be regulated and reduced, but not eliminated," to quote another Progressive Policy Institute paper, self-proclaimed environmentalists will call for an end to "widespread antagonism toward corporations and a suspicion that anything supported by business was bad for the environment." Market solutions are offered as the only alternative to the "inefficient," "centralized," "command-and-control" regulations of the past, in language closely mirroring the rhetoric of Cold War anti-communism.

While specific technology-based standards can be criticized as inflexible and sometimes even archaic, critics choose to forget that in many cases, they were instituted by Congress as a safeguard against the widespread abuses of the Reagan-era EPA. During the Reagan years, "flexible" regulations opened the door to widely criticized—and often illegal—bending of the rules for the benefit of politically favored corporations, leading to the resignation of EPA administrator Anne Gorsuch Burford and a brief jail sentence for one of her more vocal legal assistants.

The anti-regulatory fervor of the present Congress is bringing a variety of other market-oriented proposals to the fore. Some are genuinely offered to further environmental protection, while others are far more cynical attempts to replace public regulations with virtual blank checks for polluters. Some have proposed a direct charge for pollution, modeled after the comprehensive pollution taxes that have proved popular in Western Europe. Writers as diverse as Supreme Court Justice Stephen Breyer, American Enterprise Institute economist Robert Hahn and environmental business guru Paul Hawken have defended pollution taxes as an ideal market-oriented approach to controlling pollution. Indeed, unlike tradeable credits, taxes might help reduce pollution beyond regulatory levels, as they encourage firms to control emissions as much as possible. With credits, there is no reduction in pollution below the threshold established in legislation. (If many companies were to opt for substantial new emissions controls, the market would soon be glutted and the allowances would rapidly become valueless.) And taxes would work best if combined with vigilant grassroots activism that makes industries accountable to the communities in which they operate. However, given the rapid dismissal of Bill Clinton's early plan for an energy tax, it is most likely that any pollution tax proposal would be immediately dismissed by Congressional ideologues as an outrageous new government intervention into the marketplace.

Air pollution is not the only environmental problem that free marketeers are proposing to solve with the invisible hand. Pro-development interests in Congress have floated various schemes to replace the Endangered Species Act with a system of voluntary incentives, conservation easements and other schemes through which landowners would be compensated by the government to protect critical habitat. While these proposals are being debated in Congress, the Clinton administration has quietly changed the rules for administering the Act in a manner that encourages voluntary compliance and offers some of the very same loopholes that anti-environmental advocates have sought. This, too, is being offered in the name of cooperation and "market environmentalism."

Debates over the management of publicly-owned lands have inspired far more outlandish "free market" schemes. "Nearly all environmental problems are rooted in society's failure to adequately define property rights for some resource," economist Randal O'Toole has written, suggesting a need for "property rights for owls and salmon" developed to "protect them from pollution." O'Toole initially gained the attention of environmentalists in the Pacific Northwest for his detailed studies of the inequities of the U.S. Forest Service's long-term subsidy programs for logging on public lands. Now he has proposed dividing the National Forest system into individual units, each governed by its users and operated on a for-profit basis, with a portion of user fees allocated for such needs as the protection of biological diversity. Environmental values, from clean water to recreation to scenic views, should simply be allocated their proper value in the marketplace, it is argued, and allowed to out-compete unsustainable resource extraction. Other market advocates have suggested far more sweeping transfers of federal lands to the states, an idea seen by many in the West as a first step toward complete privatization.

Market enthusiasts like O'Toole repeatedly overlook the fact that ecological values are far more subjective than the market value of timber and minerals removed from public lands. Efforts to quantify these values are based on various sociological methods, market analysis and psychological studies. People are asked how much they would pay to protect a resource, or how much money they would accept to live without it, and their answers are compared with the prices of everything from wilderness expeditions to vacation homes. Results vary widely depending on how questions are asked, how knowledgeable respondents are, and what assumptions are made in the analysis. Environmentalists are rightfully appalled by such efforts as a recent Resources for the Future study designed to calculate the value of human lives lost due to future toxic exposures. Outlandish absurdities like property rights for owls arouse similar skepticism.

The proliferation of such proposals—and their increasing credibility in Washington—suggest the need for a renewed debate over the relationship between ecological values and those of the free market. For many environmental economists, the processes of capitalism, with a little fine tuning, can be made to serve the needs of environmental protection. For many activists, however, there is a fundamental contradiction between the interconnected nature of ecological processes and an economic system which not only reduces everything to isolated commodities, but seeks to manipulate those commodities to further the single, immutable goal of maximizing individual gain. An ecological economy may need to more closely mirror natural processes in their stability, diversity, long time frame, and the prevalence of cooperative, symbiotic interactions over the more extreme forms of competition that thoroughly dominate today's economy. Ultimately, communities of people need to reestablish social control over economic markets and relationships, restoring an economy which, rather than being seen as the engine of social progress, is instead, in the words of economic historian Karl Polanyi, entirely "submerged in social relationships."

Whatever economic model one proposes for the long-term future, it is clear that the current phase of corporate consolidation is threatening the integrity of the earth's living ecosystems—and communities of people who depend on those ecosystems—as never before. There is little room for consideration of ecological integrity in a global economy where a few ambitious currency traders can trigger the collapse of a nation's currency, its food supply, or a centuries-old forest ecosystem before anyone can even begin to discuss the consequences. In this kind of world, replacing our society's meager attempts to restrain and regulate corporate excesses with market mechanisms can only further the degradation of the natural world and threaten the health and well-being of all the earth's inhabitants.

POSTSCRIPT

Can Pollution Rights Trading Effectively Control Environmental Problems?

Does pollution rights trading give major corporate polluters too much power to control and manipulate the market for emission credits? This is one of the key issues that continues to inspire developing countries to withhold their endorsement of the greenhouse gas emissions trading provisions of the Kyoto Protocol. The evidence that Tokar cites, which is primarily based on short-term experience with trading in sulfur dioxide pollution credits, does not appear to fully justify the broad generalizations he makes about the inherent perils in market-based regulatory plans. Recent assessments of the Acid Rain Program by the EPA and such organizations as the Environmental Defense Fund are more positive. So is the corporate world: In "Economic Man, Cleaner Planet," *The Economist* (September 29, 2001), it is asserted that economic incentives have proved very useful and that "market forces are only just beginning to make inroads into green policymaking." T. H. Tietenberg, *Emissions Trading: Principles and Practice*, 2nd ed. (RFF Press, 2006), notes that emissions trading has a definite niche in pollution control policies. Ruth Greenspan Bell, "The Kyoto Placebo," *Issues in Science and Technology* (Winter 2006), notes that "Though heavily promoted by the World Bank, U.S.-style environmental trading has yet to be tested on a global scale and has never been successfully deployed on a national level in the developing world." There is more to be gained by helping developing nations gain regulatory skills ("What to Do about Climate Change," *Foreign Affairs*, May/June 2006).

The position of those who are ideologically opposed to pollution rights is concisely stated in Michael J. Sandel's op-ed piece "It's Immoral to Buy the Right to Pollute," *The New York Times* (December 15, 1997). In "Selling Air Pollution," *Reason* (May 1996), Brian Doherty supports the concept of pollution rights trading but argues that the kind of emission cap imposed in the case of sulfur dioxide is an inappropriate constraint on what he believes should be a completely free-market program. Richard A. Kerr, in "Acid Rain Control: Success on the Cheap," *Science* (November 6, 1998), contends that emissions trading has greatly reduced acid rain and that the annual cost has been about a tenth of the $10 billion initially forecast. According to Barry D. Solomon and Russell Lee, in "Emissions Trading Systems and Environmental Justice," *Environment* (October 2000), "a significant part of the opposition to emissions trading programs is a perception that they do little to reduce environmental injustice and can even make it worse." However, Byron Swift, in "Allowance Trading and Potential Hot Spots—Good News From the Acid Rain Program," *Environment Reporter* (May 12, 2000), argues that the success of the EPA's emission trading program has not led to the creation of pollution "hot spots" as feared by some critics.

ISSUE 7

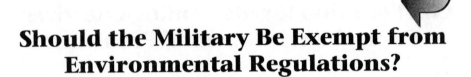

Should the Military Be Exempt from Environmental Regulations?

YES: Benedict S. Cohen, from "Impact of Military Training on the Environment," Testimony before the Senate Committee on Environment and Public Works (April 2, 2003)

NO: Jamie Clark, from "Impact of Military Training on the Environment," Testimony before the Senate Committee on Environment and Public Works (April 2, 2003)

ISSUE SUMMARY

YES: Benedict S. Cohen argues that environmental regulations interfere with military training and other "readiness" activities, and that though the U.S. Department of Defense will continue "to provide exemplary stewardship of the lands and natural resources in our trust," those regulations must be revised to permit the military to do its job without interference.

NO: Jamie Clark argues that reducing the Department of Defense's environmental obligations is dangerous because both people and wildlife would be threatened with serious, irreversible, and unnecessary harm.

Most of us have heard of "scorched earth" wars, in which an army destroys forests and farms in order to deny the enemy their benefit. We have surely seen the images of a Europe laid waste by World War II. More recently, we may recall, the Gulf War saw oil deliberately released to flood desert sands and the waters of the Persian Gulf. Enough smoke poured from burning oil wells to threaten both local climate change and human health. See, for example, Randy Thomas, "Eco War," *Earth Island Journal* (Spring 1991), B. Ruben, "Gulf Smoke Screens," *Environmental Action* (July/August 1991), and Jeffrey L. Lange, David A. Schwartz, Bradley N. Doebbeling, Jack M. Heller, and Peter S. Thorne, "Exposures to the Kuwait Oil Fires and Their Association with Asthma and Bronchitis among Gulf War Veterans," *Environmental Health Perspectives* (November 2002). Weaponry can have environmental effects by destroying dams, by physically destroying plants and animals, by causing erosion, and by disseminating toxic materials; see Henryk Bem and Firyal Bou-Rabee, "Environmental and Health Consequences of Depleted Uranium Use in the 1991 Gulf War," *Environment International* (March 2004).

The environmental impact of war would seem impossible to deny. But even after the Gulf War, the U.S. Department of Defense tried to suppress satellite photos showing the extent of the damage; see Shirley Johnston, "Gagged on Smoke," *Earth Island Journal* (Summer 1991). And for many years, it insisted that nuclear war was survivable, until researchers made it clear that even a small nuclear war would produce a "nuclear winter" that would probably destroy civilization, if not the human species. See T. Rueter and T. Kalil, "Nuclear Strategy and Nuclear Winter," *World Politics* (July 1991), and Carl Sagan and Richard Turco, *A Path Where No Man Thought: Nuclear Winter and Its Implications* (Random House, 1990).

Preparations for war may also have serious environmental impacts. Puerto Rico's island of Vieques was long a bomb depot and bombing range for the U.S. Navy. Local residents protested vigorously and documented heavy-metal contamination of the local ecosystem. After the Navy left the island, it "has continued to deny that it has been anything but an excellent environmental steward in Vieques." See Shane DuBow and Scott S. Warren, "Vieques on the Verge," *Smithsonian* (January 2004).

In 2002, the U.S. Congress, through the Readiness and Range Preservation Initiative, granted the Department of Defense a temporary exemption to the Migratory Bird Treaty Act that allowed the "incidental taking" of endangered birds during bombing and other training on military lands. In 2003, the Department of Defense asked Congress for additional exemptions from environmental regulations, specifically the Clean Air Act, Marine Mammal Protection Act, Endangered Species Act, Migratory Bird Treaty Act, and federal toxic waste laws. Paul Mayberry, deputy undersecretary of defense for readiness, said the exemptions were justified because many environmental restrictions were putting the nation's military readiness at stake. See "Pentagon Seeks Clarity in Environmental Laws Affecting Ranges," Agency Group 09, FDCH Regulatory Intelligence Database (March 21, 2003). The United States Senate Committee on Environment and Public Works held a hearing on the "Impact of Military Training on the Environment" on April 2, 2003. In the following selections, Benedict S. Cohen, deputy general counsel for environment and installations, Department of Defense, argues in his testimony before the Committee that environmental regulations interfere with military training and other "readiness" activities, and that though the Department of Defense will continue "to provide exemplary stewardship of the lands and natural resources in our trust" those regulations must be revised to permit the military to do its job without interference. Jamie Clark, senior vice president for conservation programs, National Wildlife Federation, argues that reducing the Department of Defense's environmental obligations is dangerous for two reasons. First, both people and wildlife will be threatened with serious, irreversible, and unnecessary harm. Second, other federal agencies and industry sectors with important missions, using the same logic as used here by the Department of Defense, would demand similar exemptions from environmental laws.

YES

Benedict S. Cohen

Impact of Military Training on the Environment

noregu

id inteslve with training

Mr. Chairman and distinguished members of this Committee, I appreciate the opportunity to discuss with you the very important issue of sustaining our test and training capabilities, and the legislative proposal that the Administration has put forward in support of that objective. In these remarks I would like particularly to address some of the comments and criticisms offered concerning these legislative proposals.

Addressing Encroachment

We have only recently begun to realize that a broad array of encroachment pressures at our operational ranges are increasingly constraining our ability to conduct the testing and training that we must do to maintain our technological superiority and combat readiness. Given World events today, we know that our forces and our weaponry must be more diverse and flexible than ever before. Unfortunately, this comes at the same time that our ranges are under escalating demands to sustain the diverse operations required today, and that will be increasingly required in the future.

This current predicament has come about as a cumulative result of a slow but steady process involving many factors. Because external pressures are increasing, the adverse impacts to readiness are growing. Yet future testing and training needs will only further exacerbate these issues, as the speed and range of our weaponry and the number of training scenarios increase in response to real-world situations our forces will face when deployed. We must therefore begin to address these issues in a much more comprehensive and systematic fashion and understand that they will not be resolved overnight, but will require a sustained effort.

Environmental Stewardship

Before I address our comprehensive strategy, let me first emphasize our position concerning environmental stewardship. Congress has set aside 25 million acres of land—some 1.1% of the total land area in the United States. These

From Testimony before the Senate Committee on Environment and Public Works, April 2, 2003. Notes omitted.

lands were entrusted to the Department of Defense (DoD) to use efficiently and to care for properly. In executing these responsibilities we are committed to more than just compliance with the applicable laws and regulations. We are committed to protecting, preserving, and, when required, restoring, and enhancing the quality of the environment.

- We are investing in pollution prevention technologies to minimize or reduce pollution in the first place. Cleanup is far more costly than prevention.
- We are managing endangered and threatened species, and all of our natural resources, through integrated natural resource planning
- We are cleaning up contamination from past practices on our installations and are building a whole new program to address unexploded ordnance on our closed, transferring, and transferred ranges.

The American people have entrusted these 25 million acres to our care. Yet, in many cases, these lands that were once "in the middle of nowhere" are now surrounded by homes, industrial parks, retail malls, and interstate highways.

On a daily basis our installation and range managers are confronted with a myriad of challenges—urban sprawl, noise, air quality, air space, frequency spectrum, endangered species, marine mammals, and unexploded ordnance. Incompatible development outside our fence-lines is changing military flight paths for approaches and take-offs to patterns that are not militarily realistic—results that lead to negative training and potential harm to our pilots. With over 300 threatened and endangered species on DoD lands, nearly every major military installation and range has one or more endangered species, and for many species, these DoD lands are often the last refuge. Critical habitat designations for an ever increasing number of threatened or endangered species limit our access to and use of thousands of acres at many of our training and test ranges. The long-term prognosis is for this problem to intensify as new species are continually added to the threatened and endangered list.

Much too often these many encroachment challenges bring about unintended consequences to our readiness mission. This issue of encroachment is not going away. Nor is our responsibility to "train as we fight."

2003 READINESS AND RANGE PRESERVATION INITIATIVE (RRPI)

Overview

DoD's primary mission is maintaining our Nation's military readiness, today and into the future. DoD is also fully committed to high-quality environmental stewardship and the protection of natural resources on its lands. However, expanding restrictions on training and test ranges are limiting realistic preparations for combat and therefore our ability to maintain the readiness of America's military forces.

Last year, the Administration submitted to Congress an eight-provision legislative package, the Readiness and Range Preservation Initiative (RRPI). Congress enacted three of those provisions as part of the National Defense Authorization Act for Fiscal Year 2003. Two of the enacted provisions allow us to cooperate more effectively with local and State governments, as well as private entities, to plan for growth surrounding our training ranges by allowing us to work toward preserving habitat for imperiled species and assuring development and land uses that are compatible with our training and testing activities on our installations.

Under the third provision, Congress provided the Department a regulatory exemption under the Migratory Bird Treaty Act for the incidental taking of migratory birds during military readiness activities. We are grateful to Congress for these provisions, and especially for addressing the serious readiness concerns raised by recent judicial expansion of the prohibitions under the Migratory Bird Treaty Act. I am pleased to inform this Committee that as a direct result of your legislation, Air Force B-1 and B-52 bombers, forward deployed to Anderson Air Force Base, Guam, are performing dry run training exercises over the Navy's Bombing Range at Farallon de Medinilla in the Commonwealth of the Northern Mariana Islands.

Last year, Congress also began consideration of the other five elements of our Readiness and Range Preservation Initiative. These five proposals remain essential to range sustainment and are as important this year as they were last year—maybe more so. The five provisions submitted this year reaffirm the principle that military lands, marine areas, and airspace exist to ensure military preparedness, while ensuring that the Department of Defense remains fully committed to its stewardship responsibilities. These five remaining provisions:

- Authorize use of Integrated Natural Resource Management Plans in appropriate circumstances as a substitute for critical habitat designation;
- Reform obsolete and unscientific elements of the Marine Mammal Protection Act, such as the definition of "harassment," and add a national security exemption to that statute;
- Modestly extend the allowable time for military readiness activities like bed-down of new weapons systems to comply with Clean Air Act; and
- Limit regulation of munitions on operational ranges under the Comprehensive Environmental Response, Compensation, and Liability Act (CERCLA) and Resource Conservation and Recovery Act (RCRA), if and only if those munitions and their associated constituents remain there, and only while the range remains operational.

Before discussing the specific elements of our proposal, I would like to address some overarching issues. A consistent theme in criticisms of our proposal is that it would bestow a sweeping or blanket exemption for the Defense Department from the Nation's environmental laws. No element of this allegation is accurate.

First, our initiative would apply only to military readiness activities, not to closed ranges or ranges that close in the future, and not to "the routine operation of installation operating support functions, such as administrative offices, military exchanges, commissaries, water treatment facilities, storage, schools, housing, motor pools . . . nor the operation of industrial activities, or the construction or demolition of such facilities." Our initiative thus is not applicable to the Defense Department activities that have traditionally been of greatest concern to state and federal regulators. It does address only uniquely military activities—what DoD does that is unlike any other governmental or private activity. DoD is, and will remain, subject to precisely the same regulatory requirements as the private sector when we perform the same types of activities as the private sector. We seek alternative forms of regulation only for the things we do that have no private-sector analogue: military readiness activities.

Moreover, our initiative largely affects environmental regulations that don't apply to the private sector or that disproportionately impact DoD:

- Endangered Species Act "critical habitat" designation has limited regulatory consequences on private lands, but can have crippling legal consequences for military bases.
- Under the Marine Mammal Protection Act, the private sector's Incidental Take Reduction Plans give commercial fisheries the flexibility to take significant numbers of marine mammal each year, but are unavailable to DoD—whose critical defense activities are being halted despite far fewer marine mammal deaths or injuries a year.
- The Clean Air Act's "conformity" requirement applies only to federal agencies, not the private sector.

Our proposals therefore are of the same nature as the relief Congress afforded us last year under the Migratory Bird Treaty Act, which environmental groups are unable to enforce against private parties but, as a result of a 2000 circuit court decision were able and willing to enforce, in wartime, against vital military readiness activities of the Department of Defense.

Nor does our initiative "exempt" even our readiness activities from the environmental laws; rather, it clarifies and confirms existing regulatory policies that recognize the unique nature of our activities. It codifies and extends EPA's existing Military Munitions Rule; confirms the prior Administration's policy on Integrated Natural Resource Management Plans and critical habitat; codifies the prior Administration's policy on "harassment" under the Marine Mammal Protection Act; ratifies longstanding state and federal policy concerning regulation under RCRA and CERCLA of our operational ranges; and gives states and DoD temporary flexibility under the Clean Air Act. Our proposals are, again, of the same nature as the relief Congress provided us under the Migratory Bird Treaty Act last year, which codified the prior Administration's position on DoD's obligations under the Migratory Bird Treaty Act.

Ironically, the alternative proposed by many of our critics—invocation of existing statutory emergency authority—would fully exempt DoD from the waived statutory requirements for however long the exemption lasted, a more far-reaching solution than the alternative forms of regulation we propose.

Accordingly, our proposals are neither sweeping nor exemptive; to the contrary, it is our critics who urge us to rely on wholesale, repeated use of emergency exemptions for routine, ongoing readiness activities that could easily be accommodated by minor clarifications and changes to existing law.

Existing emergency authorities

As noted above, many of our critics state that existing exemptions in the environmental laws and the consultative process in 10 USC 2014 render the Defense Department's initiative unnecessary.

Although existing exemptions are a valuable hedge against unexpected future emergencies, they cannot provide the legal basis for the Nation's everyday military readiness activities.

- The Marine Mammal Protection Act, like the Migratory Bird Treaty Act the Congress amended last year, has no national security exemption.
- 10 USC 2014, which allows a delay of at most five days in regulatory actions significantly affecting military readiness, is a valuable insurance policy for certain circumstances, but allows insufficient time to resolve disputes of any complexity. The Marine Corps' negotiations with the Fish and Wildlife Service over excluding portions of Camp Pendleton from designation as critical habitat took months. More to the point, Section 2014 merely codifies the inherent ability of cabinet members to consult with each other and appeal to the President. Since it does not address the underlying statutes giving rise to the dispute, it does nothing for readiness in circumstances where the underlying statute itself—not an agency's exercise of discretion—is the source of the readiness problem. This is particularly relevant to our RRPI proposal because none of the five amendments we propose have been occasioned by the actions of state or federal regulators. Four of the five proposed amendments (RCRA, CERCLA, MMPA, and ESA), like the MBTA amendment Congress passed last year, were occasioned by private litigants seeking to overturn federal regulatory policy and compel federal regulators to impose crippling restrictions on our readiness activities. The fifth, our Clean Air Act amendment, was proposed because DoD and EPA concluded that the Act's "general conformity" provision unnecessarily restricted the flexibility of DoD, state, and federal regulators to accommodate military readiness activities into applicable air pollution control schemes. Section 2014, therefore, although useful in some circumstances, would be of no use in addressing the critical readiness issues that our five RRPI initiatives address.
- Most of the environmental statutes with emergency exemptions clearly envisage that they will be used in rare circumstances, as a last resort, and only for brief periods.
- Under these statutes, the decision to grant an exemption is vested in the President, under the highest possible standard: "the paramount interest of the United States," a standard understood to involve exceptionally grave threats to national survival. The exemptions are also

usually limited to renewable periods of a year (or in some cases as much as three years for certain requirements).

- The ESA's section 7(j) exemption process, which differs significantly from typical emergency exemptions, allows the Secretary of Defense to direct the Endangered Species Committee to exempt agency actions in the interest of national security. However, the Endangered Species Committee process has given rise to procedural litigation in the past, potentially limiting its usefulness—especially in exigent circumstances. In addition, because it applies only to agency actions rather than to ranges themselves, any exemption secured by the Department would be of limited duration and benefit: because military testing and training evolve continuously, such an exemption would lose its usefulness over time as the nature of DoD actions on the range evolved.
- The exemption authorities do not work well in addressing those degradations in readiness that result from the cumulative, incremental effects of many different regulatory requirements and actions over time (as opposed to a single major action).
- Moreover, readiness is maintained by thousands of discrete test and training activities at hundreds of locations. Many of these are being adversely affected by environmental provisions. Maintaining military readiness through use of emergency exemptions would therefore involve issuing and renewing scores or even hundreds of Presidential certifications annually.
- And although a discrete activity (e.g., a particular carrier battle group exercise) might only rarely rise to the extraordinary level of a "paramount national interest," it is clearly intolerable to allow all activities that do not individually rise to that level to be compromised or ended by overregulation.
- Finally, to allow continued unchecked degradation of readiness until an external event like Pearl Harbor or September 11 caused the President to invoke the exemption would mean that our military forces would go into battle having received degraded training, with weapons that had received degraded testing and evaluation. Only the testing and training that occurred after the emergency exemption was granted would be fully realistic and effective.

The Defense Department believes that it is unacceptable as a matter of public policy for indispensable readiness activities to require repeated invocation of emergency authority—particularly when narrow clarifications of the underlying regulatory statutes would enable both essential readiness activities and the protection of the environment to continue. Congress would never tolerate a situation in which another activity vital to the Nation, like the practice of medicine, was only permitted to go forward through the repeated use of emergency exemptions.

That having been said, I should make clear that the Department of Defense is in no way philosophically opposed to the use of national security waivers or exemptions where necessary. We believe that every environmental statute should have a well-crafted exemption, as an insurance policy, though we continue to hope that we will seldom be required to have recourse to them. . . .

Specific Proposals

This year's proposals do include some clarifications and modifications based on events since last year. Of the five, the Endangered Species Act (ESA) and Clean Air Act provisions are unchanged. Let me address the changed provisions first.

RCRA and CERCLA

The legislation would codify and confirm the longstanding regulatory policy of EPA and every state concerning regulation of munitions use on operational ranges under RCRA and CERCLA. It would confirm that military munitions are subject to EPA's 1997 Military Munitions Rule while on range, and that cleanup of operational ranges is not required so long as material stays on the range. If such material moves off range, it still must be addressed promptly under existing environmental laws. Moreover, if munitions constituents cause an imminent and substantial endangerment on range, EPA will retain its current authority to address it on range under CERCLA section 106. (Our legislation explicitly reaffirms EPA's section 106 authority.) The legislation similarly does not modify the overlapping protections of the Safe Drinking Water Act, NEPA, and the ESA against environmentally harmful activities at operational military bases. The legislation has no effect whatsoever on DoD's cleanup obligations under RCRA or CERCLA at Formerly Used Defense Sites, closed ranges, ranges that close in the future, or waste management practices involving munitions even on operational ranges (such as so-called OB/OD activities).

The core of our concern is to protect against litigation the longstanding, uniform regulatory policy that (1) use of munitions for testing and training on an operational range is not a waste management activity or the trigger for cleanup requirements, and (2) that the appropriate trigger for DoD to address the environmental consequences of such routine test and training uses involving discharge of munitions is (a) when the range closes, (b) when munitions or their elements migrate or threaten to migrate off-range, or (c) when munitions or their elements create an imminent and substantial endangerment on-range. . . .

This legislation is needed because of RCRA's broad definition of "solid waste," and because states possess broad authority to adopt more stringent RCRA regulations than EPA (enforceable both by the states and by environmental plaintiffs). EPA therefore has quite limited ability to afford DoD regulatory relief under RCRA. Similarly, the broad statutory definition of "release" under CERCLA may also limit EPA's ability to afford DoD regulatory relief. And the President's site-specific, annually renewable waiver (under a paramount national interest standard in RCRA and a national security standard in CERCLA) is inapt for the reasons discussed above. . . .

Marine Mammal Protection Act

Although I realize this Committee is not centrally concerned with the Marine Mammal Protection Act (MMPA), I would like to take a moment to discuss it for purposes of completeness. This year's MMPA proposal includes some new provisions. This year's proposal, like last year's, would amend the term "harassment" in the MMPA, which currently focuses on the mere "potential" to injure or disturb marine mammals.

Our initiative adopts verbatim a reform proposal developed during the prior Administration by the Commerce, Interior, and Defense Departments and applies it to military readiness activities. That proposal espoused a recommendation by the National Research Council (NRC) that the currently overbroad definition of "harassment" of marine mammals—which includes "annoyance" or "potential to disturb"—be focused on biologically significant effects. As recently as 1999, the National Marine Fisheries Service (NMFS) asserted that under the sweeping language of the existing statutory definition harassment "is presumed to occur when marine mammals react to the generated sounds or visual cues"—in other words, whenever a marine mammal notices and reacts to an activity, no matter how transient or benign the reaction. As the NRC study found, "If [this] interpretation of the law for level B harassment (detectable changes in behavior) were applied to shipping as strenuously as it is applied to scientific and naval activities, the result would be crippling regulation of nearly every motorized vessel operating in U.S. waters."

Under the prior Administration, NMFS subsequently began applying the NRC's more scientific, effects-based definition. But environmental groups have challenged this regulatory construction as inconsistent with the statute. As you may know, the Navy and the National Oceanic and Atmospheric Administration suffered an important setback last year involving a vital anti-submarine warfare sensor—SURTASS LFA, a towed array emitting low-frequency sonar that is critical in detecting ultra-quiet diesel-electric submarines while they are still at a safe distance from our vessels. In the SURTASS LFA litigation environmental groups successfully challenged the new policy as inconsistent with the sweeping statutory standard, putting at risk NMFS' regulatory policy, clearly substantiating the need to clarify the existing statutory definition of harassment that we identified in our legislative package last year. . . .

The last change we are proposing, a national security exemption process, also derives from feedback the Defense Department received from environmental advocates last year after we submitted our proposal, as I discussed above. Although DoD continues to believe that predicating essential military training, testing, and operations on repeated invocations of emergency authority is unacceptable as a matter of public policy, we do believe that every environmental statute should have such authority as an insurance policy. The comments we received last year highlighted the fact that the MMPA does not currently contain such emergency authority, so this year's submission does include a waiver mechanism. Like the Endangered Species Act, our proposal would allow the Secretary of Defense, after conferring with the Secretaries of Commerce or Interior, as appropriate, to waive MMPA provisions for actions or categories of actions when required by national security. This provision is not a substitute for the other clarifications we have proposed to the MMPA, but rather a failsafe mechanism in the event of emergency.

The only substantive changes are those described above. The reason that the text is so much more extensive than last year's version is that last year's version was drafted as a freestanding part of title 10—the Defense Department title—rather than an amendment to the text of the MMPA itself. This year, because we were making several changes, we concluded that as a drafting matter we should include our changes in the MMPA itself. That necessitated a lot more language, largely just reciting existing MMPA language that we are not otherwise modifying.

The environmental impacts of our proposed reforms would be minimal. Although our initiative would exclude transient, biologically insignificant effects from regulation, the MMPA would remain in full effect for biologically significant effects—not only death or injury but also disruption of significant activities. The Defense Department could neither harm marine mammals nor disrupt their biologically significant activities without obtaining authorization from FWS or NMFS, as appropriate.

Nor does our initiative depart from the precautionary premise of the MMPA. The Precautionary Principle holds that regulators should proceed conservatively in the face of scientific uncertainty over environmental effects. But our initiative embodies a conservative, science-based approach validated by the National Research Council. By defining as "harassment" any readiness activities that "injure or have the significant potential to injure," or "disturb *or are* likely to disturb," our initiative includes a margin of safety fully consistent with the Precautionary Principle. The alternative is the existing grossly overbroad, unscientific definition of harassment, which sweeps in any activity having the "potential to disturb." As the National Research Council found, such sweeping overbreadth is unscientific and not mandated by the Precautionary Principle. . . .

The Defense Department already exercises extraordinary care in its maritime programs: all DoD activities worldwide result in fewer than 10 deaths or injuries annually (as opposed to 4800 deaths annually from commercial fishing activities). And DoD currently funds much of the most significant research on marine mammals, and will continue this research in future.

Although the environmental effects of our MMPA reforms will be negligible, their readiness implications are profound. Application of the current hair-trigger definition of "harassment" has profoundly affected both vital R&D efforts and training. Navy operations are expeditionary in nature, which means world events often require planning exercises on short notice. To date, the Navy has been able to avoid the delay and burden of applying for a take permit only by curtailing and/or dumbing down training and research/testing. For six years, the Navy has been working on research to develop a suite of new sensors and tactics (the Littoral Advanced Warfare Development Program, or LWAD) to reduce the threat to the fleet posed by ultraquiet diesel submarines operating in the littorals and shallow seas like the Persian Gulf, the Straits of Hormuz, the South China Sea, and the Taiwan Strait. These submarines are widely distributed in the world's navies, including "Axis of Evil" countries such as Iran and North Korea and potentially hostile great powers. In the 6 years that the program has operated, over 75% of the tests have been impacted by environmental considerations. In the last 3 years, 9 of 10 tests have been affected. One was cancelled entirely, and 17 different projects have been scaled back.

Endangered Species Act

Our Endangered Species Act provision is unchanged from last year. The legislation would confirm the prior Administration's decision that an Integrated Natural Resources Management Plan (INRMP) may in appropriate circumstances obviate the need to designate critical habitat on military installations. These plans for conserving natural resources on military property, required by the Sikes Act, are developed in cooperation with state wildlife agencies, the U.S. Fish and Wildlife Service, and the public. In most cases they offer comparable or better protection for the species because they consider the base's environment holistically, rather than using a species-by-species analysis. The prior Administration's decision that INRMPs may adequately provide for appropriate endangered species habitat management is being challenged in court by environmental groups, who cite Ninth Circuit caselaw suggesting that other habitat management programs provided an insufficient basis for the Fish and Wildlife Service to avoid designating Critical Habitat. These groups claim that no INRMP, no matter how protective, can ever substitute for critical habitat designation. This legislation would confirm and insulate the Fish and Wildlife Service's policy from such challenges.

Both the prior and current Administrations have affirmed the use of INRMPs as a basis for possible exclusion from critical habitat. Such plans are required to provide for fish and wildlife management, land management, forest management, and fish and wildlife-oriented recreation; fish and wildlife habitat enhancement; wetland protection, enhancement, and restoration; establishment of specific natural resource management goals, objectives, and timeframes; and enforcement of natural resource laws and regulations. And unlike the process for designation of critical habitat, INRMPs assure a role for state regulators. Furthermore, INRMPs must be reviewed by the parties on a regular basis, but not less than every five years, providing a continuing opportunity for FWS input.

By contrast, in 1999, the Fish and Wildlife Service stated in a Notice of Proposed Rulemaking that "we have long believed that, in most circumstances, the designation of 'official' critical habitat is of little additional value for most listed species, yet it consumes large amounts of conservation resources. [W]e have long believed that separate protection of critical habitat is duplicative for most species."

Our provision does not automatically eliminate critical habitat designation, precisely because under the Sikes Act, the statute giving rise to INRMPs, the Fish & Wildlife Service is given approval authority over those elements of the INRMP under its jurisdiction. This authority guarantees the Fish & Wildlife Service the authority to make a case-by-case determination concerning the adequacy of our INRMPs as a substitute for critical habitat designation. And if the Fish & Wildlife Service does not approve the INRMP, our provision will not apply to protect the base from critical habitat designation.

Our legislation explicitly requires that the Defense Department continue to consult with the Fish and Wildlife Service and the National Marine Fisheries Service under Section 7 of the Endangered Species Act (ESA); the other provisions of the ESA, as well as other environmental statutes such as the National Environmental Policy Act, would continue to apply, as well.

The Defense Department's proposal has vital implications for readiness. Absent this policy, courts, based on complaints filed by environmental litigants, compelled the Fish and Wildlife Service to re-evaluate "not prudent" findings for many critical habitat determinations, and as a result FWS proposed to designate over 50% of the 12,000-acre Marine Corps Air Station (MCAS) Miramar and over 56% of the 125,000-acre Marine Corps Base (MCB) Camp Pendleton. Prior to adoption of this policy, 72% of Fort Lewis and 40% of the Chocolate Mountains Aerial Gunnery Range were designated as critical habitat for various species, and analogous habitat restrictions were imposed on 33% of Fort Hood. These are vital installations.

Unlike Sikes Act INRMPs, critical habitat designation can impose rigid limitations on military use of bases, denying commanders the flexibility to manage their lands for the benefit of both readiness and endangered species.

Clean Air Act General Conformity Amendment

Our Clean Air Act amendment is unchanged since last year. The legislation would provide more flexibility for the Defense Department in ensuring that emissions from its military training and testing are consistent with State Implementation Plans under the Clean Air Act by allowing DoD and the states a slightly longer period to accommodate or offset emissions from military readiness activities.

The Clean Air Act's "general conformity" requirement, applicable only to federal agencies, has repeatedly threatened deployment of new weapons systems and base closure/realignment despite the fact that relatively minor levels of emissions were involved. . . .

Conclusion

In closing Mr. Chairman, let me emphasize that modern warfare is a "come as you are" affair. There is no time to get ready. We must be prepared to defend our country wherever and whenever necessary. While we want to train as we fight, in reality our soldiers, sailors, airmen and Marines fight as they train. The consequences for them, and therefore for all of us, could not be more momentous.

DoD is committed to sustaining U.S. test and training capabilities in a manner that fully satisfies that military readiness mission while also continuing to provide exemplary stewardship of the lands and natural resources in our trust. . . .

Jamie Clark  **NO**

Impact of Military Training on the Environment

. . . Prior to arriving at the National Wildlife Federation in 2001, I served for 13 years at the U.S. Fish and Wildlife Service, with the last 4 years as the Director of the agency. Prior to that, I served as Fish and Wildlife Administrator for the Department of the Army, Natural and Cultural Resources Program Manager for the National Guard Bureau, and Research Biologist for U.S. Army Medical Research Institute. I am the daughter of a U.S. Army Colonel, and lived on or near military bases throughout my entire childhood.

Based on this experience, I am very familiar with the Defense Department's long history of leadership in wildlife conservation. On many occasions during my tenures at FWS and the Defense Department, DOD rolled up its sleeves and worked with wildlife agency experts to find a way to comply with environmental laws and conserve imperiled wildlife while achieving military preparedness objectives.

The Administration now proposes in its Readiness and Range Preservation Initiative that Congress scale back DOD's responsibilities to conserve wildlife and to protect people from the hazardous pollution that DOD generates. This proposal is both unjustified and dangerous. It is unjustified because DOD's longstanding approach of working through compliance issues on an installation-by-installation basis works. As DOD itself has acknowledged, our armed forces are as prepared today as they ever have been in their history, and this has been achieved without broad exemptions from environmental laws.

The DOD proposal is dangerous because, if Congress were to broadly exempt DOD from its environmental protection responsibilities, both people and wildlife would be threatened with serious, irreversible and unnecessary harm. Moreover, other federal agencies and industry sectors with important missions, using the same logic as used here by DOD, would line up for their own exemptions from environmental laws.

My expertise is in the Endangered Species Act (ESA), so I would like to focus my testimony on why exempting the Defense Department from key provisions of the ESA would be a serious mistake. I will rely on my fellow witnesses to explain why the proposed exemptions from other environmental and public health and safety laws is similarly unwise.

From Testimony before the Senate Committee on Environment and Public Works, April 2, 2003.

Concerns with the ESA Exemption

The Defense Department's proposed ESA exemption suffers from three basic flaws: it would severely weaken this nation's efforts to conserve imperiled species and the ecosystems on which all of us depend; it is unnecessary for maintaining military readiness; and it ignores the Defense Department's own record of success in balancing readiness and conservation objectives under existing law.

1. Section 2017 Removes a Key Species Conservation Tool

Section 2017 of the Administration's Readiness and Range Preservation Initiative would preclude designations of critical habitat on any lands owned or controlled by DOD if DOD has prepared an Integrated Natural Resources Management Plan (INRMP) pursuant to the Sikes Act and has provided "special management consideration or protection" of listed species pursuant to Section 3(5)(A) of the ESA.

This proposal would effectively eliminate critical habitat designations on DOD lands, thereby removing an essential tool for protecting and recovering species listed under the ESA. Of the various ESA protections, the critical habitat provision is the only one that specifically calls for protection of habitat needed for recovery of listed species. It is a fundamental tenet of biology that habitat must be protected if we ever hope to achieve the recovery of imperiled fish, wildlife and plant species.

Section 2017 would replace this crucial habitat protection with management plans developed pursuant to the Sikes Act. The Sikes Act does not require the protection of listed species or their habitats; it simply directs DOD to prepare INRMPs that protect wildlife "to the extent appropriate." Moreover, the Sikes Act provides no guaranteed funding for INRMPs and the annual appropriations process is highly uncertain. Even the best-laid management plans can go awry when the anticipated funding fails to come through. Yet, under Section 2017, even poorly designed INRMPs that allow destruction of essential habitat and put fish, wildlife or plant species at serious risk of extinction would be substituted for critical habitat protections.

Section 2017 contains one minor limitation on the substitution of INRMPs for critical habitat designations: such a substitution is allowed only where the INRMP provides "special management consideration or protection" within the meaning of Section 3(5)(A) of the ESA. Unfortunately, this limitation does nothing to ensure that INRMPs truly conserve listed species.

The term "special management consideration or protection" was never intended to provide a biological threshold that land managers must achieve in order to satisfy the ESA. The term is found in Section 3(5) of the ESA, which sets forth a two-part definition of critical habitat. Section 3(5)(A) states that critical habitat includes areas occupied by a listed species that are "essential for the conservation of the species" and "which may require special management consideration or protection." Section 3(5)(B) states that critical habitat also includes areas not currently occupied by a listed species that are simply "essential for the conservation of the species."

As this language makes clear, an ESA §3(5) finding by the U.S. Fish and Wildlife Service or National Marine Fisheries Service (Services) that a parcel of land "may require special management consideration or protection" is not the same as finding that it is already receiving adequate protection. Such a finding simply highlights the importance of a parcel of land to a species, and it should lead to designation of that land as critical habitat. See *Center for Biological Diversity v. Norton*, 240 F. Supp. 2d 1090 (D. Ariz. 2003) (rejecting, as contrary to plain meaning of ESA, defendant's interpretation of "special management consideration or protection" as providing a basis for substituting a U.S. Forest Service management plan for critical habitat protection). By allowing DOD to substitute INRMPs for critical habitat designations whenever it unilaterally makes a finding of "special management consideration or protection," Section 2017 significantly weakens the ESA.

Section 2017 is also problematic because it would eliminate many of the ESA Section 7 consultations that have stimulated DOD to "look before it leaps" into a potentially harmful training exercise. As a result of Section 7 consultations, DOD and the Services have routinely developed what is known as "workarounds," strategies for avoiding or minimizing harm to listed species and their habitats while still providing a rigorous training regimen.

Section 2017 purports to retain Section 7 consultations. However, the duty to consult only arises when a proposed federal action would potentially jeopardize a listed species or adversely modify or destroy its critical habitat. By removing critical habitat designations on lands owned or controlled by DOD, Section 2017 would eliminate one of the two possible justifications for initiating a consultation, reducing the likelihood that consultations will take place. This would mean that DOD and the Services would pay less attention to species concerns and would be less effective in conserving imperiled species and maintaining the sustainability of the land.

The reductions in species protection proposed by DOD would have major implications for our nation's rich natural heritage. DOD manages approximately 25 million acres of land on more than 425 major military installations. These lands are home to at least 300 federally listed species. Without the refuge provided by these bases, many of these species would slide rapidly toward extinction. These installations have played a crucial role in species conservation and must continue to do so.

2. The ESA Exemption Is Not Necessary to Maintain Military Readiness

The ESA already has the flexibility needed for the Defense Department to balance military readiness and species conservation objectives. Three key provisions provide this flexibility. First, under the consultation provision of Section 7(a)(2) of the Act, DOD is provided with the opportunity to develop solutions in tandem with the Services to avoid unnecessary harm to listed species from military activities. Typically, the Services conclude, after informal consultation, that the proposed action will not adversely affect a listed species or its designated critical habitat or, after formal consultation, that it will not likely jeopardize a listed species or destroy or adversely modify its critical habitat. See, e.g., U.S. Army Environmental Center, Installation Summaries from the FY 2001 Survey of Threatened and Endangered Species on Army Lands (August 2002) at 9 (noting successful conclusion of 282 informal consultations and 36 formal consultations, with no "jeopardy" biological opinions). In both informal and formal consultations, the Services either will recommend that the action go forward without changes, or it will work with DOD to design "work arounds" for avoiding and minimizing harm to the species and its habitat. In either case, DOD accomplishes its readiness objectives while achieving ESA compliance.

Second, under Section 4(b)(2) of the ESA, the Services are authorized to exclude any area from critical habitat designation if they determine that the benefits of exclusion outweigh the benefits of specifying the area. (An exception is made for when the Services find that failure to designate an area as critical habitat will result in the extinction of a species—a finding that the Services have never made.) In making this decision, the Services must consider "the economic impact, and any other relevant impact" of the critical habitat designation. DOD has recently availed itself of this provision to convince the U.S. Fish and Wildlife Service to exclude virtually all of the habitat at Camp Pendleton—habitat deemed critical to five listed species in proposed rulemakings—from final critical habitat designations. Thus, for situations where the Section 7(a)(2) consultation procedures place undue burdens on readiness activities, DOD already has a tool for working with the Services on excluding land from critical habitat designation. Attached to my testimony is a factsheet that shows how the Services have worked cooperatively with DOD on these exclusions, and another factsheet showing the importance of maintaining the Services' role in evaluating proposed exclusions.

Third, under Section 7(j) of the ESA an exemption "shall" be granted for an activity if the Secretary of Defense finds the exemption is necessary for reasons of national security. To this date, DOD has never sought an exemption under Section 7(j)—highlighting the fact that other provisions of the ESA have provided DOD with all the flexibility it needs to reconcile training needs with species conservation objectives.

Where there are site-specific conflicts between training needs and species conservation needs, the ESA provides these three mechanisms for resolving them in a manner that allows DOD to achieve its readiness objectives. Granting DOD a nationwide ESA exemption, which would apply in many places where no irreconcilable conflicts between training needs and conservation needs have arisen, would be harmful to imperiled species and totally unnecessary to achieve readiness objectives.

a. DOD Has Misstated the Law Regarding Its Ability to Continue with a Cooperative, Case-by-Case Approach to Critical Habitat Designations

DOD has stated that the ESA exemption is necessary because a recent court ruling in Arizona would prevent DOD from taking the cooperative, case-by-case approach to critical habitat designations that was developed when I served as Director of the Fish and Wildlife Service. This description of the court ruling is inaccurate—the ruling clearly allows DOD to continue the cooperative, case-by-case approach if it wishes.

The court ruling at issue is entitled *Center for Biological Diversity v. Norton*, 240 F. Supp. 2d 1090 (D. Ariz. 2003). In this case, FWS excluded San Carlos Apache tribal lands from a critical habitat designation pursuant to ESA §4(b)(2) because the tribal land management plan was adequate and the benefits of exclusion outweighed the benefits of inclusion. The federal district court upheld the exclusion as within FWS's broad authority under ESA §4(b)(2). At the same time, the court held that lands could not legitimately be excluded from a critical habitat designation on the basis of the "special management" language in ESA §3(5).

Under the court's reasoning, FWS continues to have the broad flexibility to exclude DOD lands from a critical habitat designation on the basis of a satisfactory INRMP and the benefits to military training that the exclusion would provide. The ruling simply clarifies that such exclusions must be carried out pursuant to ESA §4(b)(2) rather than ESA §3(5). Thus, DOD's assertion that the Center for Biological Diversity ruling prevents it from working with FWS to secure exclusions of DOD lands from critical habitat designations is inaccurate.

b. DOD's Anecdotes Do Not Demonstrate That the ESA Has Reduced Readiness

The DOD has offered a series of misleading anecdotes describing difficulties it has encountered in balancing military readiness and conservation objectives. Before Congress moves forward with any exemption legislation, the appropriate Congressional committees should get a more complete picture of what is really happening at DOD installations.

Some of DOD's anecdotes are simply unpersuasive on their face, such as DOD's repeated assertion that environmental laws have prevented the armed services from learning how to dig foxholes and that troops abroad have been put at greater risk as a result. There is simply no evidence that environmental laws have ever prevented foxhole digging. Moreover, given its vast and varied landholdings and the many management options available, the Defense Department certainly can find places on which troops can learn to dig foxholes without encountering endangered species or other environmental issues.

Other anecdotes have simply disregarded the truth. For example, DOD and its allies have repeatedly argued that more than 50 percent of Camp Pendleton may not be available for training due to critical habitat designations. In fact, only five species have been proposed for critical habitat designations at Camp Pendleton. In each of these five instances, DOD raised concerns about impacts to military readiness, and in each instance, FWS worked closely with DOD to craft a solution. FWS ultimately excluded virtually all of the habitats for the five listed species on Camp Pendleton from critical habitat designations—even though FWS had earlier found that these habitats were essential to the conservation of the species. As a result of FWS's exclusion decisions, less than one percent of the training land at Camp Pendleton, and less than 4 percent of all of Camp Pendleton, is designated critical habitat. (Most of the critical habitat designated at Camp Pendleton is non-training land leased to San Onofre State Park, agricultural operations, and others. DOD's repeated suggestion that more than 50 percent of Camp Pendleton is at risk of being rendered off-limits to training due to critical habitat is simply inaccurate.

DOD also has argued that training opportunities and expansion plans at Fort Irwin have been thwarted by the desert tortoise. Yet just two weeks ago this official line was contradicted by the reality on the ground. In an article dated March 21, 2003, Fort Irwin spokesman Army Maj. Michael Lawhorn told the Barstow Desert Dispatch that he is unaware of any environmental regulations that interfere with troops' ability to train there. He also said there isn't any environmental law that hinders the expansion. . . .

These examples of misleading anecdotes highlight the need for Congress to look behind the reasons that are being put forward by DOD as the basis for weakening environmental laws. DOD uses the anecdotes in an attempt to demonstrate that conflicts between military readiness and species conservation objectives are irreconcilable. However, solutions to these conflicts are within reach if DOD is willing to invest sufficient time and energy into finding them. DOD has vast acres of land on which to train and vast stores of creativity and expertise among its land managers. With careful inventorying and planning, DOD can find a proper balance.

Has DOD made the necessary effort to inventory and plan for its training needs? In June 2002, the General Accounting Office issued a report entitled "Military Training: DOD Lacks a Comprehensive Plan to Manage Encroachment on Training Ranges," suggesting that the answer is no. The GAO found:

- DOD has not fully defined its training range requirements and lacks information on training resources available to the Services to meet those requirements, and that problems at individual installations may therefore be overstated.
- The Armed Services have never assessed the overall impacts of encroachment on training.
- DOD's readiness reports show high levels of training readiness for most units. In those few instances of when units reported lower training readiness, DOD officials rarely cited lack of adequate training ranges, areas or airspace as the cause.
- DOD officials themselves admit that population growth around military installations is responsible for past and present encroachment problems.
- The Armed Services' own readiness data do not show that environmental laws have significantly affected training readiness.

Ten months after the issuance of the GAO report, DOD still has not produced evidence that environmental laws are at fault for any of the minor gaps in readiness that may exist. EPA Administrator Whitman confirmed this much at a recent hearing. At a February 26, 2003, Senate Environment and Public Works Committee hearing on EPA's budget, EPA Administrator Whitman stated that she was "not aware of any particular area where environmental protection regulations are preventing the desired training."

To this date, DOD has not provided Congress with the most basic facts about the impacts of ESA critical habitat requirements on its readiness activities. Out of DOD's 25 million acres of training land, how many acres are designated critical habitat? At which installations? Which species? In what ways have the critical habitat designations limited readiness activities? What efforts did DOD make to alert FWS to these problems and to negotiate resolutions? Without answers to these most basic questions, Congress cannot fairly conclude that the ESA is at fault for any readiness gaps or that a sweeping ESA exemption is warranted.

3. DOD Has Worked Successfully with the Services to Balance Readiness and Species Conservation Objectives

The third reason why enacting DOD's proposed ESA changes would be a mistake is because the current approach—developing solutions at the local level, rather than relying on broad, national exemptions—has worked. My experience at both FWS and DOD has shown me that solutions developed at the local level are sometimes difficult to arrive at, but they are almost always more intelligent and long-lasting than one-size-fits-all solutions developed at the national level.

Allow me to provide a few brief examples. At the Marine Corps Base at Camp Lejeune in North Carolina, every colony tree of the endangered red-cockaded woodpecker is marked on a map, and Marines are trained to operate their vehicles as if those mapped locations are land mines. Here is the lesson that Major General David M. Mize, the Commanding General at Camp Lejeune, has drawn from this experience:

> "Returning to the old myth that military training and conservation are mutually exclusive; this notion has been repeatedly and demonstrably debunked. In the overwhelming majority of cases, with a good plan along with common sense and flexibility, military training and the conservation and recovery of endangered species can very successfully coexist."
>
> "Military installations in the southeast are contributing to red-cockaded woodpecker recovery while sustaining our primary mission of national military readiness."
>
> "I can say with confidence that the efforts of our natural resource managers and the training community have produced an environment in which endangered species management and military training are no longer considered mutually exclusive, but are compatible."

These sentiments, which I share, were relayed by Major General Mize just eight weeks ago at a National Defense University symposium sponsored by the U.S. Army Forces Command (FORSCOM) and others. At that symposium, representatives of Camp Lejeune Marine Corps Base, Eglin Air Force Base, Fort Bragg Army Base, Fort Stewart Army Base, Camp Blanding Training Center in Florida, the U.S. Army Environmental Center, and other Defense facilities—some of the most heavily utilized training bases in the country—heralded the success that Defense Department installations have had in furthering endangered species conservation while maintaining military readiness.

On the Mokapu Peninsula of Marine Corps Base Hawaii, the growth of non-native plants, which can decrease the reproductive success of endangered waterbirds, is controlled through annual "mud-ops" maneuvers by Marine Corps Assault Vehicles. Just before the onset of nesting season, these 26 ton vehicles are deployed in plow-like maneuvers that break the thick mats of invasive plants, improving nesting and feeding opportunities while also giving drivers valuable practice in unusual terrain. . . .

These success stories highlight a major trend that I believe has been missed by those promoting the DOD exemptions. In recent years, DOD has increasingly recognized the importance of sustainability because it meets several importance objectives at once. Sustainable use of the land helps DOD achieve not only compliance with environmental laws, but also long-term military readiness and cost-effectiveness goals. For example, by operating tanks so that they avoid the threatened desert tortoise, DOD prevents erosion, a problem that is extremely difficult and costly to remedy. If DOD abandons its commitment to environmental compliance, it will incur greater long-term costs for environmental remediation and will sacrifice land health and military readiness.

A November 2002 policy guidance issued by the then-Secretary of the Navy to the Chief of Naval Operations and the Commandant of the Marine Corps suggests that certain members of DOD's leadership are indeed willing to abandon the sustainability goal. The policy guidance on its face seems fairly innocuous—it purports to centralize at the Pentagon all decisionmaking on proposed critical habitat designations and other ESA actions. However, the Navy Secretary's cover memo makes clear that its purpose is also to discourage any negotiation of solutions to species conservation challenges by Marines or Navy personnel in the field, lest these locally-developed "win-win" solutions undercut DOD's arguments on Capitol Hill that the ESA is broken. According to paragraph 2 of the cover memo, "concessions ... could run counter to the legislative relief that we are continuing to pursue with Congress."

Similar sentiments were voiced by Deputy Defense Secretary Paul Wolfowitz in his March 7, 2003, memo to the chiefs of the Army, Navy and Air Force. Deputy Secretary Wolfowitz argued that "it is time for us to give greater consideration to requesting exemptions" from environmental laws and pleaded for specific examples of instances in which environmental regulations hamper training. The implicit message is that efforts at the installation level to resolve conflicts between conservation and training objectives should be suspended, and that such conflicts instead should be reported to the Pentagon, where environmental protections will simply be overridden.

These messages to military personnel in the field mark a very unfortunate abdication of DOD's leadership in wildlife conservation. To maintain its leadership role as steward of this nation's endangered wildlife, DOD must encourage its personnel to continue developing innovative solutions and not thwart those efforts.

Conclusion

With the Iraq war ongoing and terrorism threats always present, no one can dismiss the importance of military readiness. However, there is no justification for the Defense Department to retreat from its environmental stewardship commitments at home. As base commanders have been telling us, protecting endangered species and other important natural resources is compatible with maintaining military readiness.

Surveys show that the American people today want environmental protection from the federal government, including the Defense Department, as much as ever. According to an April 2002 Zogby Poll, 85% of registered voters believe that the Defense Department should be required to follow America's environmental and public health laws and not be exempt. Americans believe that no one, including the Defense Department, should be above the law.

Congress should reject the proposed environmental exemptions in the Administration's defense authorization package. This proposal, along with the parallel proposal in the Administration's FY04 budget request that Congress cut spending on DOD's environmental programs by $400 million, are a step in the wrong direction.

DOD has a long and impressive record of balancing readiness activities with wildlife conservation. The high quality of wildlife habitats at many DOD installations provides tangible evidence of DOD's positive contribution to the nation's conservation goals. At a time when environmental challenges are growing, DOD should be challenged to move forward with this successful model and not to sacrifice any of the progress that has been made. . . .

POSTSCRIPT

Should the Military Be Exempt from Environmental Regulations?

After the April 2003 Senate committee hearing, this issue received some press attention as environmentalists and congressional Democrats prepared to oppose the Bush administration. See "War on the Environment," *The Ecologist* (May 2003), and John Stanton, "Activists, Democrats Brace for Defense Environment Showdown," *CongressDaily AM* (May 13, 2003). The Senate voted for the exemptions, and early in 2004, the issue came before the House of Representatives. Dan Miller, first assistant attorney general, Natural Resources and Environmental Section, Colorado Department of Law, testified against them before the House Committee on Energy and Commerce Subcommittee on Energy and Air Quality, saying that.

"Even read in the narrowest possible fashion, the [proposed exemptions] would hamstring state and EPA cleanup authorities at over 24 million acres of 'operational ranges,' an area the size of Maryland, Massachusetts, New Jersey, Hawaii, Connecticut and Rhode Island combined. As a practical matter, environmental regulators would likely be precluded from using RCRA [Resource Conservation and Recovery Act], CERCLA [Comprehensive Environmental Response, Compensation, and Liability Act], and related state authorities to require any investigation or cleanup of groundwater contamination on these ranges, even if the contamination had migrated off-range, polluted drinking or irrigation water supplies, and even if it posed an imminent and substantial endangerment to human health. And it is likely that DOD's amendments would be construed more broadly to exempt even more contamination from state and EPA oversight. . . . If we have learned anything in the past thirty years of environmental regulation, it is that relying on federal agencies to 'voluntarily' address environmental contamination is often fruitless. One need look no further than the approximately 130 DOD facilities on the Superfund National Priorities List, or DOD's poor record of compliance with state and federal environmental laws to see that independent, legally enforceable state oversight of federal agencies is required to achieve effective results."

In May, the House Committees on Armed Services Readiness and Energy and Commerce announced that they were not about to consider the exemption. See "Hefley: No Plans to Exempt Military From Enviro Laws," *CongressDaily AM* (May 6, 2004).

But the issue is not about to go away. The Government Accounting Office (GAO) prepared a background paper, "Military Training: DOD Approach to Managing Encroachment on Training Ranges Still Evolving," GAO-03-621T (April 2, 2003), delivered by Barry W. Holman, director,

Defense Infrastructure Issues, as testimony before the Senate Committee on Environment and Public Works. It discussed eight encroachment issues, including urban growth around military bases, air and noise pollution, unexploded ordinance and other munitions, endangered species habitat, and protected marine resources. Since urban growth is not likely to cease and the number of endangered species and protected marine resources is sure to increase, encroachment is not about to diminish. The Department of Defense, says the GAO, must better document the impact of the encroachment on training and costs; it has not yet produced required reports to Congress. So far, "workarounds" have been enough to deal with the problem, but that may not remain sufficient. E. G. Willard, Tom Zimmerman, and Eric Bee, "Environmental Law and National Security: Can Existing Exemptions in Environmental Laws Preserve DOD Training and Operational Prerogatives without New Legislation?" *Air Force Law Review* (2004), conclude that existing exemptions are not enough to support military readiness and say that more are needed. "The bottom line is that we must be able to train the way we fight, and we must be able to operate to defend the country and its interests." Paul D. Thacker, "Are Environmental Exemptions for the U.S. Military Justified?" *Environmental Science & Technology* (October 15, 2004), noted that "many critics of the administration say that the campaign is more about undermining environmental laws than protecting military readiness."

In April 2006, the Democratic Staffs of the Committee on Energy and Commerce and the Committee on Resources published a white paper (available at http://www.house.gov/commerce_democrats/DODexemptions/dod.shtml) saying that "Once again, the Department of Defense (DOD) is seeking legislation that would grant the Department exemptions from the Resource Conservation and Recovery Act (RCRA) and the Comprehensive Environmental Response Compensation and Liability Act (CERCLA). These exemptions are spelled out in the Range Readiness and Preservation Initiative (RRPI) and are being sought for a fifth successive year." The white paper notes that there are no known cases of interference by environmental regulations with military activities.

On the Internet . . .

The Arctic National Wildlife Refuge: A Special Report

This site offers a cogent review of the debate over exploiting the Arctic National Wildlife Refuge (ANWR) for oil.

http://arcticcircle.uconn.edu/ANWR/anwrindex.html

National Wilderness Preservation System

Operated by representatives of the Arthur Carhart National Wilderness Training Center, the Aldo Leopold Wilderness Research Institute, and the Wilderness Institute at the University of Montana's School of Forestry, the National Wilderness Preservation System provides information, news, and Internet links related to wilderness. It includes a database of information on all 680 wilderness areas.

**http://www.wilderness.net/
index.cfm?fuse=NWPS&sec=fastFacts**

Intergovernmental Panel on Climate Change

The Intergovernmental Panel on Climate Change (IPCC) was formed by the World Meteorological Organization (WMO) and the United Nations Environment Programme (UNEP) to assess the scientific, technical, and socio-economic information relevant for the understanding of the risk of human-induced climate change. Note that the fourth assessment report is due to be released in 2007 and will be available at this site.

http://www.ipcc.ch/

Climate Change

The United Nations Environmental Program maintains this site as a central source for substantive work and information resources with regard to climate change.

http://climatechange.unep.net/

The International Partnership for the Hydrogen Economy

The International Partnership for the Hydrogen Economy is an international institution established to accelerate the transition to a hydrogen economy.

http://www.iphe.net/

The Center for Liquefied Natural Gas

The Center for Liquefied Natural Gas's overall goals are to enhance public education and LNG acceptance by serving as a clearinghouse for LNG information; to promote efficient regulation for permitting, siting, as well as the building and operation of LNG facilities and infrastructure; and to work toward continued safe and secure operations, stressing industry guidelines and best practices.

http://www.lngfacts.org/

Federal Energy Regulatory Commission

The Federal Energy Regulatory Commission regulates and oversees energy industries in the economic and environmental interest of the American public. This page provides information about LNG.

http://www.ferc.gov/industries/gas/indus-act/lng-what.asp

PART 3

Energy Issues

*H*umans cannot live and society cannot exist without producing environmental impacts. The reason is very simple: Humans cannot live and society cannot exist without using resources (e.g., soil, water, ore, wood, space, plants, animals, oil, sunlight), and those resources come from the environment. Many of these resources (e.g., wood, oil, coal, water, wind, sunlight, uranium) have to do with energy. The environmental impacts come from what must be done to obtain these resources and what must be done to dispose of the wastes generated in the process of obtaining and using them. The issues that arise are whether and how we should obtain these resources, whether and how we should deal with the wastes, and whether alternative answers to these questions may be preferable to the answers that experts think they already have.

The five issues in this section are not the only issues related to energy, but they will serve to demonstrate the vigor of the debate that they engender.

- Should the Arctic National Wildlife Refuge Be Opened to Oil Drilling?
- Should the U.S. Be Doing More to Combat Global Warming?
- Will Hydrogen End Our Fossil-Fuel Addiction?
- Is Additional Federal Oversight Needed for the Construction of LNG Import Facilities?
- Is It Time to Revive Nuclear Power?

ISSUE 8

Should the Arctic National Wildlife Refuge Be Opened to Oil Drilling?

YES: Dwight R. Lee, from "To Drill or Not to Drill: Let the Environmentalists Decide," *The Independent Review* (Fall 2001)

NO: Jeff Bingaman, et al., from Dissenting Views on ANWR Drilling Senate Energy Committee (October 24, 2005)

ISSUE SUMMARY

YES: Professor of economics Dwight R. Lee argues that the economic and other benefits of Arctic National Wildlife Refuge (ANWR) oil are so great that even environmentalists should agree to permit drilling—and they probably would if they stood to benefit directly.

NO: The Minority Members of the Senate Energy Committee objected when the Committee approved a bill that would authorize oil and gas development in the Arctic National Wildlife Refuge. They argued that though the bill contained serious legal and environmental flaws, the greatest flaw lay in its choice of priorities: Wilderness is to be preserved, not exploited.

The birth of environmental consciousness in the United States was marked by two strong, opposing views. Late in the nineteenth century, John Muir (1838–1914) called for the preservation of natural wilderness, untouched by human activities. At about the same time, Gifford Pinchot (1865–1946) became a strong voice for conservation (not to be confused with preservation; Gifford's conservation allowed the use of nature but in such a way that it was not destroyed; his aim was "the greatest good of the greatest number in the long run"). Both views agree that nature has value; however, they disagree on the form of that value. The preservationist says that nature has value in its own right and has a right to be left alone, neither developed with houses and roads nor exploited with farms, dams, mines, and oil wells. The conservationist says that nature's value lies chiefly in the benefits it provides to human beings.

The first national parks date back to the 1870s. Parks and the national forests are managed for "multiple use" on the premise that wildlife protection, recreation, timber cutting, and even oil drilling and mining can coexist. The first "primitive areas," where all development is barred, were created by the U.S. Forest Service in the 1920s. However, pressure from commercial interests (the timber and mining industries, among others) led to the reclassification of many such areas and their opening to exploitation. In 1964 the Federal Wilderness Act provided a mechanism for designating "wilderness" areas, defined as areas "where the earth and its community of life are untrammeled by man, where man himself is a visitor who does not remain." Since then it has become clear that pesticides and other man-made chemicals are found everywhere on earth, drifting on winds and ocean currents and traveling in migrant birds even to areas without obvious human presence. Humans might not be present in these places, but their effects are. And commercial interests are just as interested in the wealth that may be extracted from these areas as they ever were. There is continual pressure to expand commercial use of national forests and parks and to open wilderness areas to exploitation.

The Arctic National Wildlife Refuge (ANWR) provides a good illustration. It is not a "wilderness" area, for it was designated a wildlife preserve in 1960 and enlarged and renamed in 1980 with the proviso that its coastal plain be evaluated for its potential value in terms of oil and gas production. In 1987 the Department of the Interior recommended that the coastal plain be opened for oil and gas exploration. In 1995 Congress approved doing so, but President Bill Clinton vetoed the legislation. In 2001, after California experienced electrical blackouts, President George W. Bush declared that opening the ANWR to oil exploitation was essential to national energy security but could not muster enough votes in Congress to make it happen. In 2003, an attempt to link the need for Arctic oil to the war in Iraq failed in the Senate. In 2004, the Bush Administration proposed once more to open the ANWR to oil drilling, and in December 2005, the House of Representatives approved a Defense bill with a provision approving ANWR drilling. The Senate immediately blocked the provision, but with gasoline prices soaring, in March 2006 Bush renewed his call for approval to exploit the ANWR for oil.

Strict preservationists still remain, but the debate over protecting wilderness areas generally centers on economic arguments. In the following selections, Dwight R. Lee argues that the economic and other benefits of Arctic National Wildlife Refuge oil are so great that drilling should be permitted. When in October 2005 the Senate Energy Committee approved a bill that would authorize oil and gas development in the Arctic National Wildlife Refuge, the Committee's minority members argued that the bill contained serious legal and environmental flaws, but its greatest flaw lay in its choice of priorities: Wilderness is to be preserved, not exploited for its energy resources.

YES

Dwight R. Lee

To Drill or Not to Drill

High prices of gasoline and heating oil have made drilling for oil in Alaska's Arctic National Wildlife Refuge (ANWR) an important issue. ANWR is the largest of Alaska's sixteen national wildlife refuges, containing 19.6 million acres. It also contains significant deposits of petroleum. The question is, Should oil companies be allowed to drill for that petroleum?

The case for drilling is straightforward. Alaskan oil would help to reduce U.S. dependence on foreign sources subject to disruptions caused by the volatile politics of the Middle East. Also, most of the infrastructure necessary for transporting the oil from nearby Prudhoe Bay to major U.S. markets is already in place. Furthermore, because of the experience gained at Prudhoe Bay, much has already been learned about how to mitigate the risks of recovering oil in the Arctic environment.

No one denies the environmental risks of drilling for oil in ANWR. No matter how careful the oil companies are, accidents that damage the environment at least temporarily might happen. Environmental groups consider such risks unacceptable; they argue that the value of the wilderness and natural beauty that would be spoiled by drilling in ANWR far exceeds the value of the oil that would be recovered. For example, the National Audubon Society characterizes opening ANWR to oil drilling as a threat "that will destroy the integrity" of the refuge (see statement at www.audubon.org/campaign/refuge).

So, which is more valuable, drilling for oil in ANWR or protecting it as an untouched wilderness and wildlife refuge? Are the benefits of the additional oil really less than the costs of bearing the environmental risks of recovering that oil? Obviously, answering this question with great confidence is difficult because the answer depends on subjective values. Just how do we compare the convenience value of using more petroleum with the almost spiritual value of maintaining the "integrity" of a remote and pristine wilderness area? Although such comparisons are difficult, we should recognize that they can be made. Indeed, we make them all the time.

We constantly make decisions that sacrifice environmental values for what many consider more mundane values, such as comfort, convenience, and material well-being. There is nothing wrong with making such sacrifices because up to some point the additional benefits we realize from sacrificing a

This article is reprinted with permission of the publisher from *The Independent Review: A Journal of Political Economy* (Fall 2001, vol. VI, no. 2, pp. 217–226). © Copyright 2003, The Independent. info@independent.org; www.independent.org.

little more environmental "integrity" are worth more than the necessary sacrifice. Ideally, we would somehow acquire the information necessary to determine where that point is and then motivate people with different perspectives and preferences to respond appropriately to that information.

Achieving this ideal is not as utopian as it might seem; in fact, such an achievement has been reached in situations very similar to the one at issue in ANWR. In this article, I discuss cases in which the appropriate sacrifice of wilderness protection for petroleum production has been responsibly determined and harmoniously implemented. Based on this discussion, I conclude that we should let the Audubon Society decide whether to allow drilling in ANWR. That conclusion may seem to recommend a foregone decision on the issue because the society has already said that drilling for oil in ANWR is unacceptable. But actions speak louder than words, and under certain conditions I am willing to accept the actions of environmental groups such as the Audubon Society as the best evidence of how they truly prefer to answer the question, To drill or not to drill in ANWR?

Private Property Changes One's Perspective

What a difference private property makes when it comes to managing multiuse resources. When people make decisions about the use of property they own, they take into account many more alternatives than they do when advocating decisions about the use of property owned by others. This straightforward principle explains why environmental groups' statements about oil drilling in ANWR (and in other publicly owned areas) and their actions in wildlife areas they own are two very different things.

For example, the Audubon Society owns the Rainey Wildlife Sanctuary, a 26,000-acre preserve in Louisiana that provides a home for fish, shrimp, crab, deer, ducks, and wading birds, and is a resting and feeding stopover for more than 100,000 migrating snow geese each year. By all accounts, it is a beautiful wilderness area and provides exactly the type of wildlife habitat that the Audubon Society seeks to preserve. But, as elsewhere in our world of scarcity, the use of the Rainey Sanctuary as a wildlife preserve competes with other valuable uses.

Besides being ideally suited for wildlife, the sanctuary contains commercially valuable reserves of natural gas and oil, which attracted the attention of energy companies when they were discovered in the 1940s. Clearly, the interests served by fossil fuels do not have high priority for the Audubon Society. No doubt, the society regards additional petroleum use as a social problem rather than a social benefit. Of course, most people have different priorities: they place a much higher value on keeping down the cost of energy than they do on bird-watching and on protecting what many regard as little more than mosquito-breeding swamps. One might suppose that members of the Audubon Society have no reason to consider such "anti-environmental" values when deciding how to use their own land. Because the society owns the Rainey Sanctuary, it can ignore interests antithetical to its own and refuse to allow drilling. Yet, precisely because the society owns the land, it has been willing to accommodate the interests of those whose priorities are different and has allowed thirty-seven wells to pump gas and oil from the Rainey Sanctuary. In return, it has received royalties of more than $25 million.

One should not conclude that the Audubon Society has acted hypocritically by putting crass monetary considerations above its stated concerns for protecting wilderness and wildlife. In a wider context, one sees that because of its ownership of the Rainey Sanctuary, the Audubon Society is part of an extensive network of market communication and cooperation that allows it to do a better job of promoting its objectives by helping others promote theirs. Consumers communicate the value they receive from additional gas and oil to petroleum companies through the prices they willingly pay for those products, and this communication is transmitted to owners of oil-producing land through the prices the companies are willing to pay to drill on that land. Money really does "talk" when it takes the form of market prices. The money offered for drilling rights in the Rainey Sanctuary can be viewed as the most effective way for millions of people to tell the Audubon Society how much they value the gas and oil its property can provide.

By responding to the price communication from consumers and by allowing the drilling, the Audubon Society has not sacrificed its environmental values in some debased lust for lucre. Instead, allowing the drilling has served to reaffirm and promote those values in a way that helps others, many of whom have different values, achieve their own purposes. Because of private ownership, the valuations of others for the oil and gas in the Rainey Sanctuary create an opportunity for the Audubon Society to purchase additional sanctuaries to be preserved as habitats for the wildlife it values. So the society has a strong incentive to consider the benefits as well as the costs of drilling on its property. Certainly, environmental risks exist, and the society considers them, but if also responsibly weighs the costs of those risks against the benefits as measured by the income derived from drilling. Obviously, the Audubon Society appraises the benefits from drilling as greater than the costs, and it acts in accordance with that appraisal.

Cooperation Between Bird-Watchers and Hot-Rodders

The advantage of private ownership is not just that it allows people with different interests to interact in mutually beneficial ways. It also creates harmony between those whose interests would otherwise be antagonistic. For example, most members of the Audubon Society surely see the large sport utility vehicles and high-powered cars encouraged by abundant petroleum supplies as environmentally harmful. That perception, along with the environmental risks associated with oil recovery, helps explain why the Audubon Society vehemently opposes drilling for oil in the ANWR as well as in the continental shelves in the Atlantic, the Pacific, and the Gulf of Mexico. Although oil companies promise to take extraordinary precautions to prevent oil spills when drilling in these areas, the Audubon Society's position is no off-shore drilling, none. One might expect to find Audubon Society members completely unsympathetic with hot-rodding enthusiasts, NASCAR racing fans, and drivers of Chevy Suburbans. Yet, as we have seen, by allowing drilling for gas and oil in the Rainey Sanctuary, the society is accommodating the interests of those with gas-guzzling lifestyles, risking the "integrity" of its prized wildlife sanctuary to make more gasoline available to those whose energy consumption it verbally condemns as excessive.

The incentives provided by private property and market prices not only motivate the Audubon Society to cooperate with NASCAR racing fans, but also motivate those racing enthusiasts to cooperate with the Audubon Society. Imagine the reaction you would get if you went to a stock-car race and tried to convince the spectators to skip the race and go bird-watching instead. Be prepared for some beer bottles tossed your way. Yet by purchasing tickets to their favorite sport, racing fans contribute to the purchase of gasoline that allows the Audubon Society to obtain additional wildlife habitat and to promote bird-watching. Many members of the Audubon Society may feel contempt for racing fans, and most racing fans may laugh at bird-watchers, but because of private property and market prices, they nevertheless act to promote one another's interests.

The Audubon Society is not the only environmental group that, because of the incentives of private ownership, promotes its environmental objectives by serving the interests of those with different objectives. The Nature Conservancy accepts land and monetary contributions for the purpose of maintaining natural areas for wildlife habitat and ecological preservation. It currently owns thousands of acres and has a well-deserved reputation for preventing development in environmentally sensitive areas. Because it owns the land, it has also a strong incentive to use that land wisely to achieve its objectives, which sometimes means recognizing the value of developing the land.

For example, soon after the Wisconsin chapter received title to 40 acres of beach-front land on St. Croix in the Virgin Islands, it was offered a much larger parcel of land in northern Wisconsin in exchange for its beach land. The Wisconsin chapter made this trade (with some covenants on development of the beach land) because owning the Wisconsin land allowed it to protect an entire watershed containing endangered plants that it considered of greater environmental value than what was sacrificed by allowing the beach to be developed.

Thanks to a gift from the Mobil Oil Company, the Nature Conservancy of Texas owns the Galveston Bay Prairie Preserve in Texas City, a 2,263-acre refuge that is home to the Attwater's prairie chicken, a highly endangered species (once numbering almost a million, its population had fallen to fewer than ten by the early 1990s). The conservancy has entered into an agreement with Galveston Bay Resources of Houston and Aspects Resources, LLC, of Denver to drill for oil and natural gas in the preserve. Clearly some risks attend oil drilling in the habitat of a fragile endangered species, and the conservancy has considered them, but it considers the gains sufficient to justify bearing the risks. According to Ray

Johnson, East County program manager for the Nature Conservancy of Texas. "We believe this could provide a tremendous opportunity to raise funds to acquire additional habitat for the Attwater's prairie chicken, one of the most threatened birds in North America." Obviously the primary concern is to protect the endangered species, but the demand for gas and oil is helping achieve that objective. Johnson is quick to point out, "We have taken every precaution to minimize the impact of the drilling on the prairie chickens and to ensure their continued health and safety."

Back to ANWR

Without private ownership, the incentive to take a balanced and accommodating view toward competing land-use values disappears. So, it is hardly surprising that the Audubon Society and other major environmental groups categorically oppose drilling in ANWR. Because ANWR is publicly owned, the environmental groups have no incentive to take into account the benefits of drilling. The Audubon Society does not capture any of the benefits if drilling is allowed, as it does at the Rainey Sanctuary; in ANWR, it sacrifices nothing if drilling is prevented. In opposing drilling in ANWR, despite the fact that the precautions to be taken there would be greater than those required of companies operating in the Rainey Sanctuary, the Audubon Society is completely unaccountable for the sacrificed value of the recoverable petroleum.

Obviously, my recommendation to "let the environmentalists decide" whether to allow oil to be recovered from ANWR makes no sense if they are not accountable for any of the costs (sacrificed benefits) of preventing drilling. I am confident, however, that environmentalists would immediately see the advantages of drilling in ANWR if they were responsible for both the costs and the benefits of that drilling. As a thought experiment about how incentives work, imagine that a consortium of environmental organizations is given veto power over drilling, but is also given a portion (say, 10 percent) of what energy companies are willing to pay for the right to recover oil in ANWR. These organizations could capture tens of millions of dollars by giving their permission to drill. Suddenly the opportunity to realize important environmental objectives by favorably considering the benefits others gain from more energy consumption would come into sharp focus. The environmentalists might easily conclude that although ANWR is an "environmental treasure," other environmental treasures in other parts of the country (or the world) are even more valuable; moreover, with just a portion of the petroleum value of the ANWR, efforts might be made to reduce the risks to other natural habitats, more than compensating for the risks to the Arctic wilderness associated with recovering that value.

Some people who are deeply concerned with protecting the environment see the concentration on "saving" ANWR from any development as misguided even without a vested claim on the oil wealth it contains. For example, according to Craig Medred, the outdoor writer for the *Anchorage Daily News* and a self-described "development-phobic wilderness lover,"

That people would fight to keep the scar of clearcut logging from the spectacular and productive rain-forests of Southeast Alaska is easily understandable to a shopper in Seattle or a farmer in Nebraska. That people would argue against sinking a few holes through the surface of a frozen wasteland, however, can prove more than a little baffling even to development-phobic, wilderness lovers like me. Truth be known, I'd trade the preservation rights to any 100 acres on the [ANWR] slope for similar rights to any acre of central California wetlands.... It would seem of far more environmental concern that Alaska's ducks and geese have a place to winter in overcrowded, overdeveloped California than that California's ducks and geese have a place to breed each summer in uncrowded and undeveloped Alaska.

— (1996, Cl)

Even a small share of the petroleum wealth in ANWR would dramatically reverse the trade-off Medred is willing to make because it would allow environmental groups to afford easily a hundred acres of central California wetlands in exchange for what they would receive for each acre of ANWR released to drilling.

We need not agree with Medred's characterization of the ANWR as "a frozen wasteland" to suspect that environmentalists are overstating the environmental amenities that drilling would put at risk. With the incentives provided by private property, environmental groups would quickly reevaluate the costs of drilling in wilderness refuges and soften their rhetoric about how drilling would "destroy the integrity" of these places. Such hyperbolic rhetoric is to be expected when drilling is being considered on public land because environmentalists can go to the bank with it. It is easier to get contributions by depicting decisions about oil drilling on public land as righteous crusades against evil corporations out to destroy our priceless environment for short-run profit than it is to work toward minimizing drilling costs to accommodate better the interests of others. Environmentalists are concerned about protecting wildlife and wilderness areas in which they have ownership interest, but the debate over any threat from drilling and development in those areas is far more productive and less acrimonious than in the case of ANWR and other publicly owned wilderness areas.

The evidence is overwhelming that the risks of oil drilling to the arctic environment are far less than commonly claimed. The experience gained in Prudhoe Bay has both demonstrated and increased the oil companies' ability to recover oil while leaving a "light footprint" on arctic tundra and wildlife. Oil-recovery operations are now sited on gravel pads providing foundations that protect the underlying permafrost. Instead of using pits to contain the residual mud and other waste from drilling, techniques are now available for pumping the waste back into the well in ways that help maintain well pressure and reduce the risks of spills on the tundra. Improvements in arctic road construction have eliminated the need for the gravel access roads used in the development of the Prudhoe Bay oil fields. Roads are now made from ocean water pumped onto the tundra, where it freezes to form a road surface. Such roads melt without a trace during the short summers. The oversize rubber tires used on the roads further minimize any impact on the land.

Improvements in technology now permit horizontal drilling to recover oil that is far from directly below the wellhead. This technique reduces further the already small amount of land directly affected by drilling operations. Of the more than 19 million acres contained in ANWR, almost 18 million acres have been set aside by Congress—somewhat more than 8 million as wilderness and 9.5 million as wildlife refuge. Oil companies estimate that only 2,000 acres would be needed to develop the coastal plain.

This carefully conducted and closely confined activity hardly sounds like a sufficient threat to justify the rhetoric of a righteous crusade to prevent the destruction of ANWR, so the environmentalists warn of a detrimental effect on arctic wildlife that cannot be gauged by the limited acreage directly affected. Given the experience at Prudhoe Bay, however, such warnings are difficult to take seriously. The oil companies have gone to great lengths and spent tens of millions of dollars to reduce any harm to the fish, fowl, and mammals that live and breed on Alaska's North Slope. The protections they have provided for wildlife at Prudhoe Bay have been every bit as serious and effective as those the Audubon Society and the Nature Conservancy find acceptable in the Rainey Sanctuary and the Galveston Bay Prairie Preserve. As the numbers of various wildlife species show, many have thrived better since the drilling than they did before.

Before drilling began at Prudhoe Bay, a good deal of concern was expressed about its effect on caribou herds. As with many wildlife species, the population of the caribou on Alaska's North Slope fluctuates (often substantially) from year to year for completely natural reasons, so it is difficult to determine with confidence the effect of development on the caribou population. It is noteworthy, however, that the caribou population in the area around Prudhoe Bay has increased greatly since that oil field was developed, from approximately 3,000 to a high of some 23,400.... Some argue that the increase has occurred because the caribou's natural predators have avoided the area—some of these predators are shot, whereas the caribou are not. But even if this argument explains some or even all of the increase in the population, the increase still casts doubt on claims that the drilling threatens the caribou. Nor has it been shown that the viability of any other species has been genuinely threatened by oil drilling at Prudhoe Bay.

Caribou Versus Humans

Although consistency in government policy may be too much to hope for, it is interesting to contrast the federal government's refusal to open ANWR with some of its other oil-related policies. While opposing drilling in ANWR, ostensibly because we should not put caribou and other Alaskan wildlife at risk for the sake of getting more petroleum, we are exposing humans to far greater risks because of federal policies motivated by concern over petroleum supplies.

For example, the United States maintains a military presence in the Middle East in large part because of the petroleum reserves there. It is doubtful that the

U.S. government would have mounted a large military action and sacrificed American lives to prevent Iraq from taking over the tiny sheikdom of Kuwait except to allay the threat to a major oil supplier. Nor would the United States have lost the nineteen military personnel in the barracks blown up in Saudi Arabia in 1996 or the seventeen killed onboard the USS Cole in a Yemeni harbor in 2000. I am not arguing against maintaining a military presence in the Middle East, but if it is worthwhile to sacrifice Americans' lives to protect oil supplies in the Middle East, is it not worthwhile to take a small (perhaps nonexistent) risk of sacrificing the lives of a few caribou to recover oil in Alaska?

Domestic energy policy also entails the sacrifice of human lives for oil. To save gasoline, the federal government imposes Corporate Average Fuel Economy (CAFE) standards on automobile producers. These standards now require all new cars to average 27.5 miles per gallon and new light trucks to average 20.5 miles per gallon. The one thing that is not controversial about the CAFE standards is that they cost lives by inducing manufacturers to reduce the weight of vehicles. Even Ralph Nader has acknowledged that "larger cars are safer—there is more bulk to protect the occupant." An interesting question is, How many lives might be saved by using more (ANWR) oil and driving heavier cars rather than using less oil and driving lighter, more dangerous cars?

It has been estimated that increasing the average weight of passenger cars by 100 pounds would reduce U.S. highway fatalities by 200 a year. By determining how much additional gas would be consumed each year if all passenger cars were 100 pounds heavier, and then estimating how much gas might be recovered from ANWR oil, we can arrive at a rough estimate of how many human lives potentially might be saved by that oil. To make this estimate, I first used data for the technical specifications of fifty-four randomly selected 2001 model passenger cars to obtain a simple regression of car weight on miles per gallon. This regression equation indicates that every additional 100 pounds decreases mileage by 0.85 miles per gallon. So 200 lives a year could be saved by relaxing the CAFE standards to allow a 0.85 miles per gallon reduction in the average mileage of passenger cars. How much gasoline would be required to compensate for this decrease of average mileage? Some 135 million passenger cars are currently in use, being driven roughly 10,000 miles per year on average (1994–95 data from U.S. Bureau of the Census 1997, 843). Assuming these vehicles travel 24 miles per gallon on average, the annual consumption of gasoline by passenger cars is 56.25 billion gallons (= 135 million × 10,000/24). If instead of an average of 24 miles per gallon the average were reduced to 23.15 miles per gallon, the annual consumption of gasoline by passenger cars would be 58.32 billion gallons (= 135 million × 10,000/23.15). So, 200 lives could be saved annually by an extra 2.07 billion gallons of gas. It is estimated that ANWR contains from 3 to 16 billion barrels of recoverable petroleum. Let us take the midpoint in this estimated range, or 9.5 billion barrels. Given that on average each barrel of petroleum is refined into 19.5 gallons of gasoline, the ANWR oil could be turned into 185.25 billion additional gallons of gas, or enough to save 200 lives a year for almost ninety years (185.25/2.07 = 89.5). Hence, in total almost 18,000 lives could be saved by opening up ANWR to drilling and using the fuel made available to compensate for increasing the weight of passenger cars.

I claim no great precision for this estimate. There may be less petroleum in ANWR than the midpoint estimate indicates, and the study I have relied on may have overestimated the number of lives saved by heavier passenger cars. Still, any reasonable estimate will lead to the conclusion that preventing the recovery of ANWR oil and its use in heavier passenger cars entails the loss of thousands of lives on the highways. Are we willing to bear such a cost in order to avoid the risks, if any, to ANWR and its caribou?

Conclusion

I am not recommending that ANWR actually be given to some consortium of environmental groups. In thinking about whether to drill for oil in ANWR, however, it is instructive to consider seriously what such a group would do if it owned ANWR and therefore bore the costs as well as enjoyed the benefits of preventing drilling. Those costs are measured by what people are willing to pay for the additional comfort, convenience, and safety that could be derived from the use of ANWR oil. Unfortunately, without the price communication that is possible only by means of private property and voluntary exchange, we cannot be sure what those costs are or how private owners would evaluate either the costs or the benefits of preventing drilling in ANWR. However, the willingness of environmental groups such as the Audubon Society and the Nature Conservancy to allow drilling for oil an environmentally sensitive land they own suggests strongly that their adamant verbal opposition to drilling in ANWR is a poor reflection of what they would do if they owned even a small fraction of the ANWR territory containing oil.

 **NO**

ANWR Minority Views

Dissenting Views Of Senators Bingaman, Dorgan, Wyden, Johnson, Feinstein, Cantwell, Corzine, and Salazar

The Arctic National Wildlife Refuge has long stirred deep emotions and strong passions. To some it is the most promising place to look for oil in the [nation]. To others it is "the Last Great Wilderness," a vast and beautiful natural wonder, which deserves permanent protection for its wildlife, scenic, and recreational values. These two viewpoints are held with equal passion by roughly equal parts of the Senate and the nation as a whole. For a quarter of a century, neither side has been able to enact legislation either opening the area to oil and gas development or permanently preserving it as wilderness.

We come down squarely in favor of preserving the Arctic National Wildlife Refuge. Opening the Refuge to oil and gas development will do little to meet our energy needs and nothing to reduce our energy prices. Not one drop of oil will come from the Refuge for ten years. And even at its peak production—twenty years from now—it will reduce our reliance on imports by only 4 percent. We believe we should tap alternative sources of oil and gas and develop alternative energy technologies, rather than sacrifice the Refuge's unique wildlife and wilderness values.

Years ago, Senator Clinton P. Anderson said that our willingness to protect wilderness areas showed "that we are still a rich Nation, tending our resources as we should—not a people in despair searching every last nook and cranny of our land for . . . a barrel of oil." We believe we still are a rich Nation, rich in untapped oil and gas resources in other areas that can be developed consistent with environmental protection, and rich in the intellectual capital needed to develop new alternative energy technologies. We do not believe we need to sacrifice our wildlife refuges or other environmentally sensitive areas to fuel our cars or heat our homes.

But even if the day should come when we do need to exploit the Arctic National Wildlife Refuge for oil, we should approach the task with care. If we must open the Refuge to oil and gas development, we should do so in accordance with existing mineral leasing laws and regulations, existing environmental protections, and existing rules of administrative procedure

From ANWR Minority Views from the Senate Energy Committee, October 24, 2005.

and judicial review. We should, in short, afford the Arctic Refuge no less protection than current law affords any other refuge or public land that is open to oil and gas development. In addition, we should ensure that any oil that comes from the Arctic Refuge goes to Americans, and is not sold overseas; and that the Federal Treasury receives the full amount of royalties and bonus bids that we are promised. Regrettably, the legislation recommended by a majority of the Committee fails in every one of these respects.

1. The Mineral Leasing Act and rules. Oil and gas leasing on the public lands, including wildlife refuges, is currently conducted under the Mineral Leasing Act of 1920 and regulations adopted by the Bureau of Land Management under that Act. Among other things, they require minimum royalties, maximum lease sizes, and various performance standards and environmental protections. Oil and gas development on wildlife refuges can only take place with the concurrence of the Fish and Wildlife Service and subject to stipulations prescribed by the Fish and Wildlife Service that protect wildlife.

    It is unclear from the legislation recommended by the Committee whether any of the mineral leasing laws or regulations will apply to the leasing program for the Arctic Refuge. The legislation directs the Secretary to administer an "environmentally sound" leasing program in the refuge's Coastal Plain, but does not explicitly require it to be in accordance with the existing statutory and regulatory framework. The Secretary is directed to administer the program through "regulations, lease terms, conditions, restrictions, prohibitions, stipulations, and other provisions" of the Secretary's choosing that will ensure that the leasing program is "carried out in a manner that will ensure the receipt of fair market value by the public for the mineral resources to be leased." The legislation leaves it up to the Secretary, and ultimately the courts, to decide whether existing mineral leasing regulations and procedures will still apply or whether the new system is meant to supersede it.

2. The "compatibility" determination. Under current law, the Secretary of the Interior may permit oil and gas development in a national wildlife refuge if it is "compatible" with the purposes for which the refuge was established. If oil and gas development can take place in the Arctic Refuge without harm to the wildlife populations and habitats it was established to protect, as proponents believe, the Secretary should make the compatibility determination. Instead, the legislation recommended by the Committee "deems" the leasing program to be compatible and absolves the Secretary of any responsibility for determining whether it is or is not, in fact, compatible.

3. NEPA compliance. The National Environmental Policy Act of 1969 requires federal agencies contemplating a major federal action to prepare an environmental impact statement. The requirement is a continuing one. Agencies must supplement environmental impact statements if they make substantial changes in their proposed action or if there are significant new circumstances or information bearing on the proposed action or its impacts. The Department of the Interior last prepared an environmental impact statement on oil and gas development in ANWR in 1987, 18 years ago. In 1992, a federal court

held that significant new information was available that required the Department to supplement the 1987 environmental impact statement. If a court thought that a supplement was required 13 years ago, one must surely be needed today. Yet the legislation recommended by the Committee "deems" the 1987 statement to satisfy the requirements of NEPA "with respect to prelease activities," including the development of the regulations establishing the new leasing program.

NEPA also requires that environmental impact statements consider reasonable alternatives to a proposed action. Consideration of alternatives is said to be "the heart of the environmental impact statement." The courts have said this requirement is governed by a "rule of reason," and common sense, "bounded by some notion of feasibility." The legislation recommended by the Committee waives even this common sense requirement, and limits the Secretary to consideration of only her "preferred action for leasing and a single leasing alternative."

4. Judicial review. Under current law, a person harmed by agency action is entitled to judicial review. A person can bring suit either in the District of Columbia or where he resides. The reviewing court is empowered to "decide all relevant questions of law," and set aside agency actions found to be arbitrary, capricious, or unsupported by substantial evidence. Review is generally limited to the administrative record compiled by the agency when it made its decision, though the court may sometimes look beyond the record in NEPA cases to make sure the decision maker adequately considered environmental impacts and alternatives.

   The legislation recommended by the Committee restricts the right of judicial review. It does so by requiring anyone seeking to challenge the Secretary's actions to file suit in the District of Columbia Circuit, by limiting judicial review of the Secretary's decision to conduct a lease sale and the environmental analysis of that decision solely to whether the Secretary has complied with the new legislation, and by strictly limiting review to the administrative record. It is unclear to what extent the narrow scope of review imposed by the new legislation will be read to preempt to the broad scope of review afforded under the Administrative Procedure Act. The Committee leaves that question to the courts.

5. Roads, pipelines, and other rights-of-way. Current law provides a comprehensive process for approving rights-of-ways for roads, pipelines, airstrips, and other transportation and utility systems in conservation system units in Alaska. The principal purpose of this process was to provide access to and from resource development areas, but to do so in an orderly way that would avoid or minimize harm to the environment. The legislation recommended by the Committee exempts all rights-of-way for the exploration, development, production, or transportation of oil and gas on the Coastal Plain from this process.

6. 2,000-acre limitation. The legislation recommended by the Committee restricts the surface acreage covered by oil and gas production within the Arctic Refuge to a maximum of 2,000 acres on the

Coastal Plain. This limitation is cited by leasing proponents as evidence that oil and gas development will occupy a tiny footprint on the Coastal Plain. This provision contains so many loopholes, however, that exploration and development activities could impact much of the Coastal Plain. There is no requirement that the developed surface lands be contiguous or even consolidated. The limitation only applies literally to the actual ground covered, ignoring the effect on nearby lands. As example of the hollowness of the limitation, it only counts the area occupied by the footings of a pipeline support structure towards the acreage limitation, even though the pipeline itself may run many miles across the Coastal Plain.

7. Alaska Native lands. Over 100,000 acres (over 150 square miles) that are within both the boundaries of the Refuge and the definition of the Coastal Plain are owned by Alaska Natives. The surface of over 90,000 of these acres is owned by the Kaktovik Inupiat Corporation, and their subsurface is owned by the Arctic Slope Regional Corporation [ASRC]. The remaining approximately 10,000 acres are owned by individual Alaska Natives. Under a 1983 agreement . . ., oil and gas development on the Arctic Slope Regional Corporation's lands is prohibited "until Congress authorizes such activities on Refuge lands within the coastal plain or on ASRC Lands, or both." Enactment of the legislation recommended by the Committee will plainly "authorize such activities on Refuge lands within the coastal plain," enabling the Arctic Slope Regional Corporation and the individual Alaska Native owners to develop oil and gas resources on their lands within the Refuge. However, the Alaska Native lands are within the defined Coastal Plain covered by the leasing program, and thus are subject to the overall 2,000 acre limitation. But including the Native lands within this limitation appears to abrogate the Arctic Slope Regional Corporation's right to develop all of its lands under its 1983 contract . . .

   Moreover, it is unclear what effect the Committee recommendation will have on revenues derived from oil and gas development on Native lands within the Refuge. The legislation plainly states that all receipts derived from oil and gas development on the Coastal Plain, which is defined to include the Native lands, are to be divided equally between the State of Alaska and the U.S. Treasury, though neither would have any right to revenues derived from oil and gas development on Native lands.

8. Division of receipts. The Congressional Budget Office estimates that oil and gas leasing in the Arctic Refuge will generate $5 billion in bonus, rental, and royalty receipts. Under current law, 90 percent of those receipts ($4.5 billion) are to be paid to the State of Alaska, and the remaining 10 percent ($500 million) to the U.S. Treasury. The legislation recommended by the Committee changes the allocation in current law, to permit the Federal Government to retain 50 percent ($2.5 billion) of the receipts.

   This change is necessary for the Committee to meet its reconciliation instructions. The State of Alaska contends that any such reduction of its share of receipts violates the Alaska Statehood Act and apparently intends to challenge it in court. The Committee believes that Congress has the power to reduce the State's share and that the provision will ultimately be upheld by the courts. If, however, the State should prevail, opening the Arctic Refuge to oil and gas leasing may produce only one-fifth of the receipts the Committee expects.

9. Exports. The Mineral Leasing Act of 1920 generally prohibits the export of oil transported through pipelines over rights-of-way over public lands unless the President finds the export to be in the national interest. A 1995 amendment to the Mineral Leasing Act makes an exception for oil transported through the Trans-Alaska Pipeline. Oil moved through the Trans-Alaska Pipeline can be exported unless the President finds the export to not be in the national interest. Thus, since oil produced from the Arctic Refuge is likely to be transported through the Trans-Alaska Pipeline, it can be exported rather than used here in the [U.S.].

During consideration of the Committee recommendation, Democrats offered a series of amendments to address many of the problems in the legislation. Regrettably, all of them were rejected. Although we continue to oppose oil and gas development in the Arctic National Wildlife Refuge, adoption of our amendments would have at least ensured that development would have proceeded in accordance with the laws governing oil and gas development in other wildlife refuges, ensured that the Treasury would receive 50 percent of the receipts, and ensured that the oil would be used in the United States.

Our efforts to improve the legislation were thwarted by the fact that it is being considered under the extraordinary rules for budget reconciliation measures. All of our amendments were met with the argument that they might either reduce the amount of estimated receipts or delay their collection, or that they might not have any effect on the receipts at all, and thus would be extraneous. Such arguments only serve to remind us why important policy legislation of this sort should not be considered under the strictures of the reconciliation process.

The Senate's quarter-century debate over the future of the Arctic National Wildlife Refuge has never been about money. It has been, and continues to be, about two different sets of priorities: whether we should sacrifice a pristine wilderness to exploit the energy resources it may contain; or whether we should forego a much-needed energy resource to save a remote and frigid wilderness. Such a fundamental, deep-seated, philosophical controversy requires the deliberative process of the Senate.

Many years ago, when he introduced the bill that would become the Wilderness Act of 1964, Senator Anderson compared our wilderness areas to "our museums and our art galleries." They were "part of our cultural resource as well as our natural heritage," he said. "We should regard them as such and cherish them." We would all do well to keep Senator Anderson's words in mind as we consider the future of the Arctic National Wildlife Refuge.

POSTSCRIPT

Should the Arctic National Wildlife Refuge Be Opened to Oil Drilling?

Those who see in nature only values that can be expressed in human terms are well represented by Jonah Goldberg, who, in "Ugh, Wilderness! The Horror of 'ANWR,' the American Elite's Favorite Hellhole," *National Review* (August 6, 2001), describes the ANWR as so bleak and desolate that development can only improve it. On the other hand, Adam Kolton, testifying before the House Committee on Resources on July 11, 2001, in opposition to the National Energy Security Act of 2001 (NESA), presented the coastal plain as "the site of one of our continent's most awe-inspiring wildlife spectacles" and, thus, deserving of protection from exploitation. Kennan Ward, in *The Last Wilderness: Arctic National Wildlife Refuge* (Wildlight Press, 2001), describes a realm where human impact is still minimal and wilderness endures. John G. Mitchell, in "Oil Field or Sanctuary?" *National Geographic* (August 2001), is more balanced in his appraisal but sides with Amory B. Lovins and L. Hunter Lovins, "Fool's Gold in Alaska" *Foreign Affairs*, (July/August 2001), concluding that better alternatives to developing the ANWR exist.

In "ANWR Oil: An Alternative to War Over Oil," *American Enterprise* (June 2002), Walter J. Hickle, former U.S. secretary of the interior and twice the governor of Alaska, writes, "[T]he issue is not going to go away. Given our continuing precarious dependence on overseas oil suppliers ranging from . . . the Saudis to Venezuela's Castro-clone Hugo Chavez, sensible Americans will continue to press Congress in the months and years ahead to unlock America's great Arctic energy storehouse." The recent rise in the prices of oil and gasoline renews the point: The issue is not about to go away. In fact, it is gaining urgency from growing awareness that oil production may have already passed its peak, meaning that year by year the amount of oil available to the market will decline; see the "Peak Oil Forum" essays in *World Watch* (January/February 2006). On May 25, 2006, the House of Representatives voted to open a portion of the Arctic National Wildlife Refuge to oil drilling. The Senate was expected to address the issue later in the year.

Similar debate has centered on mineral exploitation in the American Southwest. President Clinton created the Grand Staircase–Escalante

National Monument by executive order to protect an important part of Utah's remaining wilderness, but opposition remains. See T. H. Watkins, *The Redrock Chronicles: Saving Wild Utah* (Johns Hopkins University Press, 2000). For a survey of the wilderness system created by the 1964 Wilderness Act, see John G. Mitchell and Peter Essick, "Wilderness: America's Land Apart," *National Geographic* (November 1998).

ISSUE 9

Should the U.S. Be Doing More to Combat Global Warming?

YES: Jerald L. Schnoor, from "Global Warming: A Consequence of Human Activities Rivaling Earth's Biogeochemical Processes," *Human and Ecological Risk Assessment* (December 2005)

NO: Bush Administration, from "Executive Summary: Global Climate Change Policy Book" (February 2002)

ISSUE SUMMARY

YES: Jerald L. Schnoor, co-director of the Center for Global and Regional Environmental Research at the University of Iowa, argues that global warming is real, human activities are to blame, and stabilizing atmospheric carbon dioxide levels within the next century will require drastic action.

NO: The Bush administration's plan for dealing with global warming insists that short-term economic health must come before reducing emissions of greenhouse gases. It is more useful to reduce "greenhouse gas intensity" or emissions per dollar of economic activity than to reduce total emissions.

Scientists have known for more than a century that carbon dioxide and other "greenhouse gases" (including water vapor, methane, and chlorofluorocarbons) help prevent heat from escaping the earth's atmosphere. In fact, it is this "greenhouse effect" that keeps the earth warm enough to support life. Yet there can be too much of a good thing. Ever since the dawn of the industrial age, humans have been burning vast quantities of fossil fuels, releasing the carbon they contain as carbon dioxide. Because of this, some estimate that by the year 2050, the amount of carbon dioxide in the air will be double what it was in 1850. By 1982 an increase was apparent. Less than a decade later, many researchers were saying that the climate had already begun to warm. The scientific consensus is that the global climate is warming and will continue to warm (see Naomi Oreskes, "The Scientific Consensus on Climate Change," *Science* [December 3, 2004]). There is less agreement on just how much it will warm or what the impact of the warming will be on human (and other) life.

The debate has been heated. The June 1992 issue of *The Bulletin of the Atomic Scientists* carries two articles on the possible consequences of the greenhouse effect. In "Global Warming: The Worst Case," Jeremy Leggett says that although there are enormous uncertainties, a warmer climate will release more carbon dioxide, which will warm the climate even further. As a result, soil will grow drier, forest fires will occur more frequently, plant pests will thrive, and methane trapped in the world's seabeds will be released and will increase global warming much further—in effect, there will be a "runaway greenhouse effect." Leggett also hints at the possibility that polar ice caps will melt and raise sea levels by hundreds of feet.

Taking the opposing view, in "Warming Theories Need Warning Label," S. Fred Singer emphasizes the uncertainties in the projections of global warming and their dependence on the accuracy of the computer models that generate them, and he argues that improvements in the models have consistently shrunk the size of the predicted change. There will be no catastrophe, he argues, and money spent to ward off the climate warming would be better spent on "so many pressing—and real—problems in need of resources." Singer remains skeptical of what he still calls the "global warming scare"; see Stephen Goode, "Singer Cool on Global Warming," *Insight* (April 27–May 10, 2004).

Global warming, says the UN Environment Programme, will do some $300 billion in damage each year to the world economy by 2050. In March 2001 President George W. Bush announced that the United States would not take steps to reduce greenhouse emissions—called for by the international treaty negotiated in 1997 in Kyoto, Japan—because such reductions would harm the American economy (the U.S. Senate has not ratified the Kyoto treaty). Since the Intergovernmental Panel on Climate Change (IPCC) had just released its third report saying that past forecasts were, in essence, too conservative, Bush's stance provoked immense outcry.

According to the IPCC (see *Climate Change 2001* [IPCC, 2001], available at http://www.ipcc.ch/), climate warming is already apparent and will get worse than previous forecasts had suggested. Sea level will rise, ice cover will shrink, rainfall patterns will change, and human activities—particularly emissions of carbon dioxide—are to blame. Yet, say writers such as Stuart Jordan, "The Global Warming Crisis," *The Humanist* (November/December 2005), "I see little evidence that the Bush Administration has given more than lip service to the problem.... There are ... things that can be done [and] The important thing is to get started doing some of these things in earnest, and not for us to stick our heads in the sand doing business as usual until the water comes in over our eyelids. If we wait long enough, it will."

In the following selections, professor Jerald L. Schnoor, co-director of the Center for Global and Regional Environmental Research at the University of Iowa, argues that global warming is real, human activities are to blame, and stabilizing atmospheric carbon dioxide levels within the next century will require drastic action, including the deployment of new energy sources. The Bush Administration's plan for dealing with global warming insists that short-term economic health must come before reducing emissions of greenhouse gases. It is more useful to reduce "greenhouse gas intensity" or emissions per dollar of economic activity, which will not interfere with continued economic growth the way reducing total emissions would surely do. Unfortunately, if greenhouse gas intensity is reduced and the economy grows, total emissions may well increase dramatically.

YES

Jerald L. Schnoor

Global Warming: A Consequence of Human Activities Rivaling Earth's Biogeochemical Processes

In J. D. Salinger's short story, *A Perfect Day for Bananafish*, the protagonist weaves the story of a fish who swims into a bananahole and eats so ravenously and grows so large that he can never again swim out. That fish could be a metaphor for our global consumer culture's addiction to oil. Our consumption of fossil fuels is unsustainable, not because we are running out of fuel anytime soon, but rather because we are facing very serious environmental consequences. We are "hooked" on an energy system totally dependent on fossil fuels without any alternative. Global warming is one manifestation of this addiction. Our planet grows warmer because of a massive disruption in the biogeochemical fluxes on earth. We are mining geologic deposits that were formed over a period of 300 million years, and we are burning them in the blink of an eye in geologic time, a few hundred years. Our materials flows for sulfur, nitrogen, and carbon rival the flows of nature.

So what? What are the damages (climate impacts) of our fossil-fuel addiction? When do they occur? Many prefer to delay taking action as long as there is no crisis. But is there? For economic reasons, the United States and Australia have decided not to ratify the Kyoto Climate Convention to limit greenhouse gases (GHGs). Such a position assumes there is no danger—or at least that there is little time-delay in the climate system once we decide to take action. But that, too, is not certain. To some extent, the world is struggling with a new question, one of intergenerational equity. What are the rights of the next 20 generations of unborn people who have no say in our treaties? Must we preserve the Earth for them as we have found it?

Many countries subscribe to the *Precautionary Principle*. They seem willing to invest, change behavior, and mitigate GHG now despite uncertainty. This commentary addresses the following questions associated with taking action: (1) How do we know that increases in greenhouse gases are due to human activities and not natural biogeochemical cycles? (2) What is the evidence that global warming is occurring? (3) What are the effects on human health and the environment? and (4) What can we do about it now?

From *Human and Ecological Risk Assessment*, December 2005, pp. 1105-1110. Copyright © 2005 by Taylor & Francis Books, Inc. Reprinted by permission.

How do we know that increases in greenhouse gases are due to human activities and not natural biogeochemical cycles? First, there is a strong environmental signal in which GHGs are increasing in the atmosphere, including carbon dioxide, methane, nitrous oxide, and others. These are caused primarily by burning of fossil fuels resulting in carbon dioxide emissions (CO_2); flooded agriculture, animal husbandry, and coal mining, resulting in the leakage of methane (CH_4); and denitrification of nitrogen-fertilizers resulting in nitrous oxide (N_2O). CO_2 is responsible for more than half of the greenhouse effect, and it is growing exponentially at ~0.4% per year. Every time you pump 10 gallons of gas into your car and burn it while driving, you exhaust 190 lbs of CO_2 into the atmosphere. Each person in the United States emits about 6 metric tons of carbon (22 metric tons of CO_2) into the atmosphere each year. Emissions from all global anthropogenic activities are more than 6–7 billion metric tons of carbon per year, and approximately half of that is accumulating in the atmosphere.

We know that the increasing global atmospheric CO_2 concentrations are caused by humans because of three separate lines of evidence. First, the increase in CO_2 concentrations began at the end of the 18th century when the industrial revolution commenced. So the timing of the signal is consistent with human activities. Second, our global emissions are sufficient to explain the amount of CO_2 accumulating in the atmosphere each year. Thus, the amount of GHG accumulation can be accounted for by human activities. Third, the temporal and spatial dynamics of CO_2 in the atmosphere, mixing from north to south, and fluctuating annually with an increase in the baseline each year, can only be explained by human emissions.

What is the evidence that global warming has already occurred? Earth is warmer than it has been for the past 1,000 years. But the average amount of warming, 1.1°F or 0.6°C, is not yet very large. *The 3rd Assessment Report* is a consensus document vetted by more than 2,500 scientists from more than 100 countries. It states that global warming has already occurred due mostly to human-caused emissions of greenhouse gases. Although the *average* amount of warming so far is not great, the temperature signal has emerged from the noise, and it is significant. More disconcerting is the forecast for the next ~50 years: an additional warming of 2°C (3.6°F).

The 11 warmest years in a 130-year record have occurred since 1990. We are losing most of our continental glaciers, rendering countries and communities that depend on glacial melt-water vulnerable. If you want to view the ancient ice at Glacier National Park, you had better hurry. It is melting fast. The recent Arctic Climate Impact Assessment Report (www.acia.uaf.edu) warns that the Arctic is warming at twice the global rate. The Arctic is 5.4°F warmer in July and August than it was just 30 years ago—we have lost 15% of arctic tundra and more than 10% of perennial sea ice since the 1980s. The World Wildlife Fund estimates that perennial sea ice is melting at 9.6% per decade. Global models (General Circulation Models) predicted 20 years ago that warming would be magnified in the Arctic because of a positive feedback loop whereby initial warming melts sea ice, which in turn decreases the albedo (reflectance) of the earth's surface, thus increasing temperatures still more.

Most climatologists worry more about the second moment (the variation and extremes in climate) than the average amount of warming. Observed increases in Earth's surface temperatures also correspond with a warmer sea. A warmer sea means more evaporation and, potentially, greater dissipation of energy by cyclones, hurricanes, tornadoes, storms, floods, and other extreme events. Financial damages paid by reinsurance agencies have escalated, and they are very concerned about global warming. But damage estimates are complicated by where people choose to live today—on the beach of barrier islands or atop the cliff overlooking the sea. Nonetheless, the World Business Council for Sustainable Development, 175 companies who consider the Triple Bottom Line (economics, environment, and social aspects) in their business decisions, have urged the U.S. government to adopt the Kyoto Climate Convention as a first step in mitigating global warming. Still, many others in industry are not convinced.

What are the effects on human health and the environment? For humans, direct health effects from heat stroke are relatively easily documented. Many thousands perished in the prolonged heat wave of summer 2003 in Europe. Peter Stott of the Hadley Climate Center estimated that GHG emissions doubled the likelihood for that high-temperature summer (Stott *et al.* 2004). According to Professor Tony McMichael from the Australian National University, approximately 160,000 people died prematurely due to climate change in 2003. McMichael says it is purely a statistical calculation, estimating the increased number of heat-related deaths (excess deaths) due to an anomaly in temperature.

Secondary air pollutants, ozone and particulate matter, have been attributed to warmer summers, and these pose health risks to elderly and asthmatic populations. Greater threats from mosquitoes, rodents, and other vectors are a concern from a warmer climate, but there is not much documentation to date. Island nations, such as the Maldives in the Indian Ocean, are concerned for their very existence, as sea level and storm surges rise.

Perhaps the greatest impacts to date have been on ecosystems (Parmesan and Galbraith 2004). Mexican jays and tree swallows have altered their breeding patterns. Plants flower earlier, many animals mate earlier than 20 years ago, and some species are migrating to higher latitudes. Ecosystem function, including carbon cycling and storage in Alaskan tundra, has been altered. The best documented changes have been reported for the migration of the red fox northward into the arctic fox's habitat, and the disappearance of Edith's checkerspot butterfly from its Mexican range and habitat success in Canada! Overall, there has been a poleward and upward elevational shift of plant and animal ranges worldwide, and a decline of some plant and animal species.

What can we do now? On February 16, 2005, the Kyoto Climate Convention came into full force and effect. This means that most developed countries (except for the United States and Australia) must meet targeted reductions in GHG emissions by 2012 compared to their 1990 emission baselines. Although many countries are falling short, implementation of the Convention is expected to reduce emissions by a few percent. It is not a perfect treaty. (I would have preferred a simple carbon tax with voucher arrangements for poor people and countries.) But any emission reductions afforded under the Kyoto Convention may not be noticeable if non-participating countries and the developing world increase their emissions too fast. Even U.S. and Australian companies have obligations under the Kyoto Convention if they want to trade with Kyoto participants. There is hope that the U.S. Congress will pass the Climate Stewardship Act, originally introduced by McCain and Liebermann in 2004, which is essentially "Kyoto-light" legislation.

The U.S. Senate voted 95–0 in July 1997 to recommend against signing the Kyoto Climate Convention. Senators opposed any treaty that would hurt the U.S. economy and fail to require developing countries to also reduce emissions (*e.g.*, China, India, Mexico, and Brazil). In the end, the United States signed the treaty but never ratified it. The United States still emits almost 25% of global GHG emissions. To stabilize the concentrations of GHGs in the atmosphere, they must become a player. Eventually, developing countries also must reduce emissions if we are to be successful in preventing global warming, but only after they have reached a certain level of income and development. Developed nations have already profited from GHG emissions that have accumulated in the atmosphere and caused the problem. Consequently, the developing world must be allowed to reach some level of prosperity before being obliged to reduce emissions. That is why I would like to see the United States ratify the Kyoto Convention and become involved with implementing and improving the treaty. Clean Development Mechanisms (CDMs) are one means whereby rich countries can invest in the developing world and "leap-frog" development problems.

Kyoto will not be enough. To stabilize CO_2 concentrations at 550 ppm (*i.e.*, two times the pre-industrial concentration) requires a 70% cut-back in global GHG emissions. That is what leads me to believe that we need a whole new way of doing business. We need to transition from the fossil fuel age.

There is room for optimism. Many people are heeding the call to action. Tony Blair in the United Kingdom has proposed a 60% reduction in carbon emissions by 2050 for the U.K. and E.U. Germany is mulling over a plan for a 40% reduction by 2020 and has even begun to remove the perverse subsidies (preference laws) on its native coal industry in the Ruhr River Valley. Canada's Climate Action Network has the ear of the government for a pledge of 50% reduction by 2030. Ontario Canada has a plan to phase out 5 of its largest coal-fired power plants by moving to wind, hydropower, and biomass.

It is doable. Our automobiles must become more efficient—we need to double the fuel efficiency of cars in the United States; the technology is already here—hybrid vehicles can serve as a wonderful transportation alternative. We can attain a 50% reduction in use of fossil fuels with energy efficiency (automobiles, appliances, buildings) and wind power alone! Leadership is needed to harvest "low-hanging fruit," such as requiring compact fluorescent rather than incandescent light bulbs, low-flow shower heads for water heater savings, heat pumps for heating and air conditioning, refillable beverage containers, and doubling the fuel efficiency of cars. There is more than enough wind in the United States to satisfy all our energy needs. Wind power is growing at 30% per year since 1995, and it is significantly cheaper than all other forms of new power with the federal tax credit currently in place. Home-grown transportation fuels (ethanol and biodiesel) have broken the energy and carbon barriers to make these technologies more renewable and sustainable. Such policies will create jobs, erase our balance of payments deficit, curtail mercury pollution, and improve our environment. Yes, we can obtain energy security while reducing greenhouse gases at the same time.

Over the next two decades, carbon emissions will be considered a pollutant. We will trade, mitigate, avoid, manage, and decrease the emission of CO_2 from fossil fuels through energy efficiency and carbon management/ sequestration. After that, we will need to find a renewable alternative to the fossil-fuel age. It may be hydrogen, but hydrogen is only an energy carrier. Hydrogen needs to be generated by some other energy system. I would prefer solar and wind to generate clean hydrogen for hydrogen fuel cells that will power our cars and homes. Until we reach that point, we may utilize fuel cells with methane, methanol, or ethanol as the energy source. Hydrogen fuel cell cars, buses, and H_2 fueling stations are already on the ground in Munich; Reykjavik; Washington, D.C.; and southern California. So although the dream is not a widespread reality, it is certainly not unthinkable.

That is why I am cautiously optimistic about the prospects for carbon trading. The Chicago Climate Exchange (www.chicagoclimatex.com) is operating as a market for voluntary trading of greenhouse gas emissions credits in anticipation of a full market. More than 70 companies and universities have joined and pledged to reduce their greenhouse gas emissions by 4% in 2006. A market in the United Kingdom has already traded hundreds of millions of dollars in carbon among countries of the European Union who ratified the Kyoto Climate Protocol. One can purchase the equivalent of a metric ton of CO_2 emission reductions for about $1.72 in the United States and $7 in the United Kingdom. The ability to remove, reduce, avoid, or mitigate a ton of CO_2 is a commodity, just as surely as pork bellies or soybeans in Iowa. Markets may be used as a means to control pollution in the post-industrial age.

This approach has already been modestly successful with the cap-and-trade program initiated under the Clean Air Act (CAA) Amendments of 1990, establishing a market for sulfur dioxide emissions trading. Once the cap-and-trade program was implemented, the cost plummeted from more than $1,000 per ton to $200 per ton. It is currently at $652 per ton as EPA is ratcheting down the cap. U.S. industry has met its obligations under the CAA at a fraction of the cost predicted.

Global warming is a consequence of the human enterprise growing too large with respect to natural cycles on Earth. We rival nature's biogeochemical fluxes for sulfur, nitrogen, and, to a lesser extent, carbon. We use a large fraction of all available freshwater; we harvest 19% of photosynthesis for agriculture (Vitousek *et al.* 1986), and we have already consumed most of the stores of several non-renewable metals including silver, zinc, mercury, thallium, cadmium, and copper.

We are not bananafish. All we need to do is let go of what we are doing, and we can swim away freely. We can transition from the fossil-fuel age. I truly believe that 100 years from now, people will look back incredulously on the fossil-fuel age. What Neanderthals! They burned most of the Earth's fossil fuels in a dozen generations, while gassing themselves in the meantime. We can do better than that.

References

Parmesan C and Galbraith H. 2004. Observed impacts of global climate change in the U.S. Pew Center on Global Climate Change, Arlington, VA, USA

Stott PA, Stone DA, and Allen MR. 2004. Human contribution to the European heatwave of 2003. Nature 432:610–14

United Nations Environment Programme, Intergovernmental Panel on Climate Change. 2001. Climate change 2001: The third assessment report. Cambridge University Press, Cambridge, UK

Vitousek P, Erlich PR, Erlich AH, and Matson P. 1986. Human appropriation of the products of photosynthesis. Available at http://dieoff.org/page83.htm

 **NO**

Global Climate Change Policy Book

Executive Summary

"Addressing global climate change will require a sustained effort, over many generations. My approach recognizes that sustained economic growth is the solution, not the problem—because a nation that grows its economy is a nation that can afford investments in efficiency, new technologies, and a cleaner environment."

—President George W. Bush

The President announced a new approach to the challenge of global climate change. This approach is designed to harness the power of markets and technological innovation. It holds the promise of a new partnership with the developing world. And it recognizes that climate change is a complex, long-term challenge that will require a sustained effort over many generations. As the President has said, "The policy challenge is to act in a serious and sensible way, given the limits of our knowledge. While scientific uncertainties remain, we can begin now to address the factors that contribute to climate change."

While investments today in science will increase our understanding of this challenge, our investments in advanced energy and sequestration technologies will provide the breakthroughs we need to dramatically reduce our emissions in the longer term. In the near term, we will vigorously pursue emissions reductions even in the absence of complete knowledge. Our approach recognizes that sustained economic growth is an essential part of the solution, not the problem. Economic growth will make possible the needed investment in research, development, and deployment of advanced technologies. This strategy is one that should offer developing countries the incentive and means to join with us in tackling this challenge together. Significantly, the President's plan will:

- **Reduce the Greenhouse Gas Intensity of the U.S. Economy by 18 Percent in the Next Ten Years.** Greenhouse gas intensity measures the ratio of greenhouse gas (GHG) emissions to economic output. This new approach focuses on reducing the growth of GHG emissions, while sustaining the economic growth needed to finance investment in new, clean energy technologies. It sets America on a path to slow

Global Climate Change Policy Book, February 2002.

the growth of greenhouse gas emissions, and—as the science justifies—to stop and then reverse that growth:

- In efficiency terms, the 183 metric tons of emissions per million dollars GDP that we emit today will be lowered to 151 metric tons per million dollars GDP in 2012.
- Existing trends and efforts in technology improvement will play a significant role. Beyond that, the President's commitment will achieve 100 million metric tons of reduced emissions in 2012 alone, with more than 500 million metric tons in cumulative savings over the entire decade.
- This goal is comparable to the average progress that nations participating in the Kyoto Protocol are required to achieve.

- **Substantially Improve the Emission Reduction Registry.** The President directed the Secretary of Energy, in consultation with the Secretary of Commerce, the Secretary of Agriculture, and the Administrator of the Environmental Protection Agency, to propose improvements to the current voluntary emission reduction registration program under section 1605(b) of the 1992 Energy Policy Act within 120 days. These improvements will enhance measurement accuracy, reliability and verifiability, working with and taking into account emerging domestic and international approaches.
- **Protect and Provide Transferable Credits for Emissions Reduction.** The President directed the Secretary of Energy to recommend reforms to ensure that businesses and individuals that register reductions are not penalized under a future climate policy, and to give transferable credits to companies that can show real emissions reductions.
- **Review Progress Toward Goal and Take Additional Action if Necessary.** If, in 2012, we find that we are not on track toward meeting our goal, and sound science justifies further policy action, the United States will respond with additional measures that may include a broad, market-based program as well as additional incentives and voluntary measures designed to accelerate technology development and deployment.
- **Increase Funding for America's Commitment to Climate Change.** The President's FY '03 budget seeks $4.5 billion in total climate spending—an increase of $700 million. This commitment is unmatched in the world, and is particularly notable given America's focus on international and homeland security and domestic economic issues in the President's FY '03 budget proposal.
- **Take Action on the Science and Technology Review.** The Secretary of Commerce and Secretary of Energy have completed their review of the federal government's science and technology research portfolios and recommended a path forward. As a result of their review, the President has established a new management structure to advance and coordinate climate change science and technology research.
  - The President has established a Cabinet-level Committee on Climate Change Science and Technology Integration to oversee this effort. The Secretary of Commerce and Secretary of Energy will lead the effort, in close coordination with the President's Science Advisor. The research effort will continue to be coordinated through the National Science and Technology Council in accordance with the Global Change Research Act of 1990.

- The President's FY '03 budget proposal dedicates $1.7 billion to fund basic scientific research on climate change and $1.3 billion to fund research on advanced energy and sequestration technologies.
- This includes $80 million in new funding dedicated to implementation of the Climate Change Research Initiative (CCRI) and the National Climate Change Technology Initiative (NCCTI) announced last June. This funding will be used to address major gaps in our current understanding of the natural carbon cycle and the role of black soot emissions in climate change. It will also be used to promote the development of the most promising "breakthrough" technologies for clean energy generation and carbon sequestration.
- **Implement a Comprehensive Range of New and Expanded Domestic Policies, Including:**
  - *Tax Incentives for Renewable Energy, Cogeneration, and New Technology.* The President's FY '03 budget seeks $555 million in clean energy tax incentives, as the first part of a $4.6 billion commitment over the next five years ($7.1 billion over the next 10 years). These tax credits will spur investments in renewable energy (solar, wind, and biomass), hybrid and fuel cell vehicles, cogeneration, and landfill gas conversion. Consistent with the National Energy Policy, the President has directed the Secretary of the Treasury to work with Congress to extend and expand the production tax credit for electricity generation from wind and biomass, to develop a new residential solar energy tax credit, and to encourage cogeneration projects through investment tax credits.
  - *Business Challenges.* The President has challenged American businesses to make specific commitments to improving the greenhouse gas intensity of their operations and to reduce emissions. Recent agreements with the semi-conductor and aluminum industries and industries that emit methane already have significantly reduced emissions of some of the most potent greenhouse gases. We will build upon these successes with new agreements, producing greater reductions.
  - *Transportation Programs.* The Administration is promoting the development of fuel-efficient motor vehicles and trucks, researching options for producing cleaner fuels, and implementing programs to improve energy efficiency. The President is committed to expanding federal research partnerships with industry, providing market-based incentives and updating current regulatory programs that advance our progress in this important area. This commitment includes expanding fuel cell research, in particular through the "FreedomCAR" initiative. The President's FY '03 budget seeks more than $3 billion in tax credits over 11 years for consumers to purchase fuel cell and hybrid vehicles. The Secretary of Transportation has asked the Congressional leadership to work with him on legislation that would authorize the Department of Transportation to reform the Corporate Average Fuel Economy (CAFE) program, fully considering the recent National Academy Sciences report, so that we can safely improve fuel economy for cars and trucks.
  - *Carbon Sequestration.* The President's FY '03 budget requests over $3 billion—a $1 billion increase above the baseline—as the first part of a

ten year (2002–2011) commitment to implement and improve the conservation title of the Farm Bill, which will significantly enhance the natural storage of carbon. The President also directed the Secretary of Agriculture to provide recommendations for further, targeted incentives aimed at forest and agricultural sequestration of greenhouse gases. The President further directed the Secretary of Agriculture, in consultation with the Environmental Protection Agency and the Department of Energy, to develop accounting rules and guidelines for crediting sequestration projects, taking into account emerging domestic and international approaches.

- **Promote New and Expanded International Policies to Complement Our Domestic Program.** The President's approach seeks to expand cooperation internationally to meet the challenge of climate change, including:
  - *Investing $25 Million in Climate Observation Systems in Developing Countries.* In response to the National Academy of Sciences' recommendation for better observation systems, the President has allocated $25 million and challenged other developed nations to match the U.S. commitment.
  - *Tripling Funding for "Debt-for-Nature" Forest Conservation Programs.* Building upon recent Tropical Forest Conservation Act (TFCA) agreements with Belize, El Salvador, and Bangladesh, the President's FY '03 budget request of $40 million to fund "debt for nature" agreements with developing countries nearly triples funding for this successful program. Under TFCA, developing countries agree to protect their tropical forests from logging, avoiding emissions and preserving the substantial carbon sequestration services they provide. The President also announced a new agreement with the Government of Thailand, which will preserve important mangrove forest in Northeastern Thailand in exchange for debt relief worth $11.4 million.
  - *Fully Funding the Global Environmental Facility.* The Administration's FY '03 budget request of $178 million for the GEF is more than $77 million above this year's funding and includes a substantial $70 million payment for arrears incurred during the prior administration. The GEF is the primary international institution for transferring energy and sequestration technologies to the developing world under the United Nations Framework Convention on Climate Change (UNFCCC).
    - *Dedicating Significant Funds to the United States Agency for International Development (USAID).* The President's FY '03 budget requests $155 million in funding for USAID climate change programs. USAID serves as a critical vehicle for transferring American energy and sequestration technologies to developing countries to promote sustainable development and minimize their GHG emissions growth.
    - *Pursue Joint Research with Japan.* The U.S. and Japan continue their High-Level Consultations on climate change issues. Later this month, a team of U.S. experts will meet with their Japanese counterparts to discuss specific projects within the various areas of climate science and technology, to identify the highest priorities for collaborative research.

- *Pursue Joint Research with Italy.* Following up on a pledge of President Bush and Prime Minister Berlusconi to undertake joint research on climate change, the U.S. and Italy convened a Joint Climate Change Research Meeting in January 2002. The delegations for the two countries identified more than 20 joint climate change research activities for immediate implementation, including global and regional modeling.
- *Pursue Joint Research with Central America.* The United States and Central American Heads of Government signed the Central American-United States of America Joint Accord (CONCAUSA) on December 10, 1994. The original agreement covered cooperation under action plans in four major areas: conservation of biodiversity, sound use of energy, environmental legislation, and sustainable economic development. On June 7, 2001, the United States and its Central American partners signed an expanded and renewed CONCAUSA Declaration, adding disaster relief and climate change as new areas for cooperation. The new CONCAUSA Declaration calls for intensified cooperative efforts to address climate change through scientific research, estimating and monitoring greenhouse gases, investing in forestry conservation, enhancing energy efficiency, and utilizing new environmental technologies.

National Goal

The President set a national goal to reduce the greenhouse gas intensity of the U.S. economy by 18 percent over the next ten years. Rather than pitting economic growth against the environment, the President has established an approach that promises real progress on climate change by tapping the power of sustained economic growth.

- **The President's Yardstick—Greenhouse Gas Intensity—Is a Better Way to Measure Progress Without Hurting Growth.** A goal expressed in terms of declining greenhouse gas intensity, measuring greenhouse gas emissions relative to economic activity, quantifies our effort to reduce emissions through conservation, adoption of cleaner, more efficient, and emission-reducing technologies, and sequestration. At the same time, an intensity goal accommodates economic growth.
- **Reducing Greenhouse Gas Intensity by 18 Percent Over the Next Ten Years Is Ambitious but Achievable.** The United States will reduce the 183 metric tons of emissions per million dollars GDP that we emit today to 151 metric tons per million dollars GDP in 2012. We expect existing trends and efforts in technology improvement to play a significant role. Beyond that, our commitment will achieve 100 million metric tons of reduced emissions in 2012 alone, with more than 500 million metric tons in cumulative savings over the entire decade.
- **Focusing on Greenhouse Gas Intensity Sets America on a Path to Slow the Growth of Greenhouse Gas Emissions, and—as the Science Justifies—to Stop and Then Reverse That Growth.** As we learn more about the science of climate change and develop new technologies to mitigate emissions, this annual decline can be accelerated. When the

annual decline in intensity equals the economic growth rate (currently, about 3% per year), emission growth will have stopped. When the annual decline in intensity exceeds the economic growth rate, emission growth will reverse. Reversing emission growth will eventually stabilize atmospheric concentrations as emissions decline.

· **As We Advance Science and Develop Technology to Substantially Reduce Greenhouse Gas Emissions in the Long Term, We Do Not Want to Risk Harming the Economy in the Short Term.** Over the past 20 years, greenhouse gas emissions have risen with economic growth, as our economy benefited from inexpensive, fossil-fuel based—and greenhouse gas emitting—energy. While new technologies promise to break this emission-economy link, a rapid reduction in emissions would be costly and threaten economic growth. Sustained economic growth is essential for any long-term solution: Prosperity is what allows us to dedicate more resources to solving environmental problems. History shows that wealthier societies demand—and can afford—more environmental protection.

· **The Intensity Based Approach Promotes Near-Term Opportunities to Conserve Fossil Fuel Use, Recover Methane, and Sequester Carbon.** Until we develop and adopt breakthrough technologies that provide safe and reliable energy to fuel our economy without emitting greenhouse gases, we need to promote more rapid adoption of existing, improved energy efficiency and renewable resources that provide cost-effective opportunities to reduce emissions. Profitable methane recovery from landfills, coal mines and gas pipelines offers another opportunity—estimated by the EPA at about 30 million tons of carbon equivalent emissions. Finally, carbon sequestration in soils and forests can provide tens of millions of tons of emission reductions at very low costs.

· **The Intensity Based Approach Advances a Serious, but Measured Mitigation Response.** The President recognizes America's responsibility to reduce emissions. At the same time, any long-term solution—one that stabilizes atmospheric concentrations of greenhouse gases at safe levels—will require the development and deployment of new technologies that are not yet cost-effective. The President's policy balances the desire for immediate reductions with the need to protect the economy and to take advantage of developing science and technology.

The President's Goal Is Ambitious and Responsible

· **Reducing Greenhouse Gas Intensity by 18 Percent Over the Next Ten Years Is Comparable to the Average Progress that Nations Participating in the Kyoto Protocol Are Required to Achieve.** Our goal translates into a 4.5 percent reduction beyond forecasts of the progress that America is expected to make based on existing programs and private activity. Forecasts of the average reductions required by nations implementing the Kyoto Protocol range from zero to 7 percent.

· **While Producing Results Similar to What the Kyoto Protocol Participants Are Required to Achieve on Average, the President's Approach Protects the Economy and Develops Institutions for a Long-Term Solution.** The focus on greenhouse gas intensity separates the goal of reducing emissions from the potential economic harm associated with a rigid emission cap. By measuring greenhouse gas emissions relative to economic activity, we have a solid yardstick against which we can measure progress as we pursue a range of programs to reduce emissions. As we develop technologies to produce more goods with fewer greenhouse gas emissions, this yardstick does not penalize economic growth.

- **Greenhouse Gas Intensity Is a More Practical Way to Discuss Goals with Developing Countries.** The close connection between economic growth, energy use and greenhouse gas emissions implies that fixed appropriate emission limits are hard to identify when economic growth is uncertain and carbon-free, breakthrough energy technologies are not yet in place. Such targets are also hard to identify for developing countries where the future rate of emissions is even more uncertain. Given its neutrality with regard to economic growth, greenhouse gas intensity solves or substantially reduces many of these problems.

Enhanced National Registry for Voluntary Emissions Reductions

The Administration will improve the current federal GHG Reduction and Sequestration Registry that recognizes greenhouse gas reductions by non-governmental organizations, businesses, farmers, and the federal, state and local governments. Registry participants and the public will have a high level of confidence in the reductions recognized by this Registry, through capture and sequestration projects, mitigation projects that increase energy efficiency and/or switch fuels, and process changes to reduce emissions of potent greenhouse gases, such as methane. An enhanced registry will promote the identification and expansion of innovative and effective ways to reduce greenhouse gases. The enhanced registry will encourage participation by removing the risk that these actions will be penalized—or inaction rewarded—by future climate policy.

- **Improve the Quality of the Current Program.** A registry is a tool for companies to publicly record their progress in reducing emissions, providing public recognition of a company's accomplishments, and a record of mitigation efforts for future policy design. This tool goes hand-in-hand with voluntary business challenges, described below, by providing a standardized, credible vehicle for reporting and recognizing progress.
  - Although businesses can already register emission reductions under section 1605(b) of the 1995 Energy Policy Act, participation has been limited.
  - The President directed the Secretary of Energy, in consultation with the Secretary of Commerce, Secretary of Agriculture, and the Administrator of the Environmental Protection Agency, to propose improvements to the current voluntary emissions reduction registration program within 120 days.
  - These improvements will enhance measurement accuracy, reliability and verifiability, working with and taking into account emerging domestic and international approaches.
- **Protect and Provide Transferable Credits for Emissions Reduction.** The President directed the Secretary of Energy to recommend reforms

to ensure that businesses and individuals that register reductions are not penalized under a future climate policy, and to give transferable credits to companies that can show real emissions reductions. These protections will encourage businesses and individuals to pursue innovative strategies to reduce or sequester greenhouse gas emissions, without the risk that future climate policy will disadvantage them.

- **Background on Current Registry Program.** The Energy Policy Act of 1992 directed the Department of Energy (with EIA as the implementing agency) to develop a program to document voluntary actions that reduce emissions of greenhouse gases or remove greenhouse gases from the atmosphere.

  - Under the Energy Policy Act, EIA was directed to issue "procedures for the accurate reporting of information on annual reductions of greenhouse gas emissions and carbon fixation achieved through any measures, including fuel switching, forest management practices, tree planting, use of renewable energy, manufacture or use of vehicles with reduced greenhouse gas emissions, appliance efficiency, methane recovery, cogeneration, chlorofluorocarbon capture and replacement, and power plant heat rate improvement."

  - In 1999, 207 companies and other organizations, representing 24 different industries or services, reported on 1,722 projects that achieved 226 million metric tons of carbon dioxide equivalent reductions—equal to 3.4 percent of national emissions. Participating companies included Clairol, AT&T, Dow Chemical, Johnson & Johnson, IBM, Motorola, Pharmacia, Upjohn, Sunoco, Southern, General Motors and DuPont.

  - EIA released a February 2002 report demonstrating that this program continues to expand. In 2000, 222 companies had undertaken 1,882 projects to reduce or sequester greenhouse gases. These achieved 269 million metric tons of carbon dioxide equivalent reductions—equal to 3.9 percent of national emissions.

  - A number of proposals to reform the existing registry—or create a new registry—have appeared in energy and/or climate policy bills introduced in the past year. The Administration will fully explore the extent to which the existing authority under the Energy Policy Act is adequate to achieve these reforms.

Progress Check in 2012

The domestic programs proposed by the President allow consumers and businesses to make flexible decisions about emission reductions rather than mandating particular control options or rigid targets. If, however, by 2012, our progress is not sufficient, and sound science justifies further action, the United States will respond with additional measures that may include a broad, market-based program, as well as additional incentives and voluntary measures designed to accelerate technology development and deployment.

POSTSCRIPT

Should the U.S. Be Doing More to Combat Global Warming?

The United Nations Conference on Environment and Development in Rio de Janeiro, Brazil, took place in 1992. High on the agenda was the problem of global warming, but despite widespread concern and calls for reductions in carbon dioxide releases, the United States refused to consider rigid deadlines or set quotas. The uncertainties seemed too great, and some thought the economic costs of cutting back on carbon dioxide might be greater than the costs of letting the climate warm.

The nations that signed the UN Framework Convention on Climate Change in Rio de Janeiro in 1992 met again in Kyoto, Japan, in December 1997 to set carbon emissions limits for the industrial nations. The United States agreed to reduce its annual greenhouse gas emissions 7 percent below the 1990 level between 2008 and 2012 "but still has not ratified the Kyoto Treaty. In November 2005, they met in Montreal, Canada, and decided to begin formal talks on mandatory post-2012 reductions in greenhouse gases. The U.S. agreed to talk but ruled out any future commitments." Ross Gelbspan, in "Rx for a Planetary Fever," *American Prospect* (May 8, 2000), blames much of that opposition on "big oil and big coal [which] have relentlessly obstructed the best-faith efforts of government negotiators." Nor do some portions of the industry seem interested in acting on their own. In May 2003 Exxon Mobil rejected proposals that it address global warming and develop renewable energy. CEO Lee Raymond, who had previously denounced the Kyoto Protocol, said the company does not "make social statements at the expense of shareholder return."

The opposition remains. Critics stress uncertainties in the data and the potential economic impacts of attempting to reduce carbon dioxide emissions. However, when the National Commission on Energy Policy published its report, *Ending the Energy Stalemate: A Bipartisan Strategy to Meet America's Energy Challenges* in December 2004 (http://bcsia.ksg.harvard.edu/BCSIA_content/documents/NCEP_KSG_4-14-05.pdf), the Energy Information Administration promptly analyzed its economic impact in *Impacts of Modeled Recommendations of the National Commission on Energy Policy* (April 2005) (http://www.eia.doe.gov/oiaf/servicerpt/bingaman/index.html) and concluded that increasing automobile fuel efficiency, encouraging alternate energy sources, and increasing oil production and clean coal technology would cost the "U.S. economy ... no more than 0.15% of GDP or about $78 per household per year, while overall GDP is projected to grow by 87%." However, there is a problem with any plan that includes increasing oil production; according to Kenneth S. Deffeyes, *Beyond Oil: The View From Hubbert's Peak* (Hill and Wang, 2005), world oil production would peak in late November 2005 and will thereafter decline despite efforts to combat the trend. See also the "Peak Oil Forum" special report in *World Watch* (January/February 2006).

There is also opposition based on the view that the methods of reducing greenhouse gas emissions called for in the Kyoto treaty are, at root, unworkable. See Frank N. Laird, "Just Say No to Greenhouse Gas Emissions Targets," *Issues in Science and Technology* (Winter 2000–2001). However, researchers have proposed a number of innovative ways to keep from adding carbon dioxide to the atmosphere. Conservation alone will be very helpful, says Amory B. Lovins, "More Profit with Less Carbon," *Scientific* American (September 2005). See also Robert H. Socolow, "Can We Bury Global Warming?" *Scientific American* (July 2005), and Jennie C. Stephens and Bob van der Zwaan, "The Case for Carbon Capture and Storage," *Issues in Science and Technology* (Fall 2005). Fred Krupp, president of Environmental Defense, in "Global Warming and the USA," *Vital Speeches of the Day* (April 15, 2003), recommends a market-based method to finding and developing innovative approaches. Thomas J. Wilbanks, et al., in "Possible Responses to Global Climate Change: Integrating Mitigation and Adaptation," *Environment* (June 2003), note that many mitigation techniques are under study around the world but that people will also have to adapt to a warming world.

In June 2002 the U.S. Environmental Protection Agency (EPA) issued its *U.S. Climate Action Report—2002* (available at http://www.epa.gov/global-warming/publications/car/index.html) to the United Nations. In it, the EPA admits for the first time that global warming is real and that human activities are most likely to blame. President George W. Bush immediately dismissed the report as "put out by the bureaucracy" and said he still opposes the Kyoto Protocol. He insists that more research is necessary before anyone can even begin to plan a proper response. Unfortunately, the latest studies warn that the consequences of global warming may be much more severe than even the IPCC has so far forecast. See, for instance, Richard A. Kerr, "Climate Modelers See Scorching Future as a Real Possibility," *Science* (January 28, 2005), and *Meeting the Climate Challenge: Recommendations of the International Climate Change Taskforce* (January 2005), which warns that the world may already be on the verge of irreversible disaster. In January 2006, the U.K.'s Department for Environment, Food and Rural Affairs (DEFRA) published "Avoiding Dangerous Climate Change" (http://www.defra.gov.uk/environment/climatechange/internat/dangerous-cc.htm), warning that immediate action is essential if catastrophe is to be avoided. In June 2006, the Supreme Court agreed to hear a suit by 12 states, several cities, and the American Samoa demanding that the EPA regulate carbon dioxide as a hazardous pollutant. The Bush administration insists that carbon dioxide is not a dangerous pollutant under federal law, and so far courts have sided with the administration. See H. Josef Hebert, "Ruling May Affect Global Warming," *Boston Globe* (June 27, 2006).

The Intergovernmental Panel on Climate Change's (IPCC) fourth report will be released in 2007.

ISSUE 10

Will Hydrogen End Our Fossil-Fuel Addiction?

YES: David L. Bodde, from "Fueling the Future: The Road to the Hydrogen Economy," Statement Presented to the Committee on Science, Subcommittee on Research and Subcommittee on Energy, U.S. House of Representatives (July 20, 2005)

NO: Michael Behar, from "Warning: The Hydrogen Economy May Be More Distant Than It Appears," *Popular Science* (January 2005)

ISSUE SUMMARY

YES: David L. Bodde argues that there is no question whether hydrogen can satisfy the nation's energy needs. The real issue is how to handle the transition from the current energy system to the hydrogen system.

NO: Michael Behar argues that the public has been misled about the prospects of the "hydrogen economy." We must overcome major technological, financial, and political obstacles before hydrogen can be a viable alternative to fossil fuels.

The 1973 oil crisis heightened awareness that the world—even if it was not yet running out of oil—was extraordinarily dependent on that fossil fuel (and therefore on supplier nations) for transportation, home heating, and electricity generation. Since the supply of oil and other fossil fuels is clearly finite, some people worried that there would come a time when demand could not be satisfied—and our dependence would leave us helpless. At the same time, we became acutely aware of the many unfortunate side-effects of fossil fuels, including air pollution, strip mines, oil spills, and more.

The 1970s saw the modern environmental movement gain momentum. The first Earth Day was in 1970. Numerous government steps were taken to deal with air pollution, water pollution, and other environmental problems. In response to the oil crisis, a great deal of public money went into developing alternative energy supplies. The emphasis was on "renewable" energy, meaning conservation, wind, solar, and fuels such as hydrogen gas (which when burned with pure oxygen produces only water vapor as exhaust). However, when the crisis passed and oil supplies were once more ample (albeit it did cost more to fill a gasoline tank), most public funding for alternative-energy research and demonstration projects vanished. What work continued was at the hands of a few enthusiasts and those corporations that saw future opportunities. In 2001, the WorldWatch Institute published Seth Dunn's *Hydrogen Futures: Toward a Sustainable Energy System*. In 2002, MIT Press published Peter Hoffman's *Tomorrow's Energy: Hydrogen, Fuel Cells, and the Prospects for a Cleaner Planet*. On the corporate side, fossil fuel companies have long been major investors in alternative energy systems.

What drives the continuing interest in hydrogen and other alternative or renewable energy systems is the continuing problems associated with fossil fuels (and the discovery of new problems such as global warming), concern about dependence and potential political instability, rising oil and gasoline prices, . . . and the growing realization that the availability of petroleum will peak in the near future. See Colin J. Campbell, "Depletion and Denial: The Final Years of Oil," *USA Today Magazine* (November 2000), Charles C. Mann, "Getting Over Oil," *Technology Review* (January/February 2002), Tim Appenzeller, "The End of Cheap Oil, "*National Geographic* (June 2004), Robert L. Hirsch, Roger H. Bezdek, and Robert M. Wendling, "Peaking Oil Production: Sooner Rather than Later?" *Issues in Science and Technology* (Spring 2005); and "The Peak Oil Forum," *World Watch* (January/February 2006).

Will that interest come to anything? There are, after all, a number of other ways to meet the need. Coal can be converted into oil and gasoline (though the air pollution and global warming problems will remain). Cars can be made more efficient (and mileage efficiency is much greater than it was in the 1970s despite the popularity of SUVs). Cars can be designed to use natural gas or battery power; "hybrid" cars use combinations of gasoline and electricity, and some are already on the market. See Mark K. Solheim, "How Green Is My Hybrid?" *Kiplinger's Personal Finance* (May 2006), and Corinna Kester, "Diesels versus Hybrids: Comparing the Environmental Costs," *World Watch* (July/August 2005). Unfortunately, so far people seem slow to give up their old-style cars; see David Welch, "If Gas Prices Are Up, Why Are Hybrid Sales Down?" *Business Week Online* (May 3, 2006).

Perhaps people are just waiting for hydrogen. Clemson University professor David L. Bodde argues that there is no question whether hydrogen can satisfy the nation's energy needs. The real issue is how to handle the transition from the current energy system to the hydrogen system. He recommends research, education, and support for entrepreneurs. Journalist Michael Behar argues that the public has been misled about the prospects of the "hydrogen economy." It will come, but first we must overcome major technological, financial, and political obstacles.

YES

David L. Bodde

Fueling the Future:
The Road to the Hydrogen Economy

The Hydrogen Transition:
A Marketplace Competition

Much thinking about the hydrogen economy concerns "what" issues—visionary descriptions of a national fuels infrastructure that would deliver a substantial fraction of goods and services with hydrogen as the energy carrier. And yet, past visions of energy futures, however desirable they might have seemed at the time, have not delivered sustained action, either from a public or private perspective. The national experience with nuclear power, synthetic fuels, and renewable energy demonstrates this well.

The difficulty arises from insufficient attention to the transition between the present and the desired future—the balance between forces that lock the energy economy in stasis and the entrepreneurial forces that could accelerate it toward a more beneficial condition. In effect, the present competes against the future, and the pace and direction of any transition will be governed by the outcome. Viewing the transition to a hydrogen economy through the lens of a competitive transition can bring a set of "how" questions to the national policy debate—questions of how policy can rebalance the competitive forces so that change prevails in the marketplace.

A Model of the Competitive Transition

The competitive battle will be fought over a half century among three competing infrastructures:[1]

- The internal combustion engine (ICE), either in a spark-ignition or compression-ignition form, and its attendant motor fuels supply chain;
- The hybrid electric vehicle (HEV), now entering the market, which achieves superior efficiency by supplementing an internal combustion engine with an electric drive system and uses the current supply chain for motor fuels; and,
- The hydrogen fuel cell vehicle (HFCV), which requires radically distinct technologies for the vehicle, for fuel production, and for fuel distribution.

Committee on Science, Subcommittee on Research and Subcommittee on Energy, U.S. House of Representatives, July 20, 2005.

Figure 1

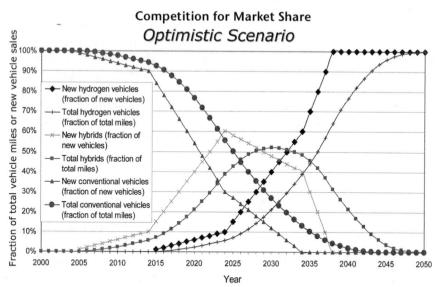

Competition for Market Share
Optimistic Scenario

Note: Complete replacement of ICE and HEV vehicles with fuel cell vehicles in 2050
Source: NRC 2004

Figure 1 shows one scenario, based on the most optimistic assumptions, of how market share could shift among the contending infrastructures (NRC 2004). Several aspects of this scenario bear special mention. First, note the extended time required for meaningful change: these are long-lived assets built around large, sunk investments. They cannot be quickly changed under the best of circumstances. Second, the road to the hydrogen economy runs smoothest through the hybrid electric vehicle. The HEV offers immediate gains in fuel economy and advanced technologies that will eventually prove useful for hydrogen fuel cell vehicles, especially battery and electric system management technologies. Although this scenario shows significant market penetration for the HEV, its success cannot be assured. The HEV might remain a niche product, despite its current popularity, if consumers conclude that the value of the fuel savings does not compensate for the additional cost of the HEV. Or, its gains in efficiency might be directed toward vehicle size and acceleration rather than fuel economy. Either circumstance would make an early hydrogen transition even more desirable.

Any transition to a HFCV fleet, however, will require overcoming a key marketplace barrier that is unique to hydrogen—widely available supplies of fuel. And to this we now turn.

The Chicken and the Egg[2]

Most analyses suggest that large-scale production plants in a mature hydrogen economy can manufacture fuel at a cost that competes well with gasoline at current prices (NRC 2004). However, investors will not build these plants and their supporting distribution infrastructure in the absence of large-scale demand. And, the demand for hydrogen will not be forthcoming unless potential purchasers of hydrogen vehicles can be assured widely available sources of fuel. Variants of this "chicken and egg" problem have limited the market penetration of other fuels, such as methanol and ethanol blends (M85 and E85) and compressed natural gas. This issue—the simultaneous development of the supply side and demand sides of the market—raises one of the highest barriers to a hydrogen transition.

Distributed Hydrogen Production for the Transition

To resolve this problem, a committee of the National Academy of Sciences (NRC 2004) recommended an emphasis on distributed production of hydrogen. In this model, the hydrogen fuel would be manufactured at dispensing stations conveniently located for consumers. Once the demand for hydrogen fuel grew sufficiently, larger manufacturing plants and logistic systems could be built to achieve scale economies. However, distributed production of hydrogen offers two salient challenges.

The first challenge is cost. . . . The "distributed technologies" offer hydrogen at a cost between 2 and 5 times the cost of the large-scale, "central station" technologies. Technological advances can mitigate, but not remove entirely, this cost disadvantage.

The second challenge concerns the environment. Carbon capture and sequestration do not appear practical in distributed production. During the opening stage of a hydrogen transition, we might simply have to accept some carbon releases in order to achieve the later benefits.

Research to Accelerate a Transition by Distributed Hydrogen Production

A study panel convened by the National Academy of Sciences (NAS) recently recommended several research thrusts that could accelerate distributed production for a transition to hydrogen (NRC 2004). These include:

- Development of a hydrogen fueling "appliance" that can be manufactured economically and used in service stations reliably and safely by relatively unskilled persons—station attendants and consumers.
- Development of an integrated, standard fueling facility that includes the above appliance as well as generation and storage equipment capable of meeting the sharply varying demands of a 24-hour business cycle.

- Advanced technologies for hydrogen production from electrolysis, essentially a fuel cell operated in reverse, to include enabling operation from intermittent energy sources, such as wind.
- Research on breakthrough technologies for small-scale reformers to produce hydrogen from fossil feedstocks.

The Department of Energy has adopted the NAS recommendations and modified its programs accordingly. It remains too early to judge progress, but in any case these technologies should receive continued emphasis as the desired transition to hydrogen nears. However, progress in research is notoriously difficult to forecast accurately. This suggests consideration be given to interim strategies that would work on the demand side of the marketplace, either to subsidize the cost of distributed hydrogen production while demand builds or to raise the cost of the competition—gasoline, and diesel fuels. Such actions would relieve the research program of the entire burden for enabling the transition.

Fundamental Research to Sustain a Hydrogen Economy

At the same time that the marketplace transition advances, several high-payoff (but also high-risk) research campaigns should be waged. These include:

- Storing hydrogen on-board vehicles at near-atmospheric pressure;
- Sequestering the carbon dioxide effluent from manufacturing hydrogen from coal;
- Sharply reducing the cost of hydrogen produced from non-coal resources, especially nuclear, photobiological, photoelectrochemical, and thin-film solar processes;
- Improving the performance and cost of fuel cells; and,
- Storing electricity on-board vehicles in batteries that provide both high-energy performance and high-power performance at reasonable cost.

On-Vehicle Hydrogen Storage

The most important long-term research challenge is to provide a more effective means of storing hydrogen on vehicles than the compressed gas or cryogenic liquid now in use. In my judgment, failure to achieve this comes closer to a complete "show-stopper" than any other possibility. I believe this true for two reasons: hydrogen leakage as the vehicle fleet ages and cost.

With regard to leakage, high-pressure systems currently store molecular hydrogen on demonstration vehicles safely and effectively. But these are new and specially built, and require trained professionals operate and maintain. What can we expect of production-run vehicles that receive the casual maintenance afforded most cars? A glance at the oil-stained pavement of any parking lot offers evidence of the leakage of heavy fluids stored in the current ICE fleet at atmospheric pressure. As high-pressure systems containing the lightest element in the universe age, we might find even greater difficulties with containment. With regard to cost, the energy losses from liquefaction and even compression severely penalize the use of hydrogen fuel, especially when it is manufactured at distributed stations.

The NAS Committee, cited earlier (NRC 2004), strongly supported an increased emphasis on game-changing approaches to on-vehicle hydrogen storage. One alternative could come from novel approaches to generating the hydrogen on board the vehicle.[3] Chemical hydrides, for example, might offer some promise here, such as the sodium borohydride system demonstrated by Daimler-Chrysler.

Carbon Sequestration

Domestic coal resources within the United States hold the potential to relieve the security burdens arising from oil dependence—but only if the environmental consequences of their use can be overcome. Further, coal offers the lowest-cost pathway to a hydrogen-based energy economy, once the transient conditions have passed. Thus, the conditions under which this resource can be used should be established as soon as possible. The prevailing assumption holds that the carbon effluent from hydrogen manufacturing can be stored as a gas (carbon dioxide, or CO_2) in deep underground formations. Yet, how long it must be contained and what leakage rates can be tolerated remain unresolved issues (Socolow 2005). Within the Department of Energy, the carbon sequestration program is managed separately from hydrogen and vehicles programs. The NAS committee recommended closer coordination between the two as well as an ongoing emphasis on carbon capture and sequestration (NRC 2004).

Producing Hydrogen Without Coal

Manufacturing hydrogen from non-fossil resources stands as an important hedge against future constraints on production from coal, or even from natural gas. And under any circumstance, the hydrogen economy will be more robust if served by production from a variety of domestic sources.

The non-fossil resource most immediately available is nuclear. Hydrogen could be produced with no CO_2 emissions by using nuclear heat and electricity in the high-temperature electrolysis of steam. Here the technology issues include the durability of the electrode and electrolyte materials, the effects of high pressure, and the scale-up of the electrolysis cell. Alternatively, a variety of thermochemical reactions could produce hydrogen with great efficiency. Here the needed research concerns higher operating temperatures (700°C to 1000°C) for the nuclear heat as well as research into the chemical cycles themselves. In both cases, the safety issues that might arise from coupling the nuclear island with a hydrogen production plant bear examination (NRC 2004).

In addition, hydrogen production from renewable sources should be emphasized, especially that avoiding the inefficiencies of the conventional chain of conversions: (1) from primary energy into electricity; (2) from electricity to hydrogen; (3) from hydrogen to electricity on-board the vehicle; (4) from electricity to mobility, which is what the customer wanted in the first place. Novel approaches to using renewable energy, such as photobiological or photoelectrochemical, should be supported strongly (NRC 2004).

Improved Fuel Cells

The cost and performance of fuel cells must improve significantly for hydrogen to achieve its full potential. To be sure, molecular hydrogen can be burned in specially designed internal combustion engines. But doing so foregoes the efficiency gains obtainable from the fuel cell, and becomes a costly and (from an energy perspective) inefficient process. The NAS Committee thought the fuel cell essential for a hydrogen economy to be worth the effort required to put it in place. They recommended an emphasis on long-term, breakthrough research that would dramatically improve cost, durability, cycling capacity, and useful life.

Improved Batteries

The battery is as important to a hydrogen vehicle as to a hybrid because it serves as the central energy management device. For example, the energy regained from regenerative braking must be stored in a battery for later reuse. Although energy storage governs the overall operating characteristics of the battery, a high rate of energy release (power) can enable the electric motor to assist the HEV in acceleration and relieve the requirements for fuel cells to immediately match their power output with the needs of the vehicle. Thus, advanced battery research becomes a key enabler for the hydrogen economy and might also expand the scope of the BEV.

Entrepreneurship for the Hydrogen Economy

For the results of DoE research to gain traction in a competitive economy, entrepreneurs and corporate innovators must succeed in bringing hydrogen-related innovations to the marketplace. In many cases, independent entrepreneurs provide the pathbreaking innovations that lead to radical improvements in performance, while established companies provide continuous, accumulating improvement.[4] The federal government, in partnership with states and universities, can become an important enabler of both pathways to a hydrogen economy.

Federal Policies Promoting Entrepreneurship

From the federal perspective, several policies could be considered to build an entrepreneurial climate on the "supply" side of the market. These include:

- Special tax consideration for investors in new ventures offering products relevant to fuel savings. The intent would be to increase the amount of venture capital available to startup companies.
- Commercialization programs might enable more entrepreneurs to bring their nascent technologies up to investment grade. . . .
- Outreach from the National Laboratories to entrepreneurs might be improved. . . .

On the demand side, any policy that increases consumer incentives to purchase fuel-efficient vehicles will provide an incentive for ongoing innovation—provided that the policy is perceived as permanent. Entrepreneurs and innovators respond primarily to opportunity; but that opportunity must be durable for the 10-year cycle required to establish a new, high-growth company.

States and Universities as Agents of Innovation/Entrepreneurship

Innovation/entrepreneurship is a contact sport, and that contact occurs most frequently and most intensely within the context of specific laboratories and specific relationships. I will use Clemson's International Center for Automotive Research (ICAR) to illustrate this principle. Most fundamentally, the ICAR is a partnership among the state of South Carolina, major automakers,[5] and their . . . suppliers. The inclusion of these suppliers will be essential for the success of ICAR or any similar research venture. This is because innovation in the auto industry has evolved toward a global, networked process, much as it has in other industries like microelectronics. The "supply chain" is more accurately described as a network, and network innovation will replace the linear model.

For these reasons, the ICAR, when fully established, will serve as a channel for research and innovation to flow into the entire cluster of auto-related companies in the Southeast United States. We anticipate drawing together and integrating the best technology from a variety of sources:

- Research performed at Clemson University and at the ICAR itself;
- Research performed at the Savannah River National Laboratory and the University of South Carolina; and,
- Relevant science and technology anywhere in the world.

Beyond research, the ICAR will include two other components of a complete innovation package: education and entrepreneur support. With regard to education, the Master of Science and PhD degrees offered through the ICAR will emphasize the integration of new technology into vehicle design, viewing the auto and its manufacturing plant as an integrated system. In addition, courses on entrepreneurship and innovation, offered through Clemson's Arthur M. Spiro Center for Entrepreneurial Leadership, will equip students with the skills to become effective agents of change within the specific context of the global motor vehicle industry.

With regard to entrepreneur support, the ICAR will host a state-sponsored innovation center to nurture startup companies that originate in the Southeast auto cluster and to draw others from around the world into that cluster. In addition, the ICAR innovation center will welcome teams from established companies seeking the commercial development of their technologies. The state of South Carolina has provided significant support through four recent legislative initiatives. The Research University Infrastructure and the Research Centers of Economic Excellence Acts build the capabilities of the state's universities; and the Venture Capital Act and Innovation Centers Act provide support for entrepreneurs.

None of these elements can suffice by itself; but taken together they combine to offer a package of technology, education, and innovation that can serve the hydrogen transition extraordinarily well.

A Concluding Observation

Revolutionary technological change of the kind contemplated here is rarely predictable and never containable. Every new technology from the computer to the airplane to the automobile carries with it a chain of social and economic consequences that reach far beyond the technology itself. Some of these consequences turn out to be benign; some pose challenges that must be overcome by future generations; but none has proven foreseeable.

For example, a hydrogen transition might bring prolonged prosperity or economic decline to the electric utility industry depending upon which path innovation takes. A pathway that leads through plug-hybrids to home appliances that manufacture hydrogen by electrolysis would reinforce the current utility business model. A pathway in which hydrogen fuel cell vehicles serve as generators for home electric energy would undermine that model. The same holds true for the coal industry. A future in which carbon sequestration succeeds will affect coal far differently from one in which it cannot be accomplished.

The only certainty is that the energy economy will be vastly different from that which we know today. It will have to be.

Reference

Socolow, Robert H. "Can We Bury Global Warming?" *Scientific American,* July 2005, pp. 49–55.

Sperling, Daniel and James D. Cannon, *The Hydrogen Transition*, Elsevier Academic Press, 2004.

U.S. National Research Council, *The Hydrogen Economy: Opportunities, Costs, Barriers, and R&D Needs*, The National Academies Press, 2004.

Notes

1. Another concept, the battery electric vehicle (BEV), offers an all-electric drivetrain with all on-board energy stored in batteries, which would be recharged from stationary sources when the vehicle is not in operation. I have not included this among the competitors because battery technology has not advanced rapidly enough for it to compete in highway markets. In contrast, BEV have proven quite successful in the personal transportation niche.

2. Alternatively framed: "Which comes first, the vehicle or the fuel?"

3. I do not include on-board reforming of fossil feedstocks, like gasoline, among these. These systems offer little gain beyond that achievable with the HEV, and most industrial proponents appear to have abandoned the idea.

4. See the Appendix: The Process of Innovation and Implications for the Hydrogen Transition for a more complete discussion.

5. BMW was the founding OEM and most significant supporter of the ICAR.

Michael Behar  **NO**

Warning: The Hydrogen Economy May Be More Distant Than It Appears

In the presidential campaign of 2004, Bush and Kerry managed to find one piece of common ground: Both spoke glowingly of a future powered by fuel cells. Hydrogen would free us from our dependence on fossil fuels and would dramatically curb emissions of air pollutants, including carbon dioxide, the gas chiefly blamed for global warming. The entire worldwide energy market would evolve into a "hydrogen economy" based on clean, abundant power. Auto manufacturers and environmentalists alike happily rode the bandwagon, pointing to hydrogen as the next big thing in U.S. energy policy. Yet the truth is that we aren't much closer to a commercially viable hydrogen-powered car than we are to cold fusion or a cure for cancer. This hardly surprises engineers, fuel cell manufacturers and policymakers, who have known all along that the technology has been hyped, perhaps to its detriment, and that the public has been misled about what Howard Coffman, editor of `fuelcell-info.com`, describes as the "undeniable realities of the hydrogen economy." These experts are confident that the hydrogen economy will arrive—someday. But first, they say, we have to overcome daunting technological, financial and political roadblocks. Herewith, our checklist of misconceptions and doubts about hydrogen and the exalted fuel cell.

1. Hydrogen Is an Abundant Fuel

True, hydrogen is the most common element in the universe; it's so plentiful that the sun consumes 600 million tons of it every second. But unlike oil, vast reservoirs of hydrogen don't exist here on Earth. Instead, hydrogen atoms are bound up in molecules with other elements, and we must expend energy to extract the hydrogen so it can be used in fuel cells. We'll never get more energy out of hydrogen than we put into it.

"Hydrogen is a currency, not a primary energy source," explains Geoffrey Ballard, the father of the modern-day fuel cell and co-founder of Ballard Power Systems, the world's leading fuel-cell developer. "It's a means of getting energy from where you created it to where you need it."

2. Hydrogen Fuel Cells Will End Global Warming

Unlike internal combustion engines, hydrogen fuel cells do not emit carbon dioxide. But extracting hydrogen from natural gas, today's primary source, does. And wresting hydrogen from water through electrolysis takes tremendous amounts of energy. If that energy comes from power plants burning fossil fuels, the end product may be clean hydrogen, but the process used to obtain it is still dirty.

Once hydrogen is extracted, it must be compressed and transported, presumably by machinery and vehicles that in the early stages of a hydrogen economy will be running on fossil fuels. The result: even more CO_2. In fact, driving a fuel cell car with hydrogen extracted from natural gas or water could produce a net increase of CO_2 in the atmosphere. "People say that hydrogen cars would be pollution-free," observes University of Calgary engineering professor David Keith. "Light-bulbs are pollution-free, but power plants are not."

In the short term, nuclear power may be the easiest way to produce hydrogen without pumping more carbon dioxide into the atmosphere. Electricity from a nuclear plant would electrolyze water—splitting H_2O into hydrogen and oxygen. Ballard champions the idea, calling nuclear power "extremely important, unless we see some other major breakthrough that none of us has envisioned."

Critics counter that nuclear power creates long-term waste problems and isn't economically competitive. An exhaustive industry analysis entitled "The Future of Nuclear Power," written last year by 10 professors from the Massachusetts Institute of Technology and Harvard University, concludes that "hydrogen produced by electrolysis of water depends on low-cost nuclear power." As long as electricity from nuclear power costs more than electricity from other sources, using that energy to make hydrogen doesn't add up.

3. The Hydrogen Economy Can Run on Renewable Energy

Perform electrolysis with renewable energy, such as solar or wind power, and you eliminate the pollution issues associated with fossil fuels and nuclear power. Trouble is, renewable sources can provide only a small fraction of the energy that will be required for a full-fledged hydrogen economy.

From 1998 to 2003, the generating capacity of wind power increased 28 percent in the U.S. to 6,374 megawatts, enough for roughly 1.6 million homes. The wind industry expects to meet 6 percent of the country's electricity needs by 2020. But economist Andrew Oswald of the University of Warwick in England calculates that converting every vehicle in the U.S. to hydrogen power would require the electricity output of a million wind turbines—enough to cover half of California. Solar panels would likewise require huge swaths of land.

Water is another limiting factor for hydrogen production, especially in the sunny regions most suitable for solar power. According to a study done by the World Resources Institute, a Washington, D.C.-based nonprofit organization, fueling a hydrogen economy with electrolysis would require 4.2 trillion gallons of water annually—roughly the amount that flows over Niagara Falls every three months. Overall, U.S. water consumption would increase by about 10 percent.

4. Hydrogen Gas Leaks Are Nothing to Worry About

Hydrogen gas is odorless and colorless, and it burns almost invisibly. A tiny fire may go undetected at a leaky fuel pump until your pant leg goes up in flames. And it doesn't take much to set compressed hydrogen gas alight. "A cellphone or a lightning storm puts out enough static discharge to ignite hydrogen," claims Joseph Romm, author of *The Hype about Hydrogen: Fact and Fiction in the Race to Save the Climate* and founder of the Center for Energy and Climate Solutions in Arlington, Virginia.

A fender bender is unlikely to spark an explosion, because carbon-fiber-reinforced hydrogen tanks are virtually indestructible. But that doesn't eliminate the danger of leaks elsewhere in what will eventually be a huge network of refineries, pipelines and fueling stations. "The obvious pitfall is that hydrogen is a gas, and most of our existing petrochemical sources are liquids," says Robert Uhrig, professor emeritus of nuclear engineering at the University of Tennessee and former vice president of Florida Power & Light. "The infrastructure required to support high-pressure gas or cryogenic liquid hydrogen is much more complicated. Hydrogen is one of those things that people have great difficulty confining. It tends to go through the finest of holes."

To calculate the effects a leaky infrastructure might have on our atmosphere, a team of researchers from the California Institute of Technology and the Jet Propulsion Laboratory in Pasadena, California, looked at statistics for accidental industrial hydrogen and natural gas leakage—estimated at 10 to 20 percent of total volume—and then predicted how much leakage might occur in an economy in which everything runs on hydrogen. Result: The amount of hydrogen in the atmosphere would be four to eight times as high as it is today.

The Caltech study "grossly overstated" hydrogen leakage, says Assistant Secretary David Garman of the Department of Energy's Office of Energy Efficiency and Renewable Energy. But whatever its volume, hydrogen added to the atmosphere will combine with oxygen to form water vapor, creating noctilucent clouds—those high, wispy tendrils you see at dawn and dusk. The increased cloud cover could accelerate global warming.

5. Cars Are the Natural First Application For Hydrogen Fuel Cells

"An economically sane, cost-effective attack on the climate problem wouldn't start with cars," David Keith says. Cars and light trucks contribute roughly 20 percent of the carbon dioxide emitted in the U.S., while power plants burning fossil fuels are responsible for more than 40 percent of CO_2 emissions. Fuel cells designed for vehicles must cope with harsh conditions and severe limitations on size and weight.

A better solution to global warming might be to hold off building hydrogen cars, and instead harness fuel cells to generate electricity for homes and businesses. Plug Power, UTC, FuelCell Energy and Ballard Power Systems already market stationary fuel-cell generators. Plug Power alone has 161 systems in the U.S., including the first fuel-cell-powered McDonald's. Collectively, however, the four companies have a peak generating capacity of about 69 megawatts, less than 0.01 percent of the total 944,000 megawatts of U.S. generating capacity.

6. The U.S. Is Committed to Hydrogen, Pouring Billions Into R&D

Consider this: President George W. Bush promised to spend $1.2 billion on hydrogen. Yet he allotted $1.5 billion to promote "healthy marriages." The monthly tab for the war in Iraq is $3.9 billion—a total of $121 billion through last September. In 2004 the Department of Energy spent more on nuclear and fossil fuel research than on hydrogen.

The federal government's FreedomCAR program, which funds hydrogen R&D in conjunction with the big three American carmakers, requires that the companies demonstrate a hydrogen-powered car by 2008—but not that they sell one.

"If you are serious about [hydrogen], you have to commit a whole lot more money," contends Guenter Conzelmann, deputy director of the Center for Energy, Environmental and Economic Systems Analysis at Argonne National Laboratory near Chicago. Conzelmann develops computer models to help the energy industry make predictions about the cost of implementing new technology. His estimate for building a hydrogen economy: more than $500 billion, and that's if 60 percent of Americans continue to drive cars with internal combustion engines.

Shell, ExxonMobil and other oil companies are unwilling to invest in production, distribution, fueling facilities and storage if there are just a handful of hydrogen cars on the road. Nor will automakers foot the bill and churn out thousands of hydrogen cars if drivers have nowhere to fill them up. Peter Devlin, head of the Department of Energy's hydrogen-production research group, says, "Our industry partners have told us that unless a fourth to a third of all refueling stations in the U.S. offer hydrogen, they won't be willing to take a chance on fuel cells."

To create hydrogen fueling stations, California governor Arnold Schwarzenegger, who drives a Hummer, has championed the Hydrogen Highway Project. His plan is to erect 150 to 200 stations—at a cost of at least $500,000 each—along the state's major highways by the end of the decade. So that's one state. Now what about the other 100,775 filling stations in the rest of the U.S.? Retrofitting just 25 percent of those with hydrogen fueling systems would cost more than $13 billion.

7. If Iceland Can Do It, So Can We

Iceland's first hydrogen fueling station is already operating on the outskirts of Reykjavík. The hydrogen, which powers a small fleet of fuel cell buses, is produced onsite from electrolyzed tap water. Meanwhile the recently formed Icelandic New Energy—a consortium that includes automakers, Royal Dutch/Shell and the Icelandic power company Norsk Hydro—is planning to convert the rest of the island nation to a hydrogen system.

Impressive, yes. But 72 percent of Iceland's electricity comes from geothermal and hydroelectric power. With so much readily available clean energy, Iceland can electrolyze water with electricity directly from the national power grid. This type of setup is impossible in the U.S., where only about 15 percent of grid electricity comes from geothermal and hydroelectric sources, while 71 percent is generated by burning fossil fuels.

Another issue is the sheer scale of the system. It could take as few as 16 hydrogen fueling stations to enable Icelanders to drive fuel cell cars anywhere in the country. At close to 90 times the size of Iceland, the U.S. would require a minimum of 1,440 fueling stations. This assumes that stations would be strategically placed to collectively cover the entire U.S. with no overlap and that everyone knows where to find the pumps.

8. Mass Production Will Make Hydrogen Cars Affordable

Simply mass-producing fuel cell cars won't necessarily slash costs. According to Patrick Davis, the former leader of the Department of Energy's fuel cell research team, "If you project today's fuel cell technologies into high-volume production—about 500,000 vehicles a year—the cost is still up to six times too high."

Raj Choudhury, operations manager for the General Motors fuel cell program, claims that GM will have a commercial fuel cell vehicle ready by 2010. Others are doubtful. Ballard says that first there needs to be a "fundamental engineering rethink" of the proton exchange membrane (PEM) fuel cell, the type being developed for automobiles, which still cannot compete with the industry standard for internal combustion engines—a life span of 15 years, or about 170,000 driving miles. Because of membrane deterioration, today's PEM fuel cells typically fail during their first 2,000 hours of operation.

Ballard insists that his original PEM design was merely a prototype. "Ten years ago I said it was the height of engineering arrogance to think that the architecture and geometry we chose to demonstrate the fuel cell in automobiles would be the best architecture and geometry for a commercial automobile," he remarks. "Very few people paid attention to that statement. The truth is that the present geometry isn't getting the price down to where it is commercial. It isn't even entering into the envelope that will allow economies of scale to drive the price down."

In the short term, conventional gasoline-burning vehicles will be replaced by gas-electric hybrids, or by vehicles that burn clean diesel, natural gas, methanol or ethanol. Only later will hydrogen cars make sense, economically and environmentally. "Most analysts think it will take several decades for hydrogen to make a large impact, assuming hydrogen technologies reach their goals," notes Joan Ogden, an associate professor of environmental science and policy at the University of California at Davis and one of the world's leading researchers of hydrogen energy.

9. Fuel Cell Cars Can Drive Hundreds of Miles on a Single Tank of Hydrogen

A gallon of gasoline contains about 2,600 times the energy of a gallon of hydrogen. If engineers want hydrogen cars to travel at least 300 miles between fill-ups—the automotive-industry benchmark—they'll have to compress hydrogen gas to extremely high pressures: up to 10,000 pounds per square inch.

Even at that pressure, cars would need huge fuel tanks. "High-pressure hydrogen would take up four times the volume of gasoline," says JoAnn Milliken, chief engineer of the Department of Energy's Office of Hydrogen, Fuel Cells and Infrastructure Technologies.

Liquid hydrogen works a bit better. GM's liquid-fueled HydroGen3 goes 250 miles on a tank roughly double the size of that in a standard sedan. But the car must be driven every day to keep the liquid hydrogen chilled to –253 degrees Celsius—just 20 degrees above absolute zero and well below the surface temperature of Pluto—or it boils off. "If your car sits at the airport for a week, you'll have an empty tank when you get back," Milliken says.

?. If Not Hydrogen, Then *What*?

The near-future prospects for a hydrogen economy are dim, concludes *The Hydrogen Economy: Opportunities, Costs, Barriers, and R&D Needs*, a major government-sponsored study published last February by the National Research Council. Representatives from ExxonMobil, Ford, DuPont, the Natural Resources Defense Council and other stakeholders contributed to the report, which urges lawmakers to legislate tougher tailpipe-emission standards and to earmark additional R&D funding for renewable energy and alternative fuels. It foresees "major hurdles on the path to achieving the vision of the hydrogen economy" and recommends that the Department of Energy "keep a balanced portfolio of R&D efforts and continue to explore supply-and-demand alternatives that do not depend on hydrogen."

Of course, for each instance where the study points out how hydrogen falls short, there are scores of advocates armed with data to show how it can succeed. Physicist Amory Lovins, who heads the Rocky Mountain Institute, a think tank in Colorado, fastidiously rebuts the most common critiques of hydrogen with an armada of facts and figures in his widely circulated white paper "Twenty Hydrogen Myths." But although he's a booster of hydrogen, Lovins is notably pragmatic. "A lot of silly things have been written both for and against hydrogen," he says. "Some sense of reality is lacking on both sides." He believes that whether the hydrogen economy arrives at the end of this decade or closer to midcentury, interim technologies will play a signal role in the transition.

The most promising of these technologies is the gas-electric hybrid vehicle, which uses both an internal combustion engine and an electric motor, switching seamlessly between the two to optimize gas mileage and engine efficiency. U.S. sales of hybrid cars have been growing steadily, and the 2005 model year saw the arrival of the first hybrid SUVs—the Ford Escape, Toyota Highlander and Lexus RX400h.

Researchers sponsored by the FreedomCAR program are also investigating ultralight materials—plastics, fiberglass, titanium, magnesium, carbon fiber—and developing lighter engines made from aluminum and ceramic materials. These new materials could help reduce vehicle power demands, bridging the cost gap between fossil fuels and fuel cells.

Most experts agree that there is no silver bullet. Instead the key is developing a portfolio of energy-efficient technologies that can help liberate us from fossil fuels and ease global warming. "If we had a wider and more diverse set of energy sources, we'd be more robust, more stable," says Jonathan Pershing, director of the Climate, Energy and Pollution Program at the World Resources Institute. "The more legs your chair rests on, the less likely it is to tip over."

Waiting for hydrogen to save us isn't an option. "If we fail to act during this decade to reduce greenhouse gas emissions, historians will condemn us," Romm writes in *The Hype about Hydrogen*. "And they will most likely be living in a world with a much hotter and harsher climate than ours, one that has undergone an irreversible change for the worse."

POSTSCRIPT

Will Hydrogen End Our Fossil-Fuel Addiction?

Hydrogen as a fuel offers definite benefits. As Joan M. Ogden notes in "Hydrogen: The Fuel of the Future?" *Physics Today* (April 2002), the technology is available, and compared to the alternatives, it "offers the greatest potential environmental and energy-supply benefits." To put hydrogen to use, however, will require massive investments in facilities for generating, storing, and transporting the gas, as well as manufacturing hydrogen-burning engines and fuel cells. Currently, large amounts of hydrogen can easily be generated by "reforming" natural gas or other hydrocarbons. Hydrolysis—splitting hydrogen from water molecules with electricity—is also possible, and in the future this may use electricity from renewable sources, such as wind or nuclear power. The basic technologies are available right now. See Thammy Evans, Peter Light, and Ty Cashman, "Hydrogen—A Little PR," *Whole Earth* (Winter 2001). Daniel Sperling, in "Updating Automotive Research," *Issues in Science and Technology* (Spring 2002), notes, "Fuel cells and hydrogen show huge promise. They may indeed prove to be the Holy Grail, eventually taking vehicles out of the environmental equation." Making that happen, however, will require research, government assistance in building a hydrogen distribution system, and incentives for both industry and car buyers. See Also Matthew L. Wald, "Questions about a Hydrogen Economy," *Scientific American* (May 2004). First steps along these lines are already visible in a few places; see Bill Keenan, "Hydrogen: Waiting for the Revolution," *Across the Board* (May/June 2004), and Annie Birdsong, "California Drives the Future of the Automobile," *World Watch* (March/April 2005). M. Z. Jacobson, W. G. Colella, and D. M. Golden, "Cleaning the Air and Improving Health with Hydrogen Fuel-Cell Vehicles," *Science* (June 24, 2005), conclude that if all onroad vehicles are replaced with fuel-cell vehicles using hydrogen generated by wind power, air pollution and human health impacts will both be reduced and overall costs will be less than for gasoline.

Joseph J. Romm, in "The Hype about Hydrogen," *Issues in Science and Technology* (Spring 2004), cautions that replacing fossil fuels with hydrogen is not something that can be done overnight. It is a process that will take decades, and for now, efforts are best bent toward reducing greenhouse gas emissions in other ways. Lacking other long-term options, Sperling and Ogden, in "The Hope for Hydrogen," *Issues in Science and Technology* (Spring 2004), say we should be working hard on the transition now. Hydrogen, they insist, "accesses a broad array of energy resources, potentially provides broader and deeper societal benefits than any other option, potentially provides large private benefits, has no natural political or economic enemies, and has a strong industrial proponent in the automotive industry."

Jeremy Rifkin, "Hydrogen: Empowering the People," *Nation* (December 23, 2002), says local production of hydrogen could mean a much more decentralized energy system. He may be right, as John A. Turner makes clear in "Sustainable Hydrogen Production," *Science* (August 13, 2004), but Henry Payne and Diane Katz, "Gas and Gasbags . . . or, the Open Road and Its Enemies," *National Review* (March 25, 2002), contend that a major obstacle to hydrogen is market mechanisms that will keep fossil fuels in use for years to come, local hydrogen production is unlikely, and adequate supplies will require that society invest heavily in nuclear power. Jim Motavalli, in "Hijacking Hydrogen," *E Magazine* (January-February 2003), worries that the fossil fuel and nuclear industries will dominate the hydrogen future. The fossil fuel industries wish to use "reforming" to generate hydrogen from coal, and the nuclear industries see hydrolysis as creating demand for nuclear power.

In January 2003, President George W. Bush proposed $1.2 billion in funding for making hydrogen-powered cars an on-the-road reality. Gregg Easterbrook, in "Why Bush's H-Car Is Just Hot Air," *New Republic* (February 24, 2003), thinks it would make much more sense to address fuel-economy standards; Bush should "leave futurism to the futurists." Peter Schwartz and Doug Randall, in "How Hydrogen Can Save America," *Wired* (April 2003), commend Bush's proposal but say that the proposed funding is not enough. We need, they say, "an Apollo-scale commitment to hydrogen power. The fate of the republic depends on it." Toward that end, Schwartz and Randall list five steps essential to making the hydrogen future real:

- Develop fuel tanks that can store hydrogen safely and in adequate quantity.
- Encourage mass production of fuel cell vehicles.
- Convert the fueling infrastructure to hydrogen.
- Increase hydrogen production.
- Mount a PR campaign.

The difficulty of the task is underlined by Robert F. Service in "The Hydrogen Backlash," *Science* (August 13, 2004) (the lead article in a special section titled "Toward a Hydrogen Economy"). Kevin Bullis, "Hydrogen Reality Check," *Technology Review* (May 5, 2006) (http://www.technologyreview.com/read_article.aspx?id=16777&ch=biztech), says any noticeable impact of hydrogen fuel cells is many years away. On the other hand, Iceland appears to be very close to a hydrogen economy right now; see Maria H. Maack and Jon Bjorn Skulason, Implementing the Hydrogen Economy," *Journal of Cleaner Production* (January 2006).

ISSUE 11

Is Additional Federal Oversight Needed for the Construction of LNG Import Facilities?

YES: Edward J. Markey, from "LNG Import Terminal and Deepwater Port Siting: Federal and State Roles," Testimony before House Committee on Government Reform, Subcommittee on Energy Policy, Natural Resources, and Regulatory Affairs (June 22, 2004)

NO: Donald F. Santa, Jr., from "LNG Import Terminal and Deepwater Port Siting: Federal and State Roles," Testimony before House Committee on Government Reform, Subcommittee on Energy Policy, Natural Resources, and Regulatory Affairs (June 22, 2004)

ISSUE SUMMARY

YES: Edward J. Markey argues that the risks—including those associated with terrorist attack—associated with LNG (liquefied natural gas) tankers and terminals are so great that additional federal regulation is essential in order to protect the public.

NO: Donald F. Santa, Jr., argues that meeting demand for energy requires public policies that "do not unreasonably limit resource and infrastructure development." The permitting process for LNG import facilities should be governed by existing Federal Energy Regulatory Commission procedures without additional regulatory impediments.

The environmental movement has, since its dawn in the 1960s, added several interesting terms to the English language. NIMBY means "Not In My Back Yard." NIMTOF means "Not in My Term of Office." BANANA means "Build Absolutely Nothing Anywhere Near Anything." The sentiments behind these acronyms can be seen in connection with many of the issues in this book. They help drive objections to landfills, nuclear power plants, nuclear and toxic waste dumps, and many more projects that pose real or potential risks to human beings and their property.

Liquefied natural gas (LNG) provokes the same feelings. Natural gas used to be considered an unwanted byproduct of oil wells. Disposing of it meant "flaring it off"; oil well towers were routinely topped by monstrous plumes of flame and smoke. After World War II, pipelines were built to carry the gas to factories, power plants, and urban areas where it could be used for its energy content. Later, the technology was developed to compress and liquefy the natural gas so it could be transported in tankers across oceans where pipelines could not go. Of course, the tankers had to arrive someplace where their gas could be transferred to tanks on trucks and trains or converted from liquid to gaseous form and put into a pipeline. These LNG terminals are essential if the gas is to be used.

But natural gas is highly flammable and accidents do happen. Many places where terminals have been proposed have objected strenuously to the associated risks. In Maine, the LNG industry has for several years sought to convince communities to accept terminals (along with local economic benefits). In Maine, several proposals have reached the point of public debate, which has included pronouncements from Canada that LNG tankers will not be permitted to pass through Canadian waters on the way to Maine terminals. Debate has been and continues to be vigorous. Critics stress the hazards (http://www.savepassamaquoddybay.org/assets.html).

When Doug Ose (R-CA) opened the July 5, 2004, meeting of the Subcommittee on Energy Policy, Natural Resources, and Regulatory Affairs of the House Committee on Government Reform, he noted that "the United States, especially California, is relying more and more on natural gas. It is the fuel-of-choice for electric power generation because it is reliable and is much cleaner than other fossil fuels. Natural gas is also used by individual citizens, and by industry, agriculture and transportation as a raw material. As a critical resource used throughout the economy, shortages in natural gas have a more profound impact than [shortages in] most other commodities. [But] North American natural gas fields are depleting at an increasing rate. Even if some new domestic natural gas comes onto the market, most experts believe that we will need even more. Pipeline imports from Canada make up about 15 percent of total U.S. consumption but, there too, experts anticipate diminishing sources. . . . I believe that increasing U.S. importation of liquefied natural gas (LNG) should be a component of the solution—either by on-shore or off-shore facilities."

In the following selections, Representative Edward J. Markey (D-MA) argues that the risks—including those associated with terrorist attack—associated with LNG tankers and terminals are so great that additional federal regulation is essential in order to protect the public. Existing regulations are not sufficient. Donald F. Santa, Jr., President of the Interstate Natural Gas Association of America, argues that meeting demand for energy requires public policies that "do not unreasonably limit resource and infrastructure development." A comprehensive energy policy must include removing barriers both to pipeline and storage infrastructure development and to enhancing gas supply by the importation of liquefied natural gas. The permitting process should be governed by existing Federal Energy Regulatory Commission procedures without additional regulatory impediments.

YES

Edward J. Markey

Testimony Before the Subcommittee on Energy Policy, Natural Resources and Regulatory Affairs

. . . LNG is an important component of the energy supply of New England, and . . . it has great potential to help the nation meet its growing need for natural gas. As Federal Reserve Chairman Alan Greenspan noted in his testimony before the Energy and Commerce Committee, one notable difference between the oil and natural gas markets in the United States is that our nation is able to obtain access to global supplies of oil via tanker. In contrast, virtually all of our natural gas supply comes from either U.S. or Canadian resources delivered via pipeline. Only a small portion of our supply comes in via tanker in the form of LNG. Increasing LNG imports is therefore one important way to help address America's increasing demand for natural gas. Obtaining access to the global natural gas supply through LNG imports is also one way of helping to reduce the current volatility in the U.S. natural gas marketplace.

The question then is where is the most appropriate place for these facilities to be sited? I would suggest to the Subcommittee that this is an issue that the Congress already considered nearly 25 years ago, based in large part about public safety concerns surrounding the siting of the Distrigas facility in a densely populated urban area, and the inherent difficulties in trying to address the consequences of an accident or an act of sabotage at this type of facility. At that time, the Congress enacted a law, which I authored, which tried to learn from the Everett experience by directing the Secretary of Transportation to consider the need for remote siting as part of the rules applicable to all new LNG importation terminals. The Secretary of Transportation, unfortunately, has chosen to largely ignore this law and has failed to comply with Congress' intent regarding what factors the Department needs to take into account in writing rules for the siting of new LNG facilities. This failure had little consequence for more than 25 years, as no new LNG importation terminals were being built. Today, however, with dozens of LNG terminals being proposed around the country, this failure can no longer be tolerated.

From Testimony before the House Committee on Government Reform, Subcommittee on Energy Policy, Natural Resources, and Regulatory Affairs, June 22, 2004.

Key Issues

As I see it, there currently are four critical issues that need to be addressed at the federal level.

First, we need to have a much better scientific and technical assessment of the consequences of a terrorist attack against an LNG tanker or LNG terminal. Such a hazard assessment is needed to better inform federal siting decisions with respect to any new LNG terminals around the nation. It is also needed to better inform state and local emergency planning and response activities with respect to existing LNG facilities.

Second, we need help from both the federal government and the facility operator to defray the costs that local governments incur in securing LNG or other critical infrastructure facilities from a terrorist attack. While Distrigas provides some funding for this purpose today, and has taken other actions to facilitate the efforts of local law enforcement to secure the facility, I believe that federal support is needed to help ensure that local firefighters are given realistic training to deal with the types of large fires or explosions that could occur, that local police departments have the resources needed to help provide security during times of elevated Homeland Security alert status, and during LNG shipments.

Third, we need to get the Transportation Department to upgrade its LNG siting regulations to comply with the Congressional intent that all future LNG terminals be remotely sited, and demand that the Department stop merely incorporating the National Fire Protection Agency Standards into its siting rules.

Fourth, we need the Coast Guard to undertake a more thorough analysis of the safety of LNG tankers, including the issues of brittle fracture and insulation flammability. . . .

Consequences of an Attack

On page 15 of the memoirs of Richard Clark, the White House's former anti-terrorism czar, and a man who served in the Clinton Administration, the first Bush Administration, and the Reagan Administration there is a disturbing passage that describes one of the discussions he had on 9/11 with Admiral James Loy, then the Commandant of the Coast Guard, as follows:

> "Jim, you have a Captain in the Port in every harbor, right." He nodded. "Can they close the harbors? I don't want anything coming in and blowing up, like the LNG in Boston." After the Millennium Terrorist Alert we had learned that al Qaeda operatives had been infiltrating Boston by coming in on liquid natural gas tankers from Algeria. We had also learned that had one of the giant tankers blown up in the harbor, it would have wiped out downtown Boston.

> "I have that authority." Loy turned and pointed at another admiral. "And I have just exercised it."

The fact that al Qaeda terrorists had come into Boston on LNG tankers was extremely disturbing to those of us who live near the Distrigas LNG facility, and it heightens the importance of ensuring that this facility, and others like it, are fully protected against terrorist attack. It also underscores the need for us to better understand the hazardous presented by such an attack. In recent months, numerous press reports have raised concerns about nature and adequacy of some of the hazard studies that were performed for the Distrigas facility shortly after the September 11th attacks.

In the fall of 2001, the Department of Energy commissioned a study by Quest Consultants, Inc. regarding public safety issues relating to the transportation of LNG to the Distrigas facility and the storage of LNG at the facility. Secretary of Transportation Mineta wrote me about the study on October 26, 2001, noting that:

> "Quest Consultants, Inc., has been hired by DOE [the Department of Energy] to perform studies related to security on vessels transporting LNG and on the onshore LNG storage tanks."

On page 10 Secretary Mineta indicated that:

> "Quest Consultants, an engineering firm, has been asked by DOE to perform a study to analyze the threat that could result from a five-meter diameter hole in an LNG tank on a vessel. Quest has performed some initial calculations to quantify the gas dispersion and fire scenarios that could follow a large release from the LNG storage tanks."

Also on page 10, Secretary Mineta further stated that in addition to actions undertaken by the Department of Transportation to enhance security at the Distrigas facility, it was his understanding that:

> "To improve security measures, DOE will work directly with the local law enforcement officials and Distrigas. MEMA [Massachusetts Emergency Management Agency] will review the studies performed by Quest and develop a plan of action. RSPA [the Department of Transportation's Research and Special Projects Administration] will be involved in the review of the onshore plant protection security features."

My office was subsequently provided with a copy of the Quest study. This Quest study, along with a study prepared for the facility operator by Lloyd's Register of Shipping, which my office was also provided, has been used by the federal government and the facility operator to reassure the Commonwealth Massachusetts about the potential danger of a fire and explosion at or near the Distrigas facility, thereby allowing the facility to reopen.

Last fall, several press reports called the accuracy of these studies into question. For example, the Quest study focused on accidents in Boston's Outer Harbor, when the most troubling public safety threats could occur in the Inner Harbor. The methodology of the study has also been called into question by numerous experts. Even the author backed away from the study's findings and conclusions. According to an October 19, 2003 article in the *Mobile Register* quotes John Cornwell, the lead scientists on the Quest study of LNG fires, as stating:

> "Some of the modeling we did for DOE—in hindsight, we should have done a more complete paper. . . . I've learned you never write anything you don't want public. We violated our own rules on that score."

The *Register* article goes on to report that Mr. Cornwell did the Quest study on short notice and that he . . . believed that it would be employed in-house by federal agencies as one of several tools used to examine LNG fire scenarios. However, according to the *Register* article:

> "In Boston, the Quest study—which has never been published in scientific journals—was apparently used by the DOE to suggest that a terrorist attack on an LNG tanker would result in only limited damage immediately around the ship. In stark contrast, published scientific studies have suggested that an LNG fire could have disastrous consequences for densely populated neighborhoods around Boston Harbor."

An article in the *Boston Herald* further suggests that the Quest study also was used by the Coast Guard to justify the resumption of LNG shipments in the months after the September 11th attacks. . . .

I wrote to the Department of Energy, the Department of Transportation, and the FERC about this study. In response, DOE acknowledged that it had commissioned the study, and reported that it had been used by DOE officials in a presentation to an interagency working group formed to assist Massachusetts following the September 11th attacks. FERC indicated that it had cited the Quest Study in the Environmental Impact Statements for four LNG terminals (The Trunkline LNG Expansion Project, the Elba Island Expansion Project, the Hackberry LNG Project, and the Freeport LNG project). DOT reported that it had used the Quest study "as a hazard assessment model that was applied specifically to the Distrigas facility" and that "the results were used to justify enhanced security procedures for vessels transporting LNG and the onshore LNG storage tanks.

All three agencies seem to have tacitly admitted the shortcomings of the Quest study in deciding to support additional LNG safety studies.

The FERC commissioned a study by ABS Consulting, which was recently released to widespread criticism from both the industry and independent experts. The ABS Study found the earlier Quest study to have several flaws, and did not recommend that it be used to analyze the consequences of a terrorist attack on an LNG tanker or terminal. While the FERC put the ABS study out for public comment, it has also indicated that it regards the ABS Study to be a final study and does not plan to request a formal peer review of this study or update it to take account of the comments that have been submitted. Both industry and expert commentary submitted to the FERC about the ABS Consulting study has been largely critical, nothing several flaws in its meth-

odology and urging that it be peer reviewed before it is used. Despite this rec-
ommendation, FERC appears to have no plans to request a peer review of the
ABS study, but has nonetheless cited the study in the Environmental Impact
Statement for the Freeport LNG project. . . .

I understand that the DOE has commissioned a study by the Sandia Laboratory, which is expected to be available later in the year. While I don't know
what is in the Sandia Study, I can only hope that it is more thorough than the
previous government-funded LNG hazard studies. I would suggest to the Subcommittee that if the EPA issued an environmental regulation based on studies with as many flaws or shortcomings at the Quest and ABS studies, the
regulated industry would be in an uproar and we would be hearing complaints about "junk science" being used to justify new regulations. Here,
when we are talking about a matter that directly affects public safety; Congress also needs to demand that the science be done right, that it be method-
ologically sound, and that it be subjected to peer review. . . .

LNG Carrier Vessel Vulnerabilities

A second issue that I would call to the Subcommittee's attention is the potential
for a terrorist attack on an LNG carrier vessel to result in failure of the cargo
containment systems. Earlier this year, my office received a copy of a letter that
Professor Jerry Havens of the University of Arkansas had sent Secretary Ridge
regarding potential LNG tanker vulnerabilities. The Department's response suggested that the concerns posed by Professor Havens regarding: 1) the suscepti-
bility of the foam insulation used on LNG carrier vessels to fire; 2) the
possibility of rupture of the LNG containment system; and, 3) the potential
for vapor pressure in the ship's LNG tanks to be elevated to levels beyond the
capacity of the relief valves are either unfounded or are already being adequately addressed.

I have written the Department to request further information about
the Department's basis for reaching such conclusions, based on contradic-
tory evidence which is readily available from the public record. Here are my
concerns:

First, the Department alleges that "foam polystyrene insulation, cited by
Professor Havens, is not used on LNG carriers precisely because it's susceptible
to melting and deformation in a fire."

This statement appears to be inaccurate. The Finnish LNG vessel manufacturer, Kvaerner Masa-Yards, reports in a sales brochure that, "the majority
of the world's present LNG fleet, including those on order, incorporate the
[the company's] Kvaerner Moss LNG tank design." This document goes on to
state that "The design of the cargo tank insulation is based on panels made of
expanded *polystyrene*." [Emphasis added]

A quick look at the Kvaerner Masa-Yard website confirms that polystyrene is still being used by the company for its LNG carrier vessels (see http:/
/www.masayards.fi/publications/pdf/LNG.pdf). This publication
describes the use of "inserts of very soft polystyrene for flexibility and fiber-

glass fibre reinforcement to absorb forces which are built up during the cooling down of the cargo tank.". . .

The Department told me that "the insulation on LNG carriers is a complex assembly of many layers" and that "each layer is tested for fire resistance, and its ability to stop the spread of a fire, before it can be used on LNG carriers in U.S. waters." I have several questions about this statement, . . . including:

1. Who in the federal government tests the insulation on LNG carriers for fire resistance?
2. Who is responsible for determining whether this insulation is acceptable for use on LNG carrier vessels operating in US waters?
3. What are the standards used by the federal government for determining whether or not the insulating materials used on LNG carrier vessels are acceptable?
4. What hazard analysis has been done to examine what would happen in the event that a fire on an LNG carrier vessel ignited the insulation or otherwise compromised it?
5. Are older ships required to be retrofitted with new insulation if they use insulating materials, like polystyrene, which have now been determined to be highly flammable? If not, why not? If so, how does the federal government verify that this has occurred?
6. In light of the post-9/11 threat, is there any plan by the Department, or by the Coast Guard, to review the safety standards applicable to LNG carriers (including fire safety standards) to determine whether they need to be upgraded to better address the threat of sabotage or terrorist attack?

In its letter, the Department stated that "the relief valve capacity of LNG carriers is designed based upon exposure to fire." This statement appears to assume that the insulation will continue to function properly. My concern is that if the insulation should fail as the result of a fire, the relief valves would not be capable of handling the increased vapor pressure that would result, since they would not allow for a sufficient flow through the valves. Professor Havens . . ., has suggested that if this were to be the case, the vessels, which are designed for only a few pounds overpressure, would be endangered.

The Department further suggests that concerns about the brittle fracture problem have been anticipated by U.S. regulations, which "require the use of a special crack-arresting steel in strategic locations throughout the vessel's hull." However, she goes on to acknowledge that "both the U.S. and international standards for LNG carriers were developed with the potential consequences posed by conventional maritime risks such as groundings, collisions, and equipment failures in mind." The Department then goes on to say that in recognition of the "new risks now possible in our post 9/11 world, the United States and the international community have responded by implementing additional operational security measures" under U.S. law and international maritime codes. My question is this: How does adoption of additional operational security measures suffice to address an issue—brittle fracture—that seems to go to the fundamental design of an LNG tanker? Might not terrorist threats require the use of additional measures to address the problem of brittle fracture of the ship's hull resulting from an LNG spill? . . .

Need for a New DOT Rulemaking

Let me now turn to . . . DOT's failure to properly exercise its authorities over LNG siting. Under a provision of the Pipeline Safety Act 1979, the Secretary of Transportation is supposed to ensure that the siting of all new LNG terminals is subject to standards which consider: 1) the kind and use of the facility; 2) existing and projected population and demographic characteristics of the location; 3) existing and proposed land use near the location; 4) natural physical aspects of the location; 5) medical, law enforcement, and fire prevention capabilities near the location that can cope with a risk caused by the facility; and 6) the need to encourage remote siting (see 49 U.S.C. 60103).

I am concerned about the nature and adequacy of the Transportation Department's efforts to carry out this authority. In the Committee report accompanying the House Energy and Commerce Committee's version of what became the Pipeline Safety Act of 1979 (H.Rept. 96-201, Part 1), the Committee noted:

> "One area of particular concern to the committee has been the failure to adopt comprehensive Federal standards regarding the sting, design, operation, and maintenance of liquefied natural gas facilities." In 1972, the industry consensus standards developed by the National Fire Protection Association were incorporated into the federal gas pipeline safety regulations, supposedly as an interim measure pending the development of comprehensive standards. Despite widespread concern over the adequacy of these interim standards and the growing importance of LNG as an energy source, the promised comprehensive standards have never been adopted. H.R. 51 addresses this problem by identifying the criteria to be considered by the Secretary in developing standards and setting firm deadlines for proposing and adopting them."

However, if you read the DOT regulations at 40 CFR Part 193, for example, you will find that the DOT's regulations still continue to largely incorporate by reference the National Fire Protection Association (NFPA) standards— specifically, NFPA Standard 59A.

Deputy Chief Joseph Flemming of Boston Fire Department, in his May 25, 2004 comments on the ABS Consulting Report, has raised some very serious concerns about the wisdom of continuing to rely on the NFPA standards. . . . [He also] notes that the NFPA Committee that made up these standards is largely comprised of representatives of the LNG industry or energy industry consultants, and that public officials—including firefighters who may have to deal with an LNG fire, are not routinely brought into discussion about what the appropriate standards should be. A quick check of the NFPA website reveals that the NFPA LNG Committee has representatives from BP Amoco, Distrigas, ExxonMobil, Weaver's Cove Energy, Keyspan, the American Gas Association, the American Petroleum Institute, the American Concrete Institute, and the Steel Plate Fabricators Association. . . .

Shortly after enactment of the 1979 Act, changes in the natural gas market place resulting from the decontrol of natural gas wellhead prices lead to the withdrawal of proposals for new LNG terminals and the shut down of all but the Everett, Massachusetts terminal. In a period when no new LNG terminals were being built, and existing ones were being shuttered, it is perhaps understandable that DOT did not take action to replace the NFPA standards with standards of its own. However, given the current resurgence of interest in LNG and the flood of new proposals to build LNG terminals, I think that DOT needs to revisit this matter now and consider revising its standards. I would also note that FERC has the legal authority to impose additional standards for LNG terminals. If DOT fails to Act, perhaps it is time for FERC to do so.

Conclusion

. . . Looking to the future, LNG is likely to become an increasing part of our energy mix. Given that fact, Congress needs to ensure that the federal government takes further steps to ensure that any future LNG terminals are sited in locations that prevent them from becoming an attractive terrorist target. Adhering to the Congressional directive that the Secretary consider the need for remote siting, looking at offshore siting alternatives, and updating the LNG siting rules so that they reflect sound science and decisions by federal agencies—as opposed to industry self regulatory bodies—is desperately needed. Finally, a more thorough examination of the potential consequences of a terrorist attack on an LNG tanker needs to be done. Perhaps the Sandia study will address this issue, but based on my experiences with the previous Quest and ABS Consulting studies, I think that the Congress needs to step up oversight in this area and demand that the studies that are being funded by the federal government are scientifically sound and subjected to a full peer review. . . .

Donald F. Santa, Jr.    **NO**

Testimony Before the Subcommittee on Energy Policy, Natural Resources and Regulatory Affairs

. . . Over the past year, LNG [liquefied natural gas] has captured the attention of the energy industry and energy policy makers. Still, the reality is that LNG is not a new product in the U.S. energy market. LNG has been utilized in various applications in this country since the Second World War. Many of our pipelines and distribution companies, for example, use LNG as a method for storing natural gas. In the 1970s, as a result of supply shortages in the U.S. interstate market, the nation developed and constructed a number of LNG importation terminals in order to supplement domestic supply with natural gas from other parts of the world. LNG's role in the domestic natural gas market was short-lived, however, once wellhead decontrol and the removal of other artificial market barriers ended the supply shortage. Imported LNG quickly became too expensive to compete against much more affordable natural gas supplies from the U.S. and Canada. Three of the four terminals that were built in the 1970s were, to a large extent, mothballed until several years ago.

Why are we again focused on LNG? It now is widely recognized that North America is experiencing a fundamental shift in the supply and demand equation for natural gas. For many years, this country had a significant excess of natural gas deliverability (what was commonly referred to as the "natural gas bubble"). This kept prices low and contributed to a shift to greater use of natural gas for electric power generation, home heating and industrial processes. Demand growth gradually eliminated this excess deliverability. Supplies now are tight and prices are considerably higher—on a sustained basis—than in previous years.

Therefore, we now must develop new natural gas supply options from multiple sources to keep pace with the still growing demand for this clean-burning fuel. INGAA agrees with the assessment that we are not running out of natural gas; rather, we are running out of places where we are permitted to explore for and produce it. Abundant natural gas resources do still exist in North America and worldwide, and can supply the market with natural gas at

From Testimony before the House Committee on Government Reform, Subcommittee on Energy Policy, Natural Resources, and Regulatory Affairs. Notes omitted.

reasonable prices, provided that public policies do not unreasonably limit resource and infrastructure development.

While it is the focus of today's hearing, LNG should not be mistaken for a "silver bullet" that alone will solve the Nation's natural gas supply problem. Our current natural gas supply challenges will not be solved *only* by expanding production in the Rocky Mountain region or the Outer Continental Shelf, or *only* by building an Alaska natural gas pipeline, or *only* by importing more LNG. In order to meet anticipated demand, we must avail ourselves of *all of these options,* and more.

An important corollary to this supply message is the critical role that pipeline and storage infrastructure play in ensuring that natural gas supply can satisfy market demand. As part of a comprehensive energy policy, removing barriers to pipeline and storage infrastructure development must go hand-in-hand with efforts to enhance gas supply.

The Existing LNG Regulatory Framework

The Federal Energy Regulatory Commission (FERC) and the U.S. Coast Guard, respectively, have the authority for the approval and siting of on-shore and off-shore LNG import terminals. Both agencies have done an excellent job in streamlining the approval process for these facilities.

The Coast Guard has demonstrated its willingness, . . . to consider off-shore terminal siting proposals expeditiously. . . .

FERC's authority to approve and site on-shore LNG terminals is pursuant to section 3(a) of the Natural Gas Act (NGA). While this statutory provision does not expressly refer to the authorization and siting of facilities for importing natural gas, the courts have made clear that this function is an integral part of authorizing natural gas imports and, therefore, is within the scope of the authority conferred by section 3(a). This was addressed by the U.S. Court of Appeals for the D.C. Circuit in the 1974 *Distrigas* decision. The court said, in part:

> . . . while imports of natural gas are a useful source of supply, their potentially detrimental effect of domestic commerce can be avoided and the interests of consumers protected only if . . . the Commission exercises with respect to them the same detailed regulatory authority that it exercises with respect to interstate commerce in natural gas. In short, we find it fully within the Commission's power, so long as that power is responsibly exercised, to impose on imports of natural gas *the equivalent of Section 7 certificate requirements both as to facilities and . . . as to sales within and without the state of importation* (emphasis added). Indeed, we think that Section 3 supplies the Commission not only with the power necessary to prevent gaps in regulation, but also with the flexibility in exercising that power.

Section 7 of the NGA empowers FERC to issue certificates of public convenience and necessity authorizing the construction and operation of interstate natural gas pipelines and storage facilities. The U.S. Department of Energy and FERC have consistently applied the *Distrigas* case's construction of section 3 of the NGA in administering this part of the law.

Mr. Chairman, without going into the extensive case law, let me state that, whenever FERC's authority under either section 3 or section 7 of the NGA has come into conflict with state law, courts have consistently held in favor of federal primacy in matters of interstate and foreign commerce. The Commerce Clause of the U.S. Constitution provides the foundation for these decisions.

While FERC has exclusive jurisdiction under the NGA over the threshold decision on whether an LNG facility or interstate pipeline can be constructed, other state and federal agencies still play a substantive role in permitting this natural gas infrastructure. There are a myriad of other state and federal permits that must be obtained before a project sponsor may begin constructing its facility. FERC's application process requires that a project sponsor list all other permits that must be obtained. And FERC's orders authorizing these facilities routinely are conditioned upon the sponsor obtaining these other authorizations.

As part of discharging its responsibilities under the National Environmental Policy Act (NEPA), FERC makes all other federal, state and local permitting agencies "participating agencies" for purposes of the comprehensive NEPA process. Apart from the NEPA process and these independent sources of authority over pipeline permitting, state agencies can, and do, participate in FERC's proceedings as intervenors in order to represent the interests of their citizens.

The industry's experience in the context of interstate natural gas pipelines has been that FERC devotes significant resources to working cooperatively with these other agencies. Furthermore, the pipeline industry's experience has been that these other sources of authority over pipeline permitting, which often are federal authorities delegated to the states, provide state agencies with considerable leverage.

Industry Concerns

Safety and Security

While regulatory certainty and permit streamlining are important to constructing new LNG terminal capacity, the most significant immediate challenge facing the industry is public perception regarding safety and security. Fear of the unknown appears to be the greatest hurdle, followed closely by the various misconceptions about LNG. Such misconceptions are difficult to overcome. All of us—industry, regulators, the Executive Branch and the Congress—have a role to play in educating the public, so that we can make informed decisions about constructing needed energy infrastructure.

Fortunately, better information is on the way. In May, FERC released a report prepared by a contractor that addressed the consequences of potential LNG spill scenarios. While the Center for LNG believes that this report needs further refinement, it still is an important step in developing a public record that will support a balanced, fact-based consideration of the safety issues associated with LNG. Within the next several weeks, the Department of Energy's Sandia National Laboratory is scheduled to complete an LNG safety and security analysis that should supplement the FERC report by addressing probability of an LNG incident. Finally, Det Norske Veritas, a private risk analysis firm, soon will be completing its own study. We hope that these studies will put to rest many of the misconceptions that have characterized some of the recent public discussion of LNG safety and security issues.

Are there risks associated with LNG? Of course there are. Still, just as with any activity, this must be placed in perspective. LNG has a long and outstanding safety record. The robust worldwide trade in LNG that takes place every day is proof that LNG can be handled safely and securely. And here in the United States, FERC and the Coast Guard, working with the Department of Transportation's Office of Pipeline Safety, can mitigate risk to an even greater extent through their safety/security regulations and enforcement. We need your help, and your leadership, in getting that message out to the public.

Approval and Siting Authority

Another set of challenges facing the industry concerns jurisdictional disputes over LNG siting authority and the potential for protracted proceedings before multiple permitting agencies. The focal point for the jurisdictional issue is the dispute between FERC and the California Public Utility Commission (CPUC) regarding the authority to site an LNG terminal in the State of California.

The jurisdictional issue has been fully adjudicated by FERC and is now ripe for judicial review. FERC has gotten it right on both the law and the policy. As already noted, the courts have interpreted the NGA to provide FERC with the authority to site an LNG import facility and to attach the necessary conditions to its determination. The facts of the California case do not include anything that we believe would cause a reviewing court to reach a conclusion at odds with the *Distrigas* decision. FERC also is on firm ground as a matter of policy. To an even greater extent than with interstate commerce, the regulation of foreign commerce clearly is a function for the federal government. The siting of facilities directly associated with foreign commerce is an obvious extension of such regulation. If this regulation were left to the states, LNG facilities almost certainly would be subject to inconsistent regulation and likely would not be constructed if they were subject to traditional public utility regulation or other burdens. The nation as a whole would suffer if the ability to enhance the capacity to import this critical source of supplemental natural gas supply were frustrated. FERC jurisdiction is important to ensuring that the larger, national public interest is served, rather than just local, parochial interests.

Some have asked whether the Congress should amend section 3 of the NGA to clarify jurisdictional boundaries. We believe that, in exercising exclusive jurisdiction over the siting of LNG import facilities, FERC is acting within the bounds of the authority already conferred by the Congress under section 3 of the NGA. Still, to the extent that such an amendment would "clear the air" and permit worthy LNG projects to proceed without what may be perceived to be a cloud over jurisdiction, such an amendment may be advisable.

Beyond this threshold jurisdictional question, we also want to draw the Subcommittee's attention to the ability of federal, state and local regulators to erect impediments to the efficient, timely construction of natural gas infrastructure already authorized by FERC. While the NGA provides FERC with the exclusive authority for determining whether such projects should be constructed, other agencies increasingly are using the jurisdictional hook provided by other laws to second guess aspects of the decisions that FERC has made following the thorough review conducted under the NGA.

As noted earlier, other state and federal agencies have an integral role to play in permitting decisions related to interstate pipeline and LNG facility construction. Our point is that fairness and administrative efficiency would be served best if these other agencies coordinate the timing of their reviews with the FERC process. The already inclusive FERC NEPA process provides a vehicle for this to occur. In that way, all of the interested federal, state and local government agencies can come together under one concurrent and comprehensive review, so that all parties have equal standing and balanced decisions can be made.

In discussing regulatory impediments to LNG import facilities, we have referred frequently to the experience with interstate pipelines. We have done so for several reasons. First, the experience with interstate pipelines provides a window on what LNG facilities likely will experience as they attempt to reach the finish line of the regulatory gauntlet that must be run before ground can be broken. Second, adequate pipeline capacity is critical to bringing new natural gas supplies to consumers, whether it be LNG or North American supply. Third, specifically with respect to LNG, import facilities must be able to interconnect with the transmission pipeline network in order for the natural gas supply to reach customers. This point is demonstrated by Dominion Resources' recent announcement of plans to increase the capacity of its Cove Point LNG terminal from 1 billion cubic feet per day ("Bcf/day") to 1.8 Bcf/day, which is dependent upon FERC approval of two associated pipelines that will move the increased supply from the terminal and into the market.

Economic Impacts

What happens if the United States is unable to construct the natural gas infrastructure that we need? Quite simply, delays in pipeline and LNG terminal construction will reduce the amount of natural gas available to consumers and thereby increase the price that they must pay. This likely will cause further job losses in industrial sectors that depend on affordable supplies of natural gas, such as chemical and fertilizer manufacturing. Because an increasing amount of electricity is generated by natural gas, electricity prices will be higher for virtually all consumers.

The INGAA Foundation, Inc. now is completing an economic analysis that quantifies some of the consumer costs associated with delays in constructing new pipeline and LNG import capacity. The preliminary results are startling. The study estimates that a two-year delay in building natural gas infrastructure (both pipelines and LNG terminals) would cost U.S. natural gas consumers in excess of $200 billion by 2020. . . . California, alone, would experience increased natural gas costs of almost $30 billion over that period. And, of course, should the end result be that certain facilities are never constructed, the economic effect would be even more severe. This INGAA Foundation study is scheduled to be published in mid-July. . . .

The bottom line is that natural gas infrastructure delays and cancellations have consequences. Every consumer will pay higher prices for natural gas, electricity and the goods produced using natural gas if we do not act to ensure that adequate LNG and pipeline capacity are constructed in time to keep supplies affordable.

Legislative Proposals

Several important provisions in H.R. 6, the pending comprehensive energy legislation, would remove impediments to building LNG and pipeline infrastructure. . . .

These provisions represent areas where changes in the statutory framework for U.S. energy policy can make a real contribution to ensuring that there is adequate LNG import and pipeline infrastructure to serve the energy needs of the nation's consumers and its economy. We continue to urge the Congress to pass this legislation.

We also wish to comment on H.R. 4413, a bill recently introduced by Representative Lee Terry that would establish clear authority for LNG terminal approval, siting, and regulation. The bill would clarify exclusive FERC authority for on-shore terminal siting decisions, and require other federal and state agencies involved in permitting to work within the FERC process and make final decisions within one year of the original application. . . . Both the Center for LNG and INGAA strongly support this legislation, and believe that it should be the model for future discussions in Congress on removing impediments to new LNG import capacity.

Conclusion

In conclusion, let me emphasize the importance of public policies that foster a positive environment for natural gas infrastructure construction and investment. These large and capital-intensive projects will be constructed only if there is a rational process for reviewing and siting these facilities. Delays and detours are costly, both to project sponsors and ultimately to consumers, and in some cases the cumulative effect can be fatal to a project. We believe that the FERC provides an appropriate and inclusive forum for authorizing on-shore LNG import terminals and that FERC has done an admirable job in discharging its responsibilities. If anything, FERC's authority in these matters should be enhanced by Congress, to send a clear message as to the national importance of building natural gas infrastructure on a timely, responsible basis. . . .

POSTSCRIPT

Is Additional Federal Oversight Needed for the Construction of LNG Import Facilities?

H. Patel, C. Caswell, and C. Durr, "North American LNG terminals: Options?" *Hydrocarbon Processing* (July 2005), stress the rising demand for natural gas, with LNG and appropriate terminals seen as key element in meeting that demand. Jerry Havens, in "Ready to Blow?" *Bulletin of the Atomic Scientists* (July/August 2003), says that the potential of LNG tankers and terminals for catastrophic explosions and fires has been understood since the 1970s, when those continental U.S. terminals now in use were approved and built. Today, an additional, major danger is that terrorists could strike at tankers or terminals. "For nearly 50 years now, all discussions of risk and probability in LNG transport have focused on how to account for human errors. The new reality is that we must now consider malicious acts as well." Havens reiterated his concerns in "LNG: Safety in Science," *Bulletin of the Atomic Scientists* (January/February 2004). For a lengthier overview of LNG, its safety record, existing legislation, industry proposals for new terminals, and risk concerns, see Paul W. Parfomak and Aaron M. Flynn, "Liquefied Natural Gas (LNG) Import Terminals: Siting, Safety and Regulation," Congressional Research Service (Order Code RL32205) (January 28, 2004).

Unfortunately, there are few (if any) locations that are so far from cities, suburbs, and protected natural environments that the risks can be ignored. Yet the need is growing. Testifying before the House Committee on Government Reform, Subcommittee on Energy Policy, Natural Resources, and Regulatory Affairs, hearing on "LNG Import Terminal and Deepwater Port Siting: Federal and State Roles" (June 22, 2004), David K. Garman, Acting Under Secretary for Energy, Science, and Environment of the U.S. Department of Energy, noted that according to the Energy Information Administration's *Annual Energy Outlook 2004* (AEO2004), demand for imported LNG will increase 54 percent between 2007 and 2010, which will require four new LNG terminals just on the Atlantic and Gulf Coasts. Additional terminals will be needed elsewhere. The debate is being played out on a local scale in communities such as Fall River, Massachusetts, where an LNG terminal has been proposed; see http://www.greenfutures.org/projects/LNG/. The local Coalition for Responsible Siting of LNG (http://www.nolng.org/) contends that the risks are unacceptable while an opposing group, Friends of LNG, argues that the risks are outweighed by the economic benefits.

The ABS report criticized by Markey was released May 13, 2004. Among other things, it noted that the data available for properly assessing risk have serious shortcomings. Available models of LNG release do not take account of the actual structure of tankers, or of wave action and currents, and there are no experimental data. Further research is essential.

The Sandia report referred to in the readings was originally scheduled for release early in 2004. But in January 2004, in response to critics who thought the study reviewed too few previous safety studies, the Department of Energy asked Sandia to increase the scope of the report. The expanded report, released in December 2004 as "Guidance on Risk Analysis and Safety Implications of a Large Liquefied Natural Gas (LNG) Spill Over Water" (see http://www.fossil.energy.gov/programs/oilgas/storage/lng/ sandia_lng_1204.pdf), said that the greatest risks are associated with intentional releases (as from terrorist attacks) but that appropriate security measures can keep these risks to acceptable levels. If they fail, the resulting fires may cause "major injuries and significant damage to structures" up to a third of a mile from the source and cause second-degree burns on people over a mile away.

On February 15, 2005, the Energy Subcommittee of the U.S. Senate's Committee on Energy and Natural Resources held a meeting on "The Future of Liquefied Natural Gas: Siting and Safety," at which Sandia's Mike Hightower summarized the conclusions of the Sandia study; see http:// energy.senate.gov/public/index.cfm?FuseAction=Hearings. Hearing&Hearing_ID=1384. He concluded that "Risk management strategies should concentrate on incident management and emergency response measures... Measures should ensure that areas of refuge are available, and community education programs should be implemented to ensure that persons know what to do in the unlikely event of a [disaster]."

Later in 2005, President Bush signed the 2005 energy bill, which "gives the industry-friendly Federal Energy Regulatory Commission final say over LNG-plant siting, overruling state concerns." David Helvarg, "A Real Energy Boom," *Sierra* (March/April 2006), says that this cost coastal states much of their ability to protect their resources. It also limited the say of local communities. Safety concerns did not stop approval of a Long Beach, CA, terminal (see "FERC Draft EIS Gives SES LNG Project Clean Bill Of Health," *Pipeline & Gas Journal* (December 2005), but the FERC does not seem likely to approve every LNG terminal proosal that comes in; safety concerns did stop a Providence, RI, project. See Stephen Barlas, "FERC Staff Recommends Against Key-Span LNG Project," *Pipeline & Gas Journal* (July 2005).

ISSUE 12

Is It Time to Revive Nuclear Power?

YES: Michael J. Wallace, from "Nuclear Power 2010 Program," Testimony before the United States Senate Committee on Energy & Natural Resources, Hearing on the Department of Energy's Nuclear Power 2010 Program (April 26, 2005)

NO: Editors of *Public Citizen*, from "The Big Blackout and Amnesia in Congress: Lawmakers Turn a Blind Eye to the Danger of Nuclear Power and the Failure of Electricity Deregulation," *Public Citizen* (September 8, 2003)

ISSUE SUMMARY

YES: Michael J. Wallace argues that because the benefits of nuclear power include energy supply and price stability, air pollution control, and greenhouse gas reduction, new nuclear power plant construction—with federal support—is essential.

NO: *Public Citizen* argues that nuclear power is too unreliable and risky to count on. We must "embrace safe, clean, sustainable energy sources."

T he technology of releasing for human use the energy that holds the atom together got off to an inauspicious start. Its first significant application was military, and the deaths associated with the Hiroshima and Nagasaki explosions have ever since tainted the technology. It did not help that for the ensuing half century, millions of people grew up under the threat of nuclear Armageddon. But almost from the beginning, nuclear physicists and engineers wanted to put nuclear energy to more peaceful uses, largely in the form of power plants. Touted in the 1950s as an astoundingly cheap source of electricity, nuclear power soon proved to be more expensive than conventional sources, largely because safety concerns caused delays in the approval process and prompted elaborate built-in precautions. Many say that safety measures have worked well when needed—Three Mile Island, often cited as a horrific example of what can go wrong with nuclear power, released very little radioactive material to the environment. The Chernobyl disaster occurred when safety measures were ignored. In both cases, human error was more to blame than the technology itself. The related issue of nuclear waste (see Issue 19) has also raised fears and added expense to the technology.

It is clear that two factors—fear and expense—impede the wide adoption of nuclear power. If both could somehow be alleviated, it might become possible to gain the benefits of the technology. Among those benefits are that nuclear power does not burn oil, coal, nor any other fuel; does not emit air pollution and thus contribute to smog and haze; does not depend on foreign sources of fuel and thus weaken national independence; and does not emit carbon dioxide. The last may be the most important benefit at a time when society is concerned about global warming, and it is the one that prompted James Lovelock, creator of the Gaia Hypothesis and an inspiration to many environmentalists, to say, "If we had nuclear power we wouldn't be in this mess now, and whose fault was it? It was [the antinuclear environmentalists']." See his autobiography, *Homage to Gaia: The Life of an Independent Scientist* (Oxford University Press, 2001). The Organisation for Economic Co-operation and Development (OECD's) Nuclear Energy Agency, in "Nuclear Power and Climate Change," (Paris, France, 1998), available at http:// www.nea.fr/html/ndd/climate/climate.pdf, found that a greatly expanded deployment of nuclear power to combat global warming was both technically and economically feasible. In 2000 Robert C. Morris published *The Environmental Case for Nuclear Power: Economic, Medical, and Political Considerations* (Paragon House). In August 2000 *USA Today Magazine* published "A Nuclear Solution to Global Warming?" "The time seems right to reconsider the future of nuclear power," say James A. Lake, Ralph G. Bennett, and John F. Kotek, in "Next-Generation Nuclear Power," *Scientific American* (January 2002). See also I. Fells, "Clean and Secure Energy for the Twenty-First Century," *Proceedings of the Institution of Mechanical Engineers, Part A—Power & Energy* (August 1, 2002). Stephen Ansolabehere, et al., "The Future of Nuclear Power," *An Interdisciplinary MIT Study* (MIT, 2003), say that greatly expanded use of nuclear power may be needed to meet future energy needs and reduce carbon emissions but reducing costs and risks will need attention. David Talbot, "Nuclear Powers Up," *Technology Review* (September 2005), notes that "While the waste problem remains unsolved, current trends favor a nuclear renaissance. Energy needs are growing. Conventional energy sources will eventually dry up. The atmosphere is getting dirtier."

In the following selections, Michael J. Wallace, executive vice president of a major energy company, argues that because the benefits of nuclear power include energy supply and price stability, air pollution control, and greenhouse gas reduction, new nuclear power plant construction is essential, and there is a clear place for federal support. The consumer advocacy group *Public Citizen* argues that nuclear power is too unreliable and risky to count on. We must "embrace safe, clean, sustainable energy sources."

YES

Michael J. Wallace

Nuclear Power 2010 Program

...Constellation Energy, a Fortune 200 company based in Baltimore, is the nation's leading competitive supplier of electricity to large and industrial customers and the nation's largest wholesale power seller. Constellation Energy also manages fuels and energy services on behalf of energy intensive industries and utilities. The company delivers electricity and natural gas through the Baltimore Gas and Electric Company (BGE), its regulated utility in Maryland. We are the owners of 107 generating units at 35 different locations in 11 states, totaling approximately 12,500 megawatts of generation capacity. In 2004, the combined revenues of the integrated energy company totaled more that $12.5 billion and we are the fastest growing Fortune 500 Company over the past two years.

Our portfolio based on electricity produced is approximately 50 percent nuclear, 35 percent coal-fired, 7 percent gas-fired and 5 percent renewables. We own and operate the Calvert Cliffs nuclear plant in Maryland, and the Nine Mile Point and Ginna nuclear stations in New York State.

Constellation is part of the NuStart consortium that is preparing an application to the NRC for a license that would allow us to build and operate a new nuclear plant. Additionally, in December 2004, we submitted a proposal to the Department of Energy (DOE) for studies that could lead to an application to the Nuclear Regulatory Commission for an Early Site Permit as part of the Nuclear Power 2010 program. So, as you can tell, we have a vested interest in the continued success of Nuclear Power 2010, and we're bullish on the future of nuclear power.

Although I am here testifying today on behalf of Constellation, this testimony is supported by our trade association, the Nuclear Energy Institute (NEI).

My statement this morning will address four major issues:

1. The strategic value of our 103 operating nuclear power plants, and the compelling need to build new nuclear plants to preserve our nation's energy security, meet our environmental goals, and sustain our economic growth.

United States Senate Committee on Energy & Natural Resources Hearing on the Department of Energy's Nuclear Power 2010 Program, April 26, 2005.

2. The critical importance of the Department of Energy's Nuclear Power 2010 program as a platform from which to launch the next generation of nuclear power plants in the United States.
3. The need to recognize that the Nuclear Power 2010 program does not address all of the challenges facing companies interested in building new nuclear power plants, and that additional joint investment initiatives by the federal government and the private sector will be necessary.
4. The urgent need for comprehensive energy legislation that squarely addresses the critical need for additional investment in our electricity and energy infrastructure, including advanced nuclear and coal-fired generating capacity, electric and natural gas transmission, and other areas. Construction of the next nuclear power plants in the United States will require some form of investment stimulus, but I know I speak for the entire electric sector when I say that the need for investment stimulus extends well beyond nuclear power. This sector is starved for investment capital, and new federal government policy initiatives are necessary to reverse that trend and place our economy and our future on a sound foundation.

The Strategic Value of Nuclear Power and the Need for New Nuclear Power Plants

The United States has 103 reactors operating today. Nuclear power represented 20 percent of U.S. electricity supply 10 years ago, and it represents 20 percent of our electricity supply today, even though we have six fewer reactors than a decade ago and even though total U.S. electricity supply has increased by 25 percent in the period.

Nuclear power has maintained its market share thanks to dramatic improvements in reliability, safety, productivity and management of our nuclear plants, which today operate, on average, at 90 percent capacity factors, year in and year out. Improved productivity at our nuclear plants satisfied 20 percent of the growth in electricity demand over the last decade.

Due, in part, to excellent plant performance, we've seen steady growth in public support for nuclear energy. The industry has monitored public opinion closely since the early 1980s and two key trends are clear: First, public favorability to nuclear energy has never been higher; and second, the spread between those who support the use of nuclear energy and those opposed is widening steadily: 80 percent of Americans think nuclear power is important for our energy future and 67 percent favor the use of nuclear energy; 71 percent favor keeping the option to build more nuclear power plants. Six in 10 Americans agree that "we should definitely build more nuclear power plants in the future." Sixty-two percent said it would be acceptable to build new plants next to a nuclear power plant already operating.

The operating nuclear plants are such valuable electric generating assets that virtually all companies are planning to renew the operating licenses for these plants, as allowed by law and Nuclear Regulatory Commission regulations, and operate for an additional 20 years beyond their initial 40-year license terms. Sixty-eight U.S. reactors have now renewed their licenses, filed their formal applications, or indicated to the Nuclear Regulatory Commission that they intend to do so. The remaining 35 reactors have not yet declared because most of them are not yet old enough to do so. We believe that virtually all U.S. nuclear plants will renew their licenses and operate for an additional 20 years. At Constellation, we are proud that our Calvert Cliffs station was the first U.S. nuclear plant to renew its license. At the time, the license renewal process was a novel concept. Today, thanks to efficient management of the process by the Nuclear Regulatory Commission, it is a stable and predictable licensing action. Ten years from now, we hope and believe that the issuance of combined construction/operating licenses for new nuclear plants—a novel concept today—will be similarly efficient and predictable.

Although it has not yet started to build new nuclear plants, the industry continues to achieve small but steady increases in generating capability—either through power uprates or the restart of shutdown nuclear capacity. The Tennessee Valley Authority is restarting Unit 1 at its Browns Ferry site in northern Alabama. This is a very complex project—fully as challenging as building a new nuclear plant—and it is on schedule and within budget at the midpoint of the project.

However, despite the impressive gains in reliability and output, there are obviously limits to how much capacity we can derive from our existing nuclear power plants. The time has come to create the business conditions under which we can build new nuclear power plants in the United States. We believe there are compelling public policy reasons for new nuclear generating capacity.

First, new nuclear power plants will continue to contribute to the fuel and technology diversity that is the core strength of the U.S. electric supply system. This diversity is at risk because today's business environment and market conditions in the electric sector make investment in large, new capital-intensive technologies difficult, particularly the advanced nuclear power plants and advanced coal-fired power plants best suited to supply baseload electricity. More than 90 percent of all new electric generating capacity added over the past five years is fueled with natural gas. Natural gas has many desirable characteristics and should be part of our fuel mix, but over-reliance on any one fuel source leaves consumers vulnerable to price spikes and supply disruptions.

Second, new nuclear power plants provide future price stability that is not available from electric generating plants fueled with natural gas. Intense volatility in natural gas prices over the last several years is likely to continue, thanks partly to unsustainable demand for natural gas from the electric sector, and subjects the U.S. economy to potential damage. Although nuclear plants are capital-intensive to build, the operating costs of nuclear power plants are stable and can dampen volatility of consumer costs in the electricity market.

Third, new nuclear plants will reduce the price and supply volatility of natural gas, thereby relieving cost pressures on other users of natural gas that have no alternative fuel source.

And finally, new nuclear power plants will play a strategic role in meeting U.S. clean air goals and the nation's goal of reducing greenhouse gas emissions. New nuclear power plants produce electricity that otherwise would be supplied by oil-, gas- or coal-fired generating capacity, and thus avoid the emissions associated with that fossil-fueled capacity.

In summary, nuclear energy represents a unique value proposition: new nuclear power plants would provide large volumes of electricity—cleanly, reliably, safely and affordably. They would provide future price stability and serve as a hedge against price and supply volatility. New nuclear plants also have valuable environmental attributes. These characteristics demonstrate why new nuclear plant construction is such an imperative in the United States.

The Critical Value of the Nuclear Power 2010 Program

As I said earlier, the Department of Energy's Nuclear Power 2010 program is an essential foundation in the joint government/industry partnership to build new nuclear power plants. This committee and, in particular, you, Mr. Chairman, deserve great credit for your leadership in ensuring adequate funding for this program in the 2005 Fiscal Year.

Nuclear Power 2010 is designed to demonstrate the various components of the new licensing system for nuclear power plants, including the process of obtaining early site permits (ESPs) and combined construction/operating licenses (COLs), sharing the cost of the detailed design and engineering work necessary to prepare COLs, and resolving generic licensing issues. This work is an essential risk-management exercise because it allows industry and the NRC staff to identify and resolve scores of technical and regulatory issues that must be settled before companies can undertake high-risk, capital-intensive construction projects like new nuclear plant construction.

The Nuclear Power 2010 program is the springboard that launched a tangible and visible industry commitment to new plant construction. The industry's commitment to Nuclear Power 2010 includes a planned investment of $650 million over the next several years on design, engineering, and licensing work, which will create a business foundation for decisions to build. Three companies have applications for early site permits under review at NRC. In addition to these three, Constellation and possibly one other company are also considering ESP applications. The industry is developing at least three applications for construction/operating licenses; the first will be filed in 2007, the second and third in 2008.

As you know, the administration has proposed $56 million for the Nuclear Power 2010 program in the 2006 fiscal year. The $56 million funding proposed for 2006 is sufficient for the ESP and COL demonstration projects already underway. It is not adequate, however, to cover more recent expressions of interest from Constellation and others, and additional resources will be needed to ensure this program is viable into the future.

It is also important to recognize that Nuclear Power 2010 is a multi-year undertaking. Certainty of future funding and program stability are a big concern for industry. However, our biggest frustration with the Nuclear Power 2010 program involves the time it has taken the DOE to award the grants. In the case of NuStart, we submitted our application in April 2004 and we were not notified that we received the grant until November 2004. As for Constellation's ESP application, we submitted it almost four months ago and have yet to hear from DOE.

To support the ESP and COL demonstration projects currently underway and future projects, we anticipate that the Department of Energy will need to significantly increase funding for Nuclear Power 2010 over FY 2006 levels.

The process of developing the first COL applications, certifying new designs and completing NRC review of the first ESP and COL applications will take some time. We are looking for ways to accelerate that process, and the Congress may be able to help there—by ensuring sufficient funding for Nuclear Power 2010 and even accelerating that funding; and by providing NRC sufficient resources to ensure that the commission has adequate manpower to conduct licensing reviews and meet aggressive but realistic schedules.

The Nuclear Power 2010 Program Does Not Address All the Challenges Facing New Nuclear Plant Construction

The Department of Energy's Nuclear Power 2010 program is a necessary, but not sufficient, step toward new nuclear plant construction. We must address other challenges as well.

Our industry is not yet at the point where we can announce specific decisions to build. We are not yet at the point where we can take a $1.5 billion to $2 billion investment decision to our boards of directors. We do yet not have fully certified designs that are competitive, for example. We do not know the licensing process will work as intended: That is why we are working systematically through the ESP and COL processes. We must identify and contain the risks to make sure that nothing untoward occurs after we start building. We cannot make a $1.5-$2 billion investment decision and end up spending twice that because the licensing process failed us.

The industry believes federal investment is necessary and appropriate to offset some of the risks I've mentioned. We recommend that the federal government's investment include the incentives identified by the Secretary of Energy Advisory Board's Nuclear Energy Task Force in its recent report. That investment stimulus includes:

1. secured loans and loan guarantees;
2. transferable investment tax credits that can be taken as money is expended during construction;
3. transferable production tax credits;

4. accelerated depreciation.

This portfolio of incentives is necessary because it's clear that no single financial incentive is appropriate for all companies, because of differences in company-specific business attributes or differences in the marketplace—namely, whether the markets they serve are open to competition or are in a regulated rate structure.

The next nuclear plants might be built as unregulated merchant plants, or as regulated rate-base projects. The next nuclear plants could be built by single entities, or by consortia of companies. Business environment and project structure have a major impact on which financial incentives work best. Some companies prefer tax-related incentives. Others expect that construction loans or loan guarantees will enable them to finance the next nuclear plants.

It is important to preserve both approaches. We must maintain as much flexibility as possible.

It's important to understand why federal investment stimulus and investment protection is necessary and appropriate.

Federal investment stimulus is necessary to offset the higher first-time costs associated with the first few nuclear plants built.

Federal investment protection is necessary to manage and contain the one type of risk that we cannot manage, and that's the risk of some kind of regulatory failure (including court challenges) that delays construction or commercial operation.

The new licensing process codified in the 1992 Energy Policy Act is conceptually sound. It allows for public participation in the process at the time when that participation is most effective—before designs and sites are approved and construction begins. The new process is designed to remove the uncertainties inherent in the Part 50 process that was used to license the nuclear plants operating today. In principle, the new licensing process is intended to reduce the risk of delay in construction and commercial operation and thus the risk of unanticipated cost increases. The goal is to provide certainty before companies begin construction and place significant investment at risk.

In practice, until the process is demonstrated, the industry and the financial community cannot be assured that licensing will proceed in a disciplined manner, without unfounded intervention and delay. Only the successful licensing and commissioning of several new nuclear plants (such as proposed by the NuStart and Dominion-led consortia) can demonstrate that the licensing issues discussed above have been adequately resolved. Industry and investor concern over these potential regulatory impediments may require techniques like the standby default coverage and standby interest coverage contained in S. 887, introduced by Senators Hagel, Craig and others.

Let me also be clear on two other important issues:

1. The industry is not seeking a totally risk-free business environment. It is seeking government assistance in containing those risks that are

beyond the private sector's control. The goal is to ensure that the level of risk associated with the next nuclear plants built in the U.S. generally approaches what the electric industry would consider normal commercial risks. The industry is fully prepared to accept construction management risks and operational risks that are properly within the private sector's control.

2. The industry's financing challenges apply largely to the first few plants in any series of new nuclear reactors. As capital costs decline to the "n^{th}-of-a-kind" range, as investors gain confidence that the licensing process operates as intended and does not represent a source of unpredictable risk, follow-on plants can be financed more conventionally, without the support necessary for the first few projects. What is needed [is] limited federal investment in a limited number of new plants for a limited period of time to overcome the financial and economic hurdles facing the first few plants built.

In summary, we believe the industry and the federal government should work together to finance the first-of-a-kind design and engineering work and to develop an integrated package of financial incentives to stimulate construction of new nuclear power plants. Any such package must address a number of factors, including the licensing/regulatory risks; the investment risks; and the other business issues that make it difficult for companies to undertake capital-intensive projects. Such a cooperative industry/government financing program is a necessary and appropriate investment in U.S. energy security.

I hope this Committee can find a place for this type of investment stimulus in the comprehensive energy legislation now being developed

In addition, I would be remiss if I did not thank the Chairman for his support for three additional programs/provisions that will assist in the construction of new nuclear power plants in the United States:

1. Sustained progress with the Yucca Mountain project is essential. This includes the funding necessary to maintain the schedule, ensure timely filing of the license application, and access to the full receipts of the Nuclear Waste Fund.
2. Renewal of the Price-Anderson Act, which provides the framework for the industry's self-funded liability insurance. I am pleased to note that this is included in the recently House-passed energy bill.
3. Updated tax treatment of decommissioning funds that would provide comparable treatment for unregulated merchant generating companies and regulated companies. This provision, included in the energy tax legislation passed recently by the House, would allow all companies to establish qualified decommissioning funds and ensure that annual contributions to those funds are treated appropriately as a deductible business expense.

The U.S. electricity business and our nation are paying the price today for our inability to strike an appropriate balance between what was expedient and easy in the short-term, and what was prudent and more difficult in the long-term. We are paying the price today for 10 to 15 years of neglect of longer-term imperatives and the oversupply of base-load generation in the 1990s.

The United States faces a critical need for investment in energy infrastructure, including the capital-intensive, long-lead-time advanced nuclear and coal-fired power plants that represent the backbone of the U.S. electricity supply system.

While some may not realize it, the United States faces an imminent energy crisis today.

Electric power sales represent three to four percent of our gross domestic product. But the other 96 to 97 percent of our $11-trillion-a-year economy depends on that three to four percent. We cannot afford to gamble with something as fundamental as energy supply, and the biggest problem we face with nuclear energy is not having enough of it.

Editors of *Public Citizen*    **NO**

The Big Blackout and Amnesia in Congress:
Lawmakers Turn a Blind Eye to the Danger of Nuclear Power and the Failure of Electricity Deregulation

T he Northeast and Midwest blackout of 2003, the largest power outage in North American history, calls attention to the chaos that deregulation has wrought on the continent's power grid, in terms of both the opportunistic, relentless profit-seeking of energy traders and the heightened vulnerability of nuclear power reactors, 21 of which were immediately shut down when the blackout hit. The blackout should serve as a wake-up call, spurring legislators to pursue an energy policy that prioritizes safe, clean, sustainable energy sources; strengthens regulation; and places the energy needs of citizens above the endless corporate quest for profits.

Sadly, though, congressional lawmakers seem to have suffered a collective blackout of their own, forgetting the spectacular failure of electricity deregulation (epitomized by the California energy crisis), which was the prime culprit of the blackout, and denying the serious risks of nuclear power. The omnibus energy bills recently passed in both houses of Congress actually *further* electricity deregulation, repeal consumer protections and, in addition, provide huge taxpayer-financed incentives for the development of more nuclear power stations. Ironically, electricity deregulation was once touted as the antidote to expensive government support of inefficient and over-budget nuclear plants such as Grand Gulf in Mississippi constructed under regulated markets; now, however, deregulation is being coupled with obscene federal subsidies for the development of new nuclear reactors.

Unfortunately, many policymakers and politicians have misidentified the cause of the blackout, ignored one of its most serious effects, and offered as a solution massive legislation that would only make the situation worse. Although there are problems with many facets of the nation's energy system, many of the deficiencies that have been highlighted since the blackout are either nonexistent (such as the alleged shortage of electricity capacity) or have been mischaracterized.

In this report, Public Citizen analyzes one of the most serious and immediately dangerous effects of the blackout: the unreliability and heightened vulnerability of nuclear power reactors. Furthermore, we trace the cause of the blackout to the chaotic effects of electricity deregulation. Finally, we consider the folly of the pending omnibus energy legislation in Congress, which completely fails to provide the most appropriate legislative prescription for the problem: the strengthening of electricity regulations and consumer protections, coupled with investment in safe, renewable and reliable electricity generation and distribution systems.

The Blackout Demonstrates the Unreliability, Vulnerability, and Danger of Nuclear Power Reactors

Unfortunately, some nuclear industry cheerleaders are opportunistically exploiting the blackout to promote further reliance on the inherently unsafe, unreliable and polluting technology of nuclear power. As usual, they espouse nuclear "solutions" to nearly every problem, while turning a blind eye to the myriad problems caused by the nuclear industry itself. Sen. Pete Domenici (R-N.M.), chairman of the Energy and Commerce committee, and a staunch supporter of nuclear power, issued a statement after the blackout in which he claimed:

> This outage clearly demonstrates how close the nation is to its energy production and distribution limit. [...] Ensuring the proper level of power to the country demands that we make trade-offs, *including ... greater use of such sources as nuclear energy ...* [Emphasis added]

In the aftermath of the recent blackout, it is important to consider the enormous risks and reliability deficiencies of nuclear power. The unique dangers of nuclear power were exacerbated by the huge power outage: 21 nuclear reactors—which are, ironically, dependent upon off-site power—were forced to shut down in the U.S. and Canada. Power loss from the grid forces nuclear power stations to resort to emergency generators for basic safety operations while in shutdown mode—a contingency operation that presents a whole host of new risks for the plant. Power outages, especially on a grand scale, put already-vulnerable nuclear facilities at an even greater risk of serious accident.

An Ever-Present Vulnerability: The Country's Nuclear Power Plants Are Disintegrating

Our country's nuclear reactors are crumbling and most efforts to repair them are akin to putting a finger in a leaking dike. Nonetheless, the U.S. Nuclear Regulatory Commission (NRC) has granted operating license renewals (for 20-year extensions to the initial 40-year license terms) to all 16 reactors that have submitted applications—even though most of these reactors have operated for fewer than 30 years. Considering the intrinsic dangers and vulnerabilities of nuclear power, and the grave consequences that can result from any malfunction—be it a technical glitch or a human error—the most reliable thing that a nuclear plant can provide is danger. Irresponsible management and regulators that cater to the industry do not help.

Consider the dangers that exist when everything is operating normally; they are only exacerbated during a blackout. Nuclear power plants operate under enormous pressure, heat and stress, in addition to the unique interactions that radiation causes within the plants' complex array of parts. This partially explains why so many U.S. reactors are perpetually at risk of a serious accident, long before their initial 40-year license term has ended. From steam generator tubes to emergency cooling pumps to reactor vessel heads (top and bottom), there is a constant supply of crises:

- The degradation and rupture of steam generator tubes at nuclear reactors has been a problem at U.S. reactors since at least 1975, when there was a spontaneous tube rupture at the 5-year-old Point Beach reactor in Wisconsin. The NRC describes steam generator tubes as serving "an important safety role because they constitute one of the primary barriers between the radioactive and non-radioactive sides of the plant. For this reason, the integrity of the tubing is essential in minimizing the leakage of water between the two 'sides' of the plant."[1] Steam generator tube rupture can "cascade," wherein a break in one tube triggers ruptures in adjacent tubes. If severe, a cascade could precipitate a nuclear meltdown at a reactor. At a 1988 conference, former NRC Commissioner Kenneth Rogers, speaking about the effects of aging U.S. nuclear plants, said: "Degradation (of the steam generator tubes) would decrease the safety margins so that, in essence, we have a 'loaded gun,' an accident waiting to happen."[2] Nonetheless, neither the industry nor the NRC has been able to adequately address the problem, and the Indian Point 2 reactor—only 35 miles from Manhattan—experienced a serious steam generator tube failure in February 2000.[3] Reactors shut down from the recent blackout that have had tube ruptures include Indian Point 2, Indian Point 3 and Ginna—all in New York. Such a rupture occurring prior to a blackout would place a heavy burden on emergency backup systems, increase the chance of meltdown and further tax plant emergency crews. At least 16 steam generator tube ruptures have occurred since the first in 1975.
- The cracking, leaking and acid-caused degradation of reactor vessels and connected components have been a known issue at nuclear reactors for at least 15 years. In March 1987, workers at the Turkey Point 4 reactor in Florida discovered that a small amount of boric acid had corroded the reactor vessel head (the "lid" of the reactor that contains the enormous radioactivity and pressure inside). Since that time, similar cracking, leaking and acid corrosion of reactors have occurred at many plants in the U.S., including Salem, San Onofre, Arkansas Nuclear One, Fort Calhoun, Calvert Cliffs, Three Mile Island, Sequoyah and Comanche Peak, among others. With both the industry and the NRC failing to adequately address the problem, a much-delayed inspection in March 2002 at Ohio's Davis-Besse plant uncovered a football-sized corrosion hole in the reactor's head. (Davis-Besse is owned and operated by FirstEnergy, the company suspected by analysts and state officials to be responsible for an initial trigger of the recent blackout. On September 8, Davis-Besse will celebrate a plant record of 570 consecutive days without producing power, at a cost of over $500 million.) The acid had bored through over 6 inches of car-

bon steel; less than a quarter inch of stainless steel was all that prevented a serious loss-of-coolant accident at the reactor—an accident that can lead to meltdown. The seriousness of this brush with disaster shook the nuclear industry worldwide.

· After years of cutting corners, ignoring problems and cutting deals with the NRC to delay necessary inspections and repairs, FirstEnergy had to bite the bullet and replace the entire reactor vessel head (the cost of which will possibly get passed on to ratepayers). Other additional problems have since been rediscovered—including the lack of a thorough "safety culture," as documented by the NRC's Inspector General in a December 2002 report—that have kept the plant shut down. On July 30, the NRC issued to the FirstEnergy Nuclear Operating Company (FENOC) an "integrated inspection report" that included a preliminary "yellow" finding, representing a problem of "substantial safety significance" (second only to a "red" finding of "high" safety significance on the NRC's color-coded scale) regarding the reactor's emergency core cooling system. The NRC cited the company with a failure to "adequately implement design control measures" to correct known problems with its emergency cooling systems. The NRC noted that metal screens that filter recirculated cooling water in the event of a loss-of-coolant accident—the type of accident that nearly occurred at Davis-Besse—could be blocked by debris that is frequently found in the emergency core cooling system. Such a blockage could lead to a core meltdown. A similar problem had plagued another type of U.S. nuclear reactor, and its potential occurrence at pressurized water reactors (PWRs) has been known for over 10 years; a structural problem at one PWR concerns all 69 PWRs like Davis-Besse.

The Blackout of 2003 Could Have Been MUCH Worse

We know that even the normal functioning of a reactor is fraught with danger, and there is a constant risk of yet another unforeseen, undiscovered or ignored problem leading to a disaster. A close, critical examination of the vulnerabilities in the emergency infrastructure of nuclear plants reveals the great danger posed by blackouts like the one just experienced.

Malfunctioning Emergency Diesel Generators
What exactly happens to a nuclear power plant when the power goes out? First, when a plant loses offsite electrical supply, it automatically shuts down or "scrams." (One engineer likened this to applying the clutch to a car moving 60 miles per hour.) It must then connect to another source of electricity to keep coolant circulating to prevent the reactor core from overheating and causing a meltdown. All nuclear power plants maintain several diesel-powered backup generators on site for use in the event of power loss, but sudden reliance on backup diesel generators is less than reassuring, as the following case studies illustrate.

In the past 12 months—from September 2002 to August 2003—there have been 15 reported instances in which emergency diesel generators have been declared inoperable. In seven cases, when such a failure brought a plant below the required number of backups, a complete shutdown of the plant was required; on four of these occasions, all backup generators failed at once. In

April 2003, the Cook nuclear power plant in western Michigan shut down when emergency water flow to all four diesel generators was blocked by "an influx of fish on the intake screens."[4] Cook also shut down in January when one of its two emergency generators was inoperable for over 72 hours.[5]

In all, four of the nine plants affected by the blackout have shut down in the past year because of problems with backup generators: New Jersey's Oyster Creek, situated between New York City and Philadelphia; Nine Mile Point in New York state; Indian Point, located on the outskirts of New York City and the subject of tremendous controversy over problematic evacuation plans; and Fermi, located only 30 miles from Detroit. On February 1, 2003, all four backup generators at Fermi were simultaneously declared inoperable when a diesel fuel spill caught fire. All four backup generators had similar fuel drain configurations, making them equally prone to such leaks and fires. The generators had to remain off-line for several hours while they were reconfigured to avert future catastrophes.

Without emergency generators, steam and battery power provide a "last chance" means to cool a reactor and stave off a meltdown. The batteries can operate for between two and eight hours; but in the recent blackout, Detroit did not see full power returned until Saturday, August 16, over 36 hours after power first went out. Had the emergency generators failed during this timeframe—as they did in the aforementioned situations—a nuclear meltdown and widespread radioactive release is rendered not at all beyond possibility.

Emergency Sirens and Evacuation Plans

If the blackout had caused a meltdown or other severe accident, it appears that many of the emergency sirens in place to alert officials and the public would not have operated because of a lack of power. In "event reports" submitted to the NRC in the hours after power was lost, the Indian Point and Ginna nuclear stations (both in New York) noted that many of their emergency sirens would have been rendered impotent due to the blackout, and at least 25 percent of the sirens covering the area around the Ginna plant were inoperable. In the case of Indian Point, the sirens in four surrounding counties—including the densely populated Westchester County, with nearly 1 million people—would have failed, leaving the region in a tragic state of ignorance in the event of a meltdown.

It is a terrible irony that power outages, which have so much potential to cause accidents at nuclear power reactors, also disable the emergency alert sirens designed to notify the public of danger. On April 4, 2003, five nuclear power stations in New York and Wisconsin reported that more than half of their emergency sirens were not working due to power outages. (Interestingly, on that same day, the operators of the Monticello nuclear power station in Minnesota reported that some of their emergency alarms were inadvertently actuated.)[6]

Problems with emergency sirens are not uncommon and are not limited to failure due to a loss of power. In fact, on the very same day of the big blackout—though completely unrelated—the operators of the Kewaunee nuclear power station reported that all 13 emergency sirens serving Kewaunee County, Wisconsin, were rendered inoperable due to a "communications problem." Nearly 70 percent of the "coverage population" would have been left in the dark if there were a serious accident at the plant.[7]

Since the beginning of this calendar year, plant operators have filed 43 event reports with the NRC regarding emergency siren problems, which range from inadvertent actuation to disablement due to power loss. More than 50 percent of the reported problems since January were due to mere power outages. Twenty of the reports in that time span cited equipment failure or malfunction.

At some plants, problems with emergency sirens are perpetual. In the past eight months, the Indian Point nuclear power station has reported 10 separate instances of siren disablement due to either power loss or equipment failure.

In the event that there is an emergency at a reactor, and even if the sirens do function properly, will the public know what to do and where to go? A disturbing example of the inadequacies of such emergency and evacuation plans can be found at the Indian Point reactors, located only 35 miles from Manhattan. There are many reasons to doubt the emergency and evacuation plans for the site. For instance, an independent study was conducted in December 2002 by James Lee Witt, a former director of the Federal Emergency Management Agency (FEMA), to evaluate evacuation plans to be used if a terrorist attack caused radioactive releases from the plant. Witt concluded that the plan was inadequate and that key aspects were simply unfixable. The area's high population density and traffic congestion would necessarily complicate evacuation for New Yorkers and residents from surrounding states, and existing plans naively assume that only persons instructed to evacuate will attempt to do so. Indian Point is situated in the most densely populated area of any U.S. nuclear power plant, and a radioactive release could affect over 20 million people.

Local emergency personnel, who would be risking their lives in the event of an accident or attack, are hardly confident that they would be able to handle the overwhelming problems that would accompany such a disaster. In May, 175 Indian Point-area first responders signed a petition to FEMA and the NRC expressing their concerns that "even [their] best efforts may not be enough to adequately protect the public health and safety of the citizens of this region." They understand that, in the event of a major accident or terrorist attack, chaos would likely reign over the densely populated region.

The four counties that are responsible for implementing the emergency evacuation plan have refused to participate in the annual emergency and evacuation plan certification process, citing concerns by their own emergency officials who doubt that they could implement the plan. The decision by Westchester, Rockland, Putnam and Orange county officials to not participate in the certification received the support of the New York State Emergency Management Office, given New York's "home rule" policy, which defers to the judgments of local municipalities in such matters. Unfortunately, both NRC and the FEMA have pressed ahead and rubber-stamped approval of the plan.

"Spent" Fuel Pools are Highly Vulnerable

A lesser-known vulnerability at nuclear plants is the so-called "spent" fuel pools. The term "spent" fuel is itself a misnomer, since the fuel is only spent in the sense that it can no longer assist in boiling the water to turn the turbines. The fuel is exhausted for that purpose, yet it is still very hot and extremely radioactive—more so when taken out of the reactor than when it is put in. When removed from the reactor core, this irradiated fuel (a more accurate name) is submerged in large pools of water—"spent" fuel pools—in a building adjacent to the reactor for cooling and storage. These buildings are typically just standard industrial constructions, built of concrete blocks and corrugated metal (much less "robust" structures than the still-questionable reactor containment structures) and are thus even more vulnerable to terrorist attacks. In the event of an attack or an accident, these structures would do little or nothing to contain radioactive releases. Depending on the amount of fuel stored in the pools, most of which are fully stocked or overloaded, such a facility has the potential to unleash a disaster at least as great as one originating at the reactor itself.

Shockingly, these fuel pools DO NOT get backup power from emergency diesel generators. When the offsite power goes out, the pool water cannot be re-circulated to prevent boiling, evaporation, exposure of the fuel rods and, ultimately, a fire and meltdown. The risk of this occurring is greatest when a "fresh" load of fuel has recently been transferred from the reactor core to the fuel pool (most reactors refuel about every 18 months). Suffice it to say that the vulnerability of irradiated fuel pools presents a grave radiation risk to the public.

Nuclear Power Plants Can't Get the Lights Back On
To get the power grid back up and running again after a blackout requires plants that have "blackstart" capability. This means that they are able to start up independently and return power to the grid. Nuclear plants are not blackstart facilities because they must rely on offsite electricity from the grid itself to power up to full capacity. Any backup power systems at nuclear facilities are devoted to keeping the reactor core cool and avoiding meltdown. Backup systems can't provide enough power to fire up the reactor itself. Due to these same concerns, nuclear plants are given first priority to receive electricity once the grid is blackstarted, eliminating their dependence on (demonstrably unreliable) emergency backup generators.

Other sources for generating electricity are typically more flexible, reliable and faster in recovering from a blackout. When a blackout hits, all power plants—nuclear and non-nuclear—go into shutdown mode to prevent further overloading or tripping of the system. The difference among plants lies in their ability to quickly restart in preparation to connect to the grid. <u>Wind farms and hydro generators can restart right away.</u> Natural gas plants might take a few hours. Coal plants can restart in eight hours. A nuclear plant that has not been damaged by a rapid "scram" shutdown triggered by the blackout needs up to 48 hours or more (and that's just for U.S. reactors; Canada's reactors fare much worse). FirstEnergy's Perry nuclear plant—located near the likely epicenter of the blackout in eastern Ohio—suffered damage in 5 percent of its reactor control rods when the reactor was rapidly shut down. Repairs to the control rods delayed the restart. While the rest of the country's grid was back up and running, the Perry plant was down and out.

Canada's Candu Reactors Candidn't
Ontario, the Canadian province affected by the blackout, has found itself regret-ting its reliance on nuclear for 36 percent of its power. Its cleverly named "Candu" reactors were designed to automatically unlink from the grid in the event of a blackout and then remain in standby mode at 60 percent power, but

that isn't what happened during the blackout. Instead, half of the province's 12 operable reactors went into full automatic shutdown, with another four requiring full manual shutdown. Only two of the reactors responded to the grid breakdown as designed, by partially reducing power. With 10 of 12 reactors down, the difficulty of cold-restarting the Candu reactors quickly became evident, as full shutdowns involve a chemical "poisoning" of the reactor process, which takes days to dissipate, allowing the reactor to power up.

Power was restored to most of the province by the following Sunday, but a state of emergency remained in place. More than a week after the blackout, five of the province's reactors were still shut down, more than 150,000 federal, provincial and municipal workers stayed at home, and the provincial government urged businesses to restrict operations and asked that heavy industries reduce consumption by half. The economic impact was severe.

The Wrong Cure: Pending Congressional Legislation Would Exacerbate the Danger of Nuclear Power and the Failure of Deregulation

The blackout has focused public and media attention on the pending omnibus energy legislation in Congress. Each house has passed an energy bill; now these bills must be reconciled in a House-Senate energy conference committee, which will convene in September. The resulting conference report would then be subject to a vote in each chamber. Unfortunately, Congress has chronically misdiagnosed the cause of the blackout; or, more likely, key members are too beholden to the energy industry to stand up to its greed and exploitation. Despite overwhelming evidence that electricity deregulation has failed and nuclear power is unsafe and unreliable—showcased spectacularly in the recent blackout—Congress obstinately continues to promote nuclear power and push for more deregulation.

California, Round Two: There's No Shortage of Electricity Capacity, Power Plants or Transmission Lines

By next summer, the United States will have a 34 percent reserve margin for electricity generation capacity, indicating a very large surplus of power plants.[8] This glut of power weakens the Bush Administration's claims that the recent electric blackouts give us a reason to build more nuclear power plants or at least keep the current, dilapidated nuclear fleet running.

The same goes for transmission capacity. At the time of the blackout, the grid was only at 75 percent capacity. Yet, shortly afterward, Secretary of Energy Spencer Abraham claimed that $50 billion in new transmission lines need to be built to relieve bottlenecks, and that consumers should pay 100 percent of the cost. Of course, Abraham didn't say that deregulation precipitated the bottlenecks and strains on the nation's electric grid in the first place. The transmission system was designed to accommodate local electricity markets, not the large, freewheeling trading of electricity and movement of power over long distances under deregulation. Sending power over a much wider area decreases efficiency and burdens a transmission system designed to serve local utilities.

And despite its proponents' claims, deregulation has been no friend to ratepayers. Prices have increased in every deregulated wholesale market: California prices shot up 1,000 percent, electricity prices in New England's wholesale market have increased by nearly 400 percent, and power prices in the Pennsylvania-Jersey-Maryland (PJM) (the mid-Atlantic regional power grid operator) deregulated market have increased as much as 250 percent under deregulation. As a result of these failures, nine states have repealed or significantly delayed their deregulation laws. But deregulation's reliance on markets for infrastructure investment has already devastated reliability.

Deregulation freed utilities from having to reinvest ratepayer money back into the transmission system, instead replacing that orderly planning with reliance on the whims of the free market. But the market—which, by definition, lacks a sound regulatory structure—has not provided utility companies with incentives to make necessary investments in transmission; this is due largely to the loopholes added to the Public Utility Holding Company Act (PUHCA) over the past decade. PUHCA, notwithstanding its corruption under the influence of big energy companies, is an essential electricity consumer protection that limits the way in which large utilities can invest ratepayer money into non-electricity assets.

But lawmakers, apparently blind to the failure of deregulation (or, more likely, too beholden to industry interests), have targeted PUHCA for elimination, which would weaken regulators' ability to protect consumers of electricity from the market forces that have wreaked havoc upon the system. For Congress, hindsight is hazy at best. This push for deregulation is compounded with efforts to revive the most expensive and least reliable form of electricity generation: nuclear power.

Congress Must Address Deregulation in the Energy Bill

Republican leaders have sought to exploit the crisis to push for regressive measures in the energy bill. Rep. Billy Tauzin (R-La.), chair of the House Energy and Commerce Committee and cochair of the conference committee, said that the blackout points to "the critical need for Congress to enact a comprehensive national energy bill this year." However, the energy bills approved by the House and Senate not only fail to address electricity reliability, they threaten to make the situation worse.

Both bills will expand deregulation by replacing state jurisdiction over power lines with corporate-controlled Regional Transmission Organizations (RTOs). These larger, multi-state markets will not result in any savings for consumers, but they will make the market even more centralized and will facilitate larger amounts of power being moved over even greater distances. Instead, Congress should be encouraging decentralized power and transmission solutions that keep the infrastructure state-regulated and at the service of local communities.

Moreover, the House bill alters the Federal Power Act's definition of "just and reasonable rates" to permit owners of transmission lines to charge consumers more for the use of transmission lines. But building more lines is not necessary to avoid further blackouts. This provision amounts to a tremendous giveaway to utilities without doing anything substantive to address reliability. Increasing the rate of return for all owners of transmission lines—both existing and proposed—across the board is no guarantee that reliability problems will be addressed. While the rule deregulates transmission by allowing owners to charge whatever price they want, it doesn't provide any guarantee that consumers will enjoy any savings in the future. This is a big difference from the current model in which transmission rates are regulated and consumers are protected.

The House bill also grants the secretary of energy the authority to overrule state concerns regarding controversial transmission line projects. It also grants the federal government the power of eminent domain to seize private land to build the transmission lines.

Only one measure included in congressional energy legislation will actually address the reliability problems with the transmission grid. Both the House and Senate bills mandate and establish enforcement of National Transmission Reliability Standards, an important response to the recent blackout.

But the potential benefit of the reliability provision is thwarted by the full repeal of the Public Utility Holding Company Act (PUHCA) in both bills. PUHCA is an important federal electricity consumer protection that limits the way that large utilities can invest ratepayer money in non-electricity assets. Repeal of PUHCA will result in a wave of mergers, with companies like ExxonMobil likely acquiring utilities. These new, complex companies will have little incentive to reinvest money into historically low-profit assets like transmission lines, and their opaque corporate structures will make it impossible for states and the federal government to truly decipher their finances. PUHCA repeal will therefore lead to an over-concentration within the electric industry, leaving a handful of companies that are largely unaccountable to consumers.

Investors Avoid Nuclear, But Congress Turns a Blind Eye to Safety and Reliability Problems and Pushes More Subsidies

Nuclear power plants have historically proven to be a dicey business investment, at best. A recent DOE report designed to help promote new nuclear plants conceded that "economic viability for a nuclear plant is difficult to demonstrate."[9] Despite a record indicating that nuclear plant owner/operators tend to prioritize production over safety, take every conceivable shortcut, and avoid essential maintenance and upgrades, the costs of construction and decommissioning alone are still daunting to potential investors.

A May 2003 Congressional Budget Office (CBO) cost analysis of Senate energy bill, S.14 (the energy bill abandoned by the Senate in favor of the energy bill from the 107th Congress)—which, like the House energy bill, would provide loan guarantees and power purchase agreements to finance half the development and construction costs of new nuclear power reactors—warned that "plant operators would default on the borrowing that financed its capital costs" for the construction of the plant. The CBO predicted the odds for such defaults to be "very high—well above 50 percent."

The push in the House and Senate energy bills to build new nuclear power plants will not address the energy problems demonstrated in the blackout; it will only expose the public to greater dangers. Yet the House and Senate energy bills pile on subsidies, including incentives for research and development and tax breaks for nuclear operators. Both bills authorize DOE's *Nuclear Power 2010* program to promote the construction of new nuclear reactors, as well as the *Generation IV* program to develop new reactor designs. The House bill provides $3.2 billion in subsidies for nuclear energy research and development and tax breaks for nuclear operators, while the Senate bill gives away $1.5 billion to the nuclear industry.

Furthermore, both bills reauthorize the Price-Anderson Act to extend federal insurance protection to potential new reactors that get built. The private insurance sector, having made its own economic analysis of nuclear power's risks, refuses to fully cover a nuclear power plant (or individual insurance customers) in case of an accident or terrorist attack. Simply put, if Price-Anderson is not reauthorized, there will be no new nuclear power plants, as no corporation would be willing to shoulder such enormous potential liabilities in the event of a catastrophic attack or accident. The lawsuits and settlements that would certainly follow such an event could easily bankrupt any reactor owner without Price-Anderson in place.

Nuclear plants can only be a lucrative investment for investors when regulatory agencies (such as the NRC) roll over and play dead (or even act as industry promoters rather than regulators) and when legislation lavishes subsidies on the nuclear industries, at ratepayer and taxpayer expense.

Conclusion: Current Energy Legislation Is a Crass Denial of the Danger of Nuclear Power and the Failure of Electricity Deregulation

The blackout is a spectacular demonstration of the unreliability of nuclear reactors and the failure of deregulation. It also highlights the shocking imprudence of congressional attempts to revive nuclear power and promote more deregulation.

The only things that nuclear plants can always be counted on to provide are radioactive waste and the risk of catastrophic accidents and radioactive releases. Nuclear plants are also an albatross on the power grids, by not contributing to post-blackout grid recovery, but requiring a first-priority input of electricity once the power grid has been recovered. When a blackout does occur, their constant, inherent dangers are multiplied as the plants depend on unreliable diesel generators to avoid catastrophic accidents. If backup systems should fail, it is only a matter of time before disaster strikes. If that should occur, reactor communities must contend with unreliable alarm sirens and inadequate, unfixable emergency and evacuation plans. The problems with nuclear reactors in times of blackouts are an extremely disturbing combination.

And electricity deregulation, which precipitated the blackout, has failed in every regard. It has resulted in higher prices for ratepayers, diminished reliability and a strained transmission system caused by chaotic energy trading. Only the energy industry and its friends in Congress have benefited from the anarchy of a deregulated electricity market.

The only energy crises that the United States faces have been created by electricity deregulation and a foolish refusal to embrace safe, clean, sustainable energy sources. Failure by Congress to pursue this path is utterly pathological, and it puts the American public at a greater risk of more blackouts, higher electricity rates and the danger of a serious accident at a nuclear power plant. Let us hope that this blackout serves to put Congress on alert to cast aside the monied interests and make consumers' access to energy its first priority.

References

1. "Fact Sheet on Steam Generator Tube Issues." *U.S. Nuclear Regulatory Commission Web Site.* February 2002. `http://www.nrc.gov/reading-rm/doc-collections/fact-sheets/steam-gen.html`

2. Smeloff, Ed. "Nuclear mishap raises questions about Indian Point's future." *Westchester Environment.* 2000. 2 March/April 2002. `http://www.fcwc.org/WEArchive/030400/030400we.htm#Nuclear`

3. "Event Notification Report for February 17, 2000." *U.S. Nuclear Regulatory Commission Web Site.* `http://www.nrc.gov/reading-rm/doc-collections/event-status/event/2000/20000217en.html`

4. "Event Notification Report for April 25, 2003." *U.S. Nuclear Regulatory Commission Web Site.* `http://www.nrc.gov/reading-rm/doc-collections/event-status/event/2003/20030425en.html`

5. "Event Notification Report for January 27, 2003." *U.S. Nuclear Regulatory Commission Web Site.* `http://www.nrc.gov/reading-rm/doc-collections/event-status/event/2003/20030127en.html`

6. "Event Notification Report for April 7, 2003." *U.S. Nuclear Regulatory Commission Web Site.* `http://www.nrc.gov/reading-rm/doc-collections/event-status/event/2003/20030407en.html`

7. "Event Notification Report for August 15, 2003." *U.S. Nuclear Regulatory Commission Web Site.* `http://www.nrc.gov/reading-rm/doc-collections/event-status/event/2003/20030815en.html`

8. Kellerman, Larry, managing director, Goldman Sachs & Co. Interview. *Project Finance NewsWire* August 2003

9. "Summary Report." *A Roadmap to Deploy New Nuclear Power Plants in the United States by 2010.* Vol. 1. *U.S. Department of Energy.* 31 Oct. 2001.

POSTSCRIPT

Is It Time to Revive Nuclear Power?

Peter Schwartz and Spencer Reiss, "Nuclear Now: How Clean, Green Atomic Energy Can Stop Global Warming," *Wired* (February 2005), express considerable optimism for the prospects of nuclear power. Robert Evans, "Nuclear Power: Back in the Game," *Power Engineering* (October 2005), reports that a number of power companies are considering new nuclear power plants. See also Eliot Marshall, "Is the Friendly Atom Poised for a Comeback?" and Daniel Clery, "Nuclear Industry Dares to Dream of a New Dawn," *Science* (August 19, 2005). Nuclear momentum is growing, says Charles Petit, "Nuclear Power: Risking a Comeback," *National Geographic* (April 2006), thanks in part to new technologies. Karen Charman, "Brave Nuclear World?" *World Watch* (May/June 2006), objects that producing nuclear fuel uses huge amounts of electricity derived from fossil fuels, so going nuclear can hardly prevent all releases of carbon dioxide (although using electricity derived from nuclear power would reduce the problem). She also notes that "Although no comprehensive and integrated study comparing the collateral and external costs of energy sources globally has been done, all currently available energy sources have them.... Burning coal—the single, largest source of air pollution in the U.S.—causes global warming, acid rain, soot, smog, and other toxic air emissions and generates waste ash, sludge, and toxic chemicals. Landscapes and ecosystems are completely destroyed by mountaintop removal mining, while underground mining imposes high fatality, injury, and sickness rates. Even wind energy kills birds, can be noisy, and, some people complain, blights landscapes."

Michael J. Wallace tells us that there are 103 nuclear reators operating in the United States today. Stephen Ansolabehere, et al., "The Future of Nuclear Power," *An Interdisciplinary MIT Study* (MIT, 2003), note that in 2000 there were 352 in the developed world as a whole, and a mere 15 in developing nations, and that even a very large increase in the number of nuclear power plants—to 1,000 to 1,500—will not stop all releases of carbon dioxide. In fact, if carbon emissions double by 2050 as expected, from 6,500 to 13,000 million metric tons per year, the 1,800 million metric tons not emitted because of nuclear power will seem relatively insignificant. Christine Laurent, in "Beating Global Warming with Nuclear Power?" *UNESCO Courier* (February 2001), notes that "For several years, the nuclear energy industry has attempted to cloak itself in different ecological robes. Its credo: nuclear energy is a formidable asset in battle against global warming because it emits very small amounts of greenhouse gases. This stance, first presented in the late 1980s when the extent of the phenomenon was still the subject of controversy, is now at the heart of policy debates over how to avoid droughts, downpours, and floods." Laurent adds that it makes more sense to focus on reducing carbon emissions by reducing energy consumption.

The debate over the future of nuclear power is likely to remain vigorous for some time to come. But as Richard A. Meserve said in a *Science* editorial ("Global Warming and Nuclear Power," *Science* [January 23, 2004]), "For those who are serious about confronting global warming, nuclear power should be seen as part of the solution. Although it is unlikely that many environmental groups will become enthusiastic proponents of nuclear power, the harsh reality is that any serious program to address global warming cannot afford to jettison any technology prematurely.... The stakes are large, and the scientific and educational community should seek to ensure that the public understands the critical link between nuclear power and climate change."

Alvin M. Weinberg, former director of the Oak Ridge National Laboratory, notes in "New Life for Nuclear Power," *Issues in Science and Technology* (Summer 2003), that to make a serious dent in carbon emission would require perhaps four times as many reactors as suggested in the MIT study. The accompanying safety and security problems would be challenging. If the challenges can be met, says John J. Taylor, retired vice president for nuclear power at the Electric Power Research Institute, in "The Nuclear Power Bargain," *Issues in Science and Technology* (Spring 2004), there are a great many potential benefits.

Environmental groups such as Friends of the Earth are adamantly opposed, saying "Those who back nuclear over renewables and increased energy efficiency completely fail to acknowledge the deadly radioactive legacy nuclear power has created and continues to create" ("Nuclear Power Revival Plan Slammed," Press Release, April 18, 2004, http://www.foe-scotland.org.uk/press/pr20040408.html).

On the Internet . . .

The Population Council

Established in 1952, the Population Council "is an international, nonprofit institution that conducts research on three fronts: biomedical, social science, and public health. This research—and the information it produces—helps change the way people think about problems related to reproductive health and population growth." Many of the Council's publications are available on-line.

http://www.popcouncil.org/

United Nations Population Division

The United Nations Population Division is responsible for monitoring and appraising a broad range of areas in the field of population. This site offers a wealth of recent data and links.

http://www.un.org/esa/population/unpop.htm

The Agriculture Network Information Center

The Agriculture Network Information Center is a guide to quality agricultural (including biotechnology) information on the Internet as selected by the National Agricultural Library, Land-Grant Universities, and other institutions.

http://www.agnic.org/

Agriculture: Genetic Resources and GMOs

The Agriculture portion of the European Union's portal web site (EUROPA) provides information on different subjects associated with the genetic base for agricultural activities.

http://www.europa.eu.int/comm/agriculture/res/index_en.htm

Earth Policy Institute

The purpose of the Earth Policy Institute is to provide a vision of what an environmentally sustainable economy will look like, a roadmap of how to get from here to there, and an ongoing assessment of this effort, of where progress is being made and where it is not.

http://www.earth-policy.org/

Seaweb

SeaWeb is a project designed to raise awareness of the world ocean and the life within it. Its people believe that as more people understand the ocean's critical role in everyday life and the future of the planet, they will take actions to conserve the ocean and the web of life it supports. It provides access to a great deal of fisheries-related information.

http://www.seaweb.org

PART 4

Food and Population

*T*o many, "sustainability" means arranging things so that the natural world—plants and animals, forests and coral reefs, fresh water and landscapes—can continue to exist more or less (mostly less) as it did before human beings multiplied, developed technology, and began to cause extinctions, air and water pollution, soil erosion, desertification, climate change, and so on. To others, "sustainability" means arranging things so that humankind can continue to survive and thrive, even keeping up its history of growth, technological development, and energy use—as if the environment and its resources were infinite.

Because the two visions of "sustainability" are logically incompatible, we must struggle to find some sort of middle ground. Must we reduce the numbers of people on the planet? Their use of technology? Their standard of living? If we do, will human well-being be lessened? If we do not, how can we continue to feed everyone? And if we wish to eat wild food such as fish, how can we keep our appetite from destroying the source?

- Do Falling Birth Rates Pose a Threat to Human Welfare?
- Is Genetic Engineering the Answer to Hunger?
- Are Marine Reserves Needed to Protect Global Fisheries?

ISSUE 13

Do Falling Birth Rates Pose a Threat to Human Welfare?

YES: Michael Meyer, from "Birth Dearth," *Newsweek* (September 27, 2004)

NO: David Nicholson-Lord, from "The Fewer the Better," *New Statesman* (November 8, 2004)

ISSUE SUMMARY

YES: Michael Meyer argues that when world population begins to decline after about 2050, economies will no longer continue to grow, government benefits will decline, young people will have to support ever more elders, and despite some environmental benefits, quality of life will suffer.

NO: David Nicholson-Lord argues that the economic problems of population decline all have straightforward solutions. A less crowded world will not suffer from the environmental ills attendant on overcrowding and will, overall, be a roomier, gentler, less materialistic place to live, with cleaner air and water.

In 1798 the British economist Thomas Malthus published his *Essay on the Principle of Population*. In it, he pointed with alarm at the way the human population grew geometrically (a hockey-stick curve of increase) and at how agricultural productivity grew only arithmetically (a straight-line increase). It was obvious, he said, that the population must inevitably outstrip its food supply and experience famine. Contrary to the conventional wisdom of the time, population growth was not necessarily a good thing. Indeed, it led inexorably to catastrophe. For many years, Malthus was something of a laughingstock. The doom he forecast kept receding into the future as new lands were opened to agriculture, new agricultural technologies appeared, new ways of preserving food limited the waste of spoilage, and the developed nations underwent a "demographic transition" from high birth rates and high death rates to low birth rates and low death rates.

Demographers initially attributed the demographic transition to increasing prosperity and predicted that as prosperity increased in countries whose populations were rapidly growing, birth rates would surely fall. Later, some scholars analyzed the historical data and concluded that the transition had actually preceded prosperity. The two views have contrasting implications for public policy designed to slow population growth—economic aid or family planning aid—but neither has worked very well. In 1994 the UN Conference on Population and Development, which was held in Cairo, Egypt, concluded that better results would follow from improving women's access to education and health care.

Should we be trying to slow or reverse population growth? In the 1968 book *The Population Bomb* (Ballantine Books), Paul R. Ehrlich warned that unrestricted population growth would lead to both human and environmental disaster. But some religious leaders oppose population control because family planning is against God's will. Furthermore, minority groups and developing nations contend that they are unfairly targeted by family planning programs.

The world's human population has grown tremendously. In Malthus's time, there were about 1 billion human beings on earth. By 1950 there were a little over 2.5 billion. In 1999 the tally passed 6 billion. By 2025 it will be over 8 billion. Statistics like these, which are presented in *World Resources 2002-2004: Decisions for the Earth, Balance, Voice, and Power* (World Resources Institute, 2003) (http://governance.wri.org/pubs_description.cfm?PubID=3764), a biennial report of the World Resources Institute in collaboration with the United Nations Environment and Development Programmes, are positively frightening. By 2050 the UN expects the world population to be about 9 billion (see *World Population Prospects: The 2004 Revision Population Database*; http://esa.un.org/unpp/; United Nations, 2005).

While global agricultural production has also increased, it has not kept up with rising demand, and—because of the loss of topsoil to erosion, the exhaustion of aquifers for irrigation water, and the high price of energy for making fertilizer (among other things)—the prospect of improvement seems exceedingly slim to many observers. Paul R. Ehrlich and Anne H. Ehrlich argue in "The Population Explosion: Why We Should Care and What We Should Do About It," *Environmental Law* (Winter 1997) that "population growth may be the paramount force moving humanity inexorably towards disaster." They therefore maintain that it is essential to reduce the impact of population in terms of both numbers and resource consumption.

Was Malthus wrong? Both environmental scientists and many economists now say that if population continues to grow, problems are inevitable. But earlier predictions of a world population of 10 or 12 billion by 2050 are no longer looking very likely. The UN's population statistics show a slowing of growth, to be followed by an actual decline in population.

Some people worry that such a decline will not be good for human welfare. Michael Meyer argues that shrinking population will mean that the economic growth which has meant constantly increasing standards of living, must come to an end, government programs (from war to benefits for the poor and elderly) will no longer be affordable, young people will have to support ever more elders, and despite some environmental benefits, quality of life will suffer. David Nicholson-Lord argues that the economic problems of population decline all have straightforward solutions. A less crowded world will not suffer from the environmental ills attendant on overcrowding and will, overall, be a roomier, gentler, less materialistic place to live, with cleaner air and water.

YES

Michael Meyer

Birth Dearth

falling Birth rate is a threat!!! (handwritten)

Everyone knows there are too many people in the world. Whether we live in Lahore or Los Angeles, Shanghai or Sao Paulo, our lives are daily proof. We endure traffic gridlock, urban sprawl and environmental depredation. The evening news brings variations on Ramallah or Darfur—images of Third World famine, poverty, pestilence, war, global competition for jobs and increasingly scarce natural resources.

Just last week the United Nations warned that many of the world's cities are becoming hopelessly overcrowded. Lagos alone will grow from 6.5 million people in 1995 to 16 million by 2015, a miasma of slums and decay where a fifth of all children will die before they are 5. At a conference in London, the U.N. Population Fund weighed in with a similarly bleak report: unless something dramatically changes, the world's 50 poorest countries will triple in size by 2050, to 1.7 billion people.

Yet this is not the full story. To the contrary, in fact. Across the globe, people are having fewer and fewer children. Fertility rates have dropped by half since 1972, from six children per woman to 2.9. And demographers say they're still falling, faster than ever. The world's population will continue to grow—from today's 6.4 billion to around 9 billion in 2050. But after that, it will go sharply into decline. Indeed, a phenomenon that we're destined to learn much more about—depopulation—has already begun in a number of countries. Welcome to the New Demography. It will change everything about our world, from the absolute size and power of nations to global economic growth to the quality of our lives.

This revolutionary transformation will be led not so much by developed nations as by the developing ones. Most of us are familiar with demographic trends in Europe, where birthrates have been declining for years. To reproduce itself, a society's women must each bear 2.1 children. Europe's fertility rates fall far short of that, according to the 2002 U.N. population report. France and Ireland, at 1.8, top Europe's childbearing charts. Italy and Spain, at 1.2, bring up the rear. In between are countries such as Germany, whose fertility rate of 1.4 is exactly Europe's average. What does that mean? If the U.N. figures are right, Germany could shed nearly a fifth of its 82.5 million people over the next 40 years—roughly the equivalent of all of east Germany, a loss of population not seen in Europe since the Thirty Years' War.

And so it is across the Continent. Bulgaria will shrink by 38 percent, Romania by 27 percent, Estonia by 25 percent. "Parts of Eastern Europe, already sparsely populated, will just empty out," predicts Reiner Klingholz, director of the Berlin Institute for Population and Development. Russia is already losing close to 750,000 people yearly. (President Vladimir Putin calls it a "national crisis.") So is Western Europe, and that figure could grow to as much as 3 million a year by midcentury, if not more.

The surprise is how closely the less-developed world is following the same trajectory. In Asia it's well known that Japan will soon tip into population loss, if it hasn't already. With a fertility rate of 1.3 children per woman, the country stands to shed a quarter of its 127 million people over the next four decades, according to U.N. projections. But while the graying of Japan (average age: 42.3 years) has long been a staple of news headlines, what to make of China, whose fertility rate has declined from 5.8 in 1970 to 1.8 today, according to the U.N.? Chinese census data put the figure even lower, at 1.3. Coupled with increasing life spans, that means China's population will age as quickly in one generation as Europe's has over the past 100 years, reports the Center for Strategic and International Studies in Washington. With an expected median age of 44 in 2015, China will be older on average than the United States. By 2019 or soon after, its population will peak at 1.5 billion, then enter a steep decline. By midcentury, China could well lose 20 to 30 percent of its population every generation.

The picture is similar elsewhere in Asia, where birthrates are declining even in the absence of such stringent birth-control programs as China's. Indeed, it's happening despite often generous official incentives to procreate. The industrialized nations of Singapore, Hong Kong, Taiwan and South Korea all report subreplacement fertility, says Nicholas Eberstadt, a demographer at the American Enterprise Institute in Washington. To this list can be added Thailand, Burma, Australia and Sri Lanka, along with Cuba and many Caribbean nations, as well as Uruguay and Brazil. Mexico is aging so rapidly that within several decades it will not only stop growing but will have an older population than that of the United States. So much for the cliche of those Mexican youths swarming across the Rio Grande? "If these figures are accurate," says Eberstadt, "just about half of the world's population lives in subreplacement countries."

There are notable exceptions. In Europe, Albania and the outlier province of Kosovo are reproducing energetically. So are pockets of Asia: Mongolia, Pakistan and the Philippines. The United Nations projects that the Middle East will double in population over the next 20 years, growing from 326 million today to 649 million by 2050. Saudi Arabia has one of the highest fertility rates in the world, 5.7, after Palestinian territories at 5.9 and Yemen at 7.2. Yet there are surprises here, too. Tunisia has tipped below replacement. Lebanon and Iran are at the threshold. And though overall the region's population continues to grow, the increase is due mainly to lower infant mortality; fertility rates themselves are falling faster than in developed countries, indicating that over the coming decades the Middle East will age far more rapidly than other regions of the world. Birthrates in Africa remain high, and despite the AIDS epidemic its population is projected to keep growing. So is that of the United States.

We'll return to American exceptionalism, and what that might portend. But first, let's explore the causes of the birth dearth, as outlined in a pair of new books on the subject. "Never in the last 650 years, since the time of the Black Plague, have birth and fertility rates fallen so far, so fast, so low, for so long, in so many places," writes the sociologist Ben Wattenberg in "Fewer: How the New Demography of Depopulation Will Shape Our Future." Why? Wattenberg suggests that a variety of once independent trends have conjoined to produce a demographic tsunami. As the United Nations reported last week, people everywhere are leaving the countryside and moving to cities, which will be home to more than half the world's people by 2007. Once there, having a child becomes a cost rather than an asset. From 1970 to 2000, Nigeria's urban population climbed from 14 to 44 percent. South Korea went from 28 to 84 percent. So-called megacities, from Lagos to Mexico City, have exploded seemingly overnight. Birth rates have fallen in inverse correlation.

Other factors are at work. Increasing female literacy and enrollment in schools have tended to decrease fertility, as have divorce, abortion and the worldwide trend toward later marriage. Contraceptive use has risen dramatically over the past decade; according to U.N. data, 62 percent of married or "in union" women of reproductive age are now using some form of nonnatural birth control. In countries such as India, now the capital of global HIV, disease has become a factor. In Russia, the culprits include alcoholism, poor public health and industrial pollution that has whacked male sperm counts. Wealth discourages childbearing, as seen long ago in Europe and now in Asia. As Wattenberg puts it, "Capitalism is the best contraception."

The potential consequences of the population implosion are enormous. Consider the global economy, as Phillip Longman describes it in another recent book, "The Empty Cradle: How Falling Birthrates Threaten World Prosperity and What to Do About It." A population expert at the New America Foundation in Washington, he sees danger for global prosperity. Whether it's real estate or consumer spending, economic growth and population have always been closely linked. "There are people who cling to the hope that you can have a vibrant economy without a growing population, but mainstream economists are pessimistic," says Longman. You have only to look at Japan or Europe for a whiff of what the future might bring, he adds. In Italy, demographers forecast a 40 percent decline in the working-age population over the next four decades—accompanied by a commensurate drop in growth across the Continent, according to the European Commission. What happens when Europe's cohort of baby boomers begins to retire around 2020? Recent strikes and demonstrations in Germany, Italy, France and Austria over the most modest pension reforms are only the beginning of what promises to become a major sociological battle between Europe's older and younger generations.

That will be only a skirmish compared with the conflict brewing in China. There market reforms have removed the cradle-to-grave benefits of the planned economy, while the Communist Party hasn't constructed an adequate social safety net to take their place. Less than one quarter of the population is covered by retirement pensions, according to CSIS. That puts the burden of elder care almost entirely on what is now a generation of only children. The one-child policy has led to the so-called 4-2-1 problem, in which each child will be potentially responsible for caring for two parents and four grandparents.

Incomes in China aren't rising fast enough to offset this burden. In some rural villages, so many young people have fled to the cities that there may be nobody left to look after the elders. And the aging population could soon start to dull China's competitive edge, which depends on a seemingly endless supply of cheap labor. After 2015, this labor pool will begin to dry up, says economist Hu Angang. China will have little choice but to adopt a very Western-sounding solution, he says: it will have to raise the education level of its work force and make it more productive. Whether it can is an open question. Either way, this much is certain: among Asia's emerging economic powers, China will be the first to grow old before it gets rich.

Equally deep dislocations are becoming apparent in Japan. Akihiko Matsutani, an economist and author of a recent best seller, "The Economy of a Shrinking Population," predicts that by 2009 Japan's economy will enter an era of "negative growth." By 2030, national income will have shrunk by 15 percent. Speculating about the future is always dicey, but economists pose troubling questions. Take the legendarily high savings that have long buoyed the Japanese economy and financed borrowing worldwide, especially by the United States. As an aging Japan draws down those assets in retirement, will U.S. and global interest rates rise? At home, will Japanese businesses find themselves competing for increasingly scarce investment capital? And just what will they be investing in, as the country's consumers grow older, and demand for the latest in hot new products cools off? What of the effect on national infrastructure? With less tax revenue in state coffers, Matsutani predicts, governments will increasingly be forced to skimp on or delay repairs to the nation's roads, bridges, rail lines and the like. "Life will become less convenient," he says. Spanking-clean Tokyo might come to look more like New York City in the 1970s, when many urban dwellers decamped for the suburbs (taking their taxes with them) and city fathers could no longer afford the municipal upkeep. Can Japanese cope? "They will have to," says Matsutani. "There's no alternative."

Demographic change magnifies all of a country's problems, social as well as economic. An overburdened welfare state? Aging makes it collapse. Tensions over immigration? Differing birthrates intensify anxieties, just as the need for imported labor rises—perhaps the critical issue for the Europe of tomorrow. A poor education system, with too many kids left behind? Better fix it, because a shrinking work force requires higher productivity and greater flexibility, reflected in a new need for continuing job training, career switches and the health care needed to keep workers working into old age.

In an ideal world, perhaps, the growing gulf between the world's wealthy but shrinking countries and its poor, growing ones would create an opportunity. Labor would flow from the overpopulated, resource-poor south to the depopulating north, where jobs would continue to be plentiful. Capital and remittance income from the rich nations would flow along the reverse path, benefiting all. Will it happen? Perhaps, but that presupposes considerable labor mobility. Considering the resistance Europeans display toward large-scale immigration from North Africa, or Japan's almost zero-immigration policy, it's hard to be optimistic. Yes, attitudes are changing. Only a decade ago, for instance, Europeans also spoke of zero immigration. Today they recognize the need and, in bits and pieces, are beginning to plan for it. But will it happen on the scale required?

A more probable scenario may be an intensification of existing tensions between peoples determined to preserve their beleaguered national identities on the one hand, and immigrant groups on the other seeking to escape overcrowding and lack of opportunity at home. For countries such as the Philippines—still growing, and whose educated work force looks likely to break out of low-status jobs as nannies and gardeners and move up the global professional ladder—this may be less of a problem. It will be vastly more serious for the tens of millions of Arab youths who make up a majority of the population in the Middle East and North Africa, at least half of whom are unemployed.

America is the wild card in this global equation. While Europe and much of Asia shrinks, the United States' indigenous population looks likely to stay relatively constant, with fertility rates hovering almost precisely at replacement levels. Add in heavy immigration, and you quickly see that America is the only modern nation that will continue to grow. Over the next 45 years the United States will gain 100 million people, Wattenberg estimates, while Europe loses roughly as many.

This does not mean that Americans will escape the coming demographic whammy. They, too, face the problems of an aging work force and its burdens. (The cost of Medicare and Social Security will rise from 4.3 percent of GDP in 2000 to 11.5 percent in 2030 and 21 percent in 2050, according to the Congressional Budget Office.) They, too, face the prospect of increasing ethnic tensions, as a flat white population and a dwindling black one become gradually smaller minorities in a growing multicultural sea. And in our interdependent era, the troubles of America's major trading partners—Europe and Japan—will quickly become its own. To cite one example, what becomes of the vaunted "China market," invested in so heavily by U.S. companies, if by 2050 China loses an estimated 35 percent of its workers and the aged consume an ever-greater share of income?

America's demographic "unipolarity" has profound security implications as well. Washington worries about terrorism and failing states. Yet the chaos of today's fragmented world is likely to prove small in comparison to what could come. For U.S. leaders, Longman in "The Empty Cradle" sketches an unsettling prospect. Though the United States may have few military competitors, the technologies by which it projects geopolitical power—from laser-guided missiles and stealth bombers to a huge military infrastructure—may gradually become too expensive for a country facing massively rising social entitlements in an era of slowing global economic growth. If the war on terrorism turns out to be the "generational struggle" that national-security adviser Condoleezza Rice says it is, Longman concludes, then the United States might have difficulty paying for it.

None of this is writ, of course. Enlightened governments could help hold the line. France and the Netherlands have instituted family-friendly policies that help women combine work and motherhood, ranging from tax credits for kids to subsidized day care. Scandinavian countries have kept birthrates up with generous provisions for parental leave, health care and part-time employment. Still, similar programs offered by the shrinking city-state of Singapore—including a state-run dating service—have done little to reverse the birth dearth. Remember, too, that such prognoses have been wrong in the past. At the cusp of the postwar baby boom, demographers predicted a sharp fall in fertility and a global birth dearth. Yet even if this generation of seers turns out to be right, as seems likely, not all is bad. Environmentally, a smaller world is almost certainly a better world, whether in terms of cleaner air or, say, the return of wolves and rare flora to abandoned stretches of the East German countryside. And while people are living longer, they are also living healthier—at least in the developed world. That means they can (and probably should) work more years before retirement.

Yes, a younger generation will have to shoulder the burden of paying for their elders. But there will be compensations. As populations shrink, says economist Matsutani, national incomes may drop—but not necessarily per capita incomes. And in this realm of uncertainty, one mundane thing is probably sure: real-estate prices will fall. That will hurt seniors whose nest eggs are tied up in their homes, but it will be a boon to youngsters of the future. Who knows? Maybe the added space and cheap living will inspire them to, well, do whatever it takes to make more babies. Thus the cycle of life will restore its balance. . . .

David Nicholson-Lord

 **NO**

The Fewer the Better

This is a story of two Britains and two futures. In the first Britain, the work culture dominates; the talk is of economic growth and dynamism and competing with the rest of the world. Labour is young, cheap and biddable, and driven by the urge to "succeed"—to make it in material and career terms, with the consumer goods and lifestyles to match. In the cities of this 24/7 society, population densities rise: so do crime, violence and antisocial behaviour. Outside the cities, urbanisation spreads, along with noise, congestion, the creep of human clutter and development. Unspoilt places are increasingly hard to find. Pollution gets steadily worse.

The second Britain is a quieter place. The age profile is older, the values less strident and materialistic. People work longer—they are not pensioned off in their fifties—but they save more and spend less, at least on ephemera and gadgets. They drink much less, too, and don't get involved in fights. Work is important but so are hobbies, family and community life. Cities are more spacious, roads emptier, the countryside more rural. The air and water are cleaner and there is hope of getting the weather back to normal because the planet is no longer warming so rapidly.

Which future would we prefer? The first—let us call it "UK plc"—with its economic engine revving at full speed? Or the second, where quality of life matters more: not so much a plc as a community enterprise, with the emphasis on community rather than enterprise? Most people would plump for the second. Yet we seem to be heading for the first.

What we do not admit is that the difference between the two futures is largely one of human numbers. Population is a subject we don't like to mention. In September Michael Howard, the Conservative leader, pointed out that, over the next 30 years, Britain's population would grow by 5.6 million—an increase of nearly 10 percent on the current 59 million. Immigration, running at an average net inflow of 158,000 a year in the past five years, accounts for 85 percent of this increase. Because population is forecast to rise, the government plans an extra 3.8 million houses in England over the next 20 years. But that plan is based on net immigration of 65,000 a year. If it continues at 158,000 a year, we will need 4.85 million new homes.

Howard went on to quote, with approval, the conclusions of the government's Community Cohesion Panel, which said in July that people "need sufficient time to come to terms with and accommodate incoming groups, regardless of their ethnic origin. The 'pace of change' . . . is simply too great . . . at present."

Alarmist? Electioneering? Playing the race card? In so far as these parts of Howard's speech were reported at all, that was how the left-liberal media interpreted them. Yet his figures understate the contribution of immigration to housing forecasts, because they ignore the changes in fertility and household formation resulting from a younger population. According to the Optimum Population Trust, a continuation of the 2001–2003 growth rate of 0.4 percent would result in a UK population of 71 million in 2050 and 100 million by the end of the century.

The implications of this bear examination. Given a population of "only" 65–66 million by mid-century, for example, we would need an extra nine or ten million houses by 2050—more than twice the numbers Howard was talking about, and an increase of nearly 50 percent on the current English housing stock. Should this worry us? Clearly many people think so; the government's housing plans have been a source of controversy ever since they were published. Examine this controversy in greater depth, and you will find a developing awareness of what ecologists call "carrying capacity": the balance (or lack of it) between a physical environment and the numbers it can support.

About all this, the environmental lobby is now silent. The last time such issues were deemed fit for public debate was in 1973, when a government population panel said Britain must accept "that its population cannot go on increasing indefinitely". The progressive-minded believe, on the one hand, in liberal multiculturalism and, on the other, in sustainability. They cannot resolve the conflict. The field has thus been abandoned to the political right.

The demographic facts are undeniable, however. Before the start of the current immigration surge in the 1990s, Britain's population, like that of many other developed countries, was heading for decline—as early as 2013, according to some forecasts. British women are having 1.7 children each, on average, above Germany (1.4) and Japan (1.3) but below the replacement level of 2.1. If this had been allowed to continue, with no immigration, we would be down to 30 million by 2120.

What would it be like to live in a country where population halved in the space of three or four generations? Environmentally, the case for population decline is unanswerable—less pollution, less strain on natural systems, greater national self-sufficiency, a reduction in fossil-fuel emissions, the freeing up of land for other species and higher-order human uses, such as wilderness. Psychologically, what the economist Fred Hirsch called "positional goods"—a view, an unspoilt beach, a piece of heritage—would be freer of the crowds and queues that now, for most people, mar them. Applied to social and economic life, this might reduce the awful sense of competitiveness that is a relatively recent feature of cultural life, for jobs, places at school or university, or entry to prized social institutions or niches.

Given the close association between crowding, densities, congestion and stress, and the greater distances available between people, we would also probably see less casual public aggression: less of the "rage" that emerged in the late 1980s. And, because young people are more likely to commit crimes, the ageing of society that would result from population decline would reinforce these trends. A Britain of 30 million people would almost certainly be a kindlier, more easygoing, more socially concerned place—exactly the sort of Britain that many readers of the *New Statesman* would like to see.

⋅❀⋅

Most of the argument so far, however, has focused on the perils of decline: economic and social stagnation, the decrease in the support ratio (of workers to pensioners), emerging labour shortages and so on. Given that all of these "problems" are either illusory, fantastical or soluble it is instructive to ask why they obsess us. Why were the Tories, for example, thinking until recently about encouraging people to have babies and why does the government still envisage no upper limit on immigration? There are two answers. First, population growth is such a feature of the past two centuries—although not of preceding ages—that it has become synonymous in our minds with progress. Second, economic growth is how politicians and economists measure national success. And having more people is the quickest and easiest way to boost gross domestic product.

Much is made, therefore, of the impact of immigration on economic growth. Yet the growth comes almost entirely from additions to the national headcount. The increased wealth per person may be as little as 0.1 percent a year, according to US research.

More important is what happens when "immigrants" become "natives". This is the central fallacy of the demographic "timebomb" argument. Immigrants eventually become pensioners, and pensioners keep living longer. The only way to preserve a support ratio regarded as optimal is thus to have permanently high levels of immigration—and a population permanently, indeed infinitely, growing. David Coleman, professor of demography at Oxford University, has calculated that to keep the support ratio between pensioners and those of working age at roughly current levels would require a UK population in 2100 of approximately 300 million and rising. He calls it "the incredible in pursuit of the implausible".

And what about the world as a whole? How are developing countries, presumably expected to provide the young immigrants to the UK and other western countries, supposed to support their own old people?

New figures from the US Population Reference Bureau suggest a world population of roughly 9.3 billion in 2050, against 6.4 billion now. Studies such as the WWF's *Living Planet Report* say that by that time, humanity's footprint will be up to 220 percent of the earth's biological capacity. We would need, in other words, another couple of planets to survive. But if we manage to control global population (and it looks increasingly likely that we can) numbers will start to decline, possibly around 2070. What is the world supposed to do then? Import extra-planetary aliens to maintain the support ratio?

Even within Britain, it is hard to make a case for labour shortages when unemployment is three times the number of vacancies and economic inactivity, notably among the over-fifties, is at an all-time high. It is also hard, morally at least, to argue that we should deliberately cream off the skilled and educated workers of poorer countries—little different from people-trafficking, according to a National Health Service overseas recruiter addressing this year's Royal College of Nursing conference—or that we should bring people in because there is nobody else to sweep our streets and clean our toilets.

⋅✦⋅

The solutions to the "problems" of population decline, in fact, lie safely within the range of realistic policy options. They include: people saving more and consuming less; governments investing more in preventative health measures, to lengthen illness-free old age; better labour productivity; a higher retirement age; drawing the economically inactive back into economic activity (with penalties for ageism); and restructuring hard-to-fill jobs to make them more attractive. Population decline creates a (relative) shortage of workers and therefore shifts power from capital to labour and raises pay rates generally, as happened after the Black Death. Isn't the left supposed to be in favour of such an outcome and against the use of immigrants to create a US-style low-wage economy?

Yet the argument is not primarily about economics. Those who advocate increases in immigration and population do so largely on the grounds that they are good for GDP. They forget, as most economists do, that they are often bad for the environment and society. Economic growth, after all, is ethically undiscriminating: the wages earned from clearing up the effects of a car crash or a pollution mishap count towards GDP in the same way as those earned from making a loaf of bread. All over the world, Britain included, population growth is generating an extraordinary range of negative effects, from climate change and resource exhaustion to the destruction of species and habitats and the poisoning of the biosphere. Deliberately boosting Britain's population, either through large-scale net immigration or by telling people to have more babies, will ultimately make it a much worse place to live in.

How big should Britain's population be? That depends which sums you do—but some calculations from the Optimum Population Trust suggest 20 million or fewer.

Before you throw up your hands in disbelief at this idea, consider the view of a liberal from another generation, John Stuart Mill, who in his *Principles of Political Economy* (1848) acknowledged the economic potential for a "great increase in population" but confessed he could see little reason for desiring it. "The density of population necessary to enable mankind to obtain . . . all the advantages both of co-operation and of social intercourse," he wrote, "has, in all the most populous countries, been attained." In 1848, the world contained just over a billion people and the population of Britain was 21 million.

POSTSCRIPT

Do Falling Birth Rates Pose a Threat to Human Welfare?

Resources and population come together in the concept of "carrying capacity," defined very simply as the size of the population that the environment can support, or "carry," indefinitely, through both good years and bad. It is not the size of the population that can prosper in good times alone, for such a large population must suffer catastrophically when droughts, floods, hurricanes, or blights arrive or the climate warms or cools. It is a long-term concept, where "long-term" means not decades or generations, nor even centuries, but millennia or more. See Mark Nathan Cohen, "Carrying Capacity," *Free Inquiry* (August/September 2004).

What is Earth's carrying capacity for human beings? It is surely impossible to set a precise figure on the number of human beings the world can support for the long run. As Joel E. Cohen discusses in *How Many People Can the Earth Support?* (W. W. Norton, 1995), estimates of Earth's carrying capacity range from under a billion to over a trillion. The precise number depends on our choices of diet, standard of living, level of technology, willingness to share with others at home and abroad, and desire for an intact physical, chemical, and biological environment, as well as on whether or not our morality permits restraint in reproduction and our political or religious ideology permits educating and empowering women. The key, Cohen stresses, is human choice, and the choices are ones we must make within the next 50 years. See also Joel E. Cohen, "Human Population Grows Up," *Scientific American* (September 2005). Phoebe Hall, "Carrying Capacity," *E Magazine* (March/April 2003), notes that even countries with large land areas and small populations, such as Australia and Canada, ca be overpopulated in terms of resource availability. The critical resource appears to be food supply; see Russell Hopfenberg, "Human Carrying Capacity Is Determined by Food Availability," *Population & Environment* (November 2003). Jared Diamond's *Collapse: How Societies Choose to Fail or Succeed* (Viking, 2005) is an excellent presentation of how past societies have run afoul of the need to match resources and population.

Andrew R. B. Ferguson, in "Perceiving the Population Bomb," *World Watch* (July/August 2001), sets the maximum sustainable human population at about 2 billion. Sandra Postel, in the Worldwatch Institute's *State of the World 1994* (W. W.Norton, 1994), says, "As a result of our population size, consumption patterns, and technology choices, we have surpassed the planet's carrying capacity. This is plainly evident by the extent to which we are damaging and depleting natural capital" (including soil and water).

If population growth is now declining and world population will actually begin to decline during this century, there is clearly hope. But the question of carrying capacity remains. Most estimates of carrying capacity put it at well below the current world population size, and it will take a long time for global population to fall far enough to reach such levels. We seem to be moving in the right direction, but it remains an open question whether our numbers will decline far enough soon enough, *i.e.*, before environmental problems become critical. On the other hand, Jeroen Van den Bergh and Piet Rietveld, "Reconsidering the Limits to World Population: Meta-analysis and Meta-prediction," *Bioscience* (March 2004), set their best estimate of human global carrying capacity at 7.7 billion, which is distinctly reassuring, at least for the environment. The end of population growth will have severe social impacts; see Wolfgang Lutz, Warren C. Sanderson, and Sergei Scherbov, eds., *The End of World Population Growth in the 21st Century: New Challenges for Human Capital Formation and Sustainable Development* (Earthscan, 2004).

ISSUE 14

Is Genetic Engineering the Answer to Hunger?

YES: Gerald D. Coleman, from "Is Genetic Engineering the Answer to Hunger?" *America* (February 21, 2005)

NO: Sean McDonagh, from "Genetic Engineering Is Not the Answer," *America* (May 2, 2005)

ISSUE SUMMARY

YES: Gerald D. Coleman argues that genetically engineered crops are useful, healthful, and nonharmful, and although caution may be justified, such crops can help satisfy the moral obligation to feed the hungry.

NO: Sean McDonagh argues that those who wish to feed the hungry would do better to address land reform, social inequality, lack of credit, and other social issues.

In the early 1970s scientists first discovered that it was technically possible to move genes—the biological material that determines a living organism's physical traits—from one organism to another and thus (in principle) to give bacteria, plants, and animals new features. Most researchers in molecular genetics were excited by the potentialities that suddenly seemed within their reach. However, a few researchers—as well as many people outside the field—were disturbed by the idea; they thought that genetic mix-and-match games might spawn new diseases, weeds, and pests. Some people even argued that genetic engineering should be banned at the outset, before unforeseeable horrors were unleashed. Researchers in support of genetic experimentation responded by declaring a moratorium on their own work until suitable safeguards (in the form of government regulations) could be devised.

A 1987 National Academy of Sciences report said that genetic engineering posed no unique hazards. And, despite continuing controversy, by 1989 the technology had developed tremendously: researchers could obtain patents for mice with artificially added genes ("transgenic" mice); firefly genes had been added to tobacco plants to make them glow (faintly) in the dark; and growth hormone produced by genetically engineered bacteria was being used to grow low-fat pork and increase milk production in cows. The growing biotechnology industry promised more productive crops that made their own fertilizer and pesticide. Proponents argued that genetic engineering was in no significant way different from traditional selective breeding. Critics argued that genetic engineering was unnatural and violated the rights of both plants and animals to their "species integrity"; that expensive, high-tech, tinkered animals gave the competitive advantage to big agricultural corporations and drove small farmers out of business; and that putting human genes into animals, plants, or bacteria was downright offensive. Trey Popp, "God and the New Foodstuffs," *Science & Spirit* (March/April 2006), discusses objections to genetically modified rice containing human genes. For a summary of events related to the development of agricultural biotechnology, see "Biotechnology Timeline: Chronology of Key Events," *International Debates* (March 2006).

In 1992 the U.S. Office of Science and Technology issued guidelines to bar regulations that are based on the assumption that genetically engineered crops pose greater risks than similar crops produced by traditional breeding methods. The result was the rapid commercial introduction of crops that were genetically engineered to make the bacterial insecticide Bt and to resist herbicides and disease, among other things. In 2003, some 70 engineered crop varieties were grown on over 68 million hectares in 18 countries. Sales of genetically engineered crop products are expected to reach $25 billion by 2010.

Skepticism about the benefits remains, but in 2000, the national academies of science of the United States, the United Kingdom, China, Brazil, India, Mexico, and the third world recognized that though the use of genetically modified crops has some worrisome potentials that deserve further research and continuing caution, those crops hold the potential to feed the world during the twenty-first century while also protecting the environment (Royal Society of London, et al., "Transgenic Plants and World Agriculture," A Report Prepared Under the Auspices of the Royal Society of London, the U.S. National Academy of Sciences, the Brazilian Academy of Sciences, the Chinese Academy of Sciences, the Indian National Science Academy, the Mexican Academy of Sciences, and the Third World Academy of Sciences, July 2000; http://fermat.nap.edu/html/transgenic/). The following selections illustrate the different current perspectives from within the Roman Catholic Church. Gerald D. Coleman argues that genetically engineered crops are useful, healthful, and nonharmful, and though caution may be justified, such crops can help satisfy the moral obligation to feed the hungry. Sean McDonagh, a priest who writes frequently on environmental matters, argues that ethical and environmental problems connected to the implementation of genetically engineered crops stand in the way of their promise. Those who wish to feed the hungry would do better to address land reform, social inequality, lack of credit, and other social issues.

YES

Gerald D. Coleman

Is Genetic Engineering the Answer to Hunger?

Both the developed and developing worlds are facing a critical moral choice in the controversial issue of genetically modified food, also known as genetically modified organisms and genetically engineered crops. Critics of these modifications speak dismissively of biotech foods and genetic pollution. On the other hand, proponents like Nina Federoff and Nancy Marie Brown, authors of *Mendel in the Kitchen: A Scientist's View of Genetically Modified Foods* (2004), promote genetically modified organisms (GMs or GMOs) as "the miracle of seed science and fertilizers."

To mark the 20th anniversary of U.S. diplomatic relations with the Holy See, the U.S. Embassy to the Holy See, in cooperation with the Pontifical Academy of Sciences, hosted a conference last fall at Rome's Gregorian University on "Feeding a Hungry World: The Moral Imperative of Biotechnology." Archbishop Renato Martino, who heads the Pontifical Council for Justice and Peace and has been a strong and outspoken proponent of GMOs, told Vatican Radio: "The problem of hunger involves the conscience of every man. For this reason the Catholic Church follows with special interest and solicitude every development in science to help the solution of a plight that affects . . . humanity."

Americans have grown accustomed, perhaps unwittingly, to GMO products. In the United States, for example, 68 percent of the soybeans, 70 percent of the cotton crop, 26 percent of corn and 55 percent of canola are genetically engineered. GMOs represent an estimated 60 percent of all American processed foods. A recent study by the National Center for Food and Agriculture found that farmers in the United States investing in biotech products harvested 5.3 billion additional pounds of crops and realized $22 billion in increased income. Most of the world's beer and cheese is made with GMOs, as are hundreds of medications. In an article published last October, James Nicholson, then U.S. Ambassador to the Vatican and an aggressive promoter of U.S. policy in Vatican circles, wrote that "millions of Americans, Canadians, Australians, Argentines and other people have been eating genetically modified food for nearly a decade—without one proven case of an illness, allergic reaction or even the hiccups. . . . Mankind has been genetically altering food throughout human history." And according to its supporters, biotechnology helps the environment by reducing the use of pesticides and tilling.

The World Health Association recently reported that more than 3.7 billion people around the world are now malnourished, the largest number in history. To this, opponents of GMOs reply that the "real problems" causing hunger, especially in the developing world, are poverty, lack of education and training, unequal land distribution and lack of access to markets. The moral point they advance is that distribution, not production, is the key to solving hunger.

Another significant moral issue relates to "intellectual property policies" and the interest of companies in licensing potentially valuable discoveries. The Rev. Giulio Albanese, head of the Missionary News Agency, insists that unless the problem of intellectual property is resolved in favor of the poor, it represents a "provocation" to developing countries: "The concern of many in the missionary world over the property rights to GM seeds . . . cannot but accentuate the dependence of the poor nations on the rich ones." In response to this concern, a proposal was made recently (reported in *Science* magazine on March 19) that research universities cooperate to seek open licensing provisions that would allow them to share their intellectual property through a "developing-country license." Universities would still retain rights for research and education and maintain negotiating power with the biotechnology and pharmaceutical industries. Catholic social ethics would support this type of proposal, since it places the good of people over amassing profit.

Three moral paths suggest themselves:

1. *Favor the use of GMOs.* Nobel Prize winner Norman Borlaug, who developed the Green Revolution wheat and rice strains, recently wrote: "Biotechnology absolutely should be part of Africa's agricultural reform. African leaders would be making a grievous error if they turn their back on it." Proponents at the Rome conference agreed, arguing that the use of GMOs decreases pesticide use, creates more nutrient-filled crops that require less water and have greater drought resistance, produces more food at a lower cost and uses less land. One small-scale South African farmer concluded, "We need this technology. We don't want always to be fed food aid. We want access to this technology so that one day we can also become commercial farmers."

   This position concludes that the use of GMOs amounts to a moral obligation.

2. *Condemn the use of GMOs.* Many Catholic bishops take an opposing stance. Perhaps the clearest statement comes from the National Conference of Bishops of Brazil and their Pastoral Land Commission. Their argument is threefold: the use of GMOs involves potential risks to human health; a small group of large corporations will be the greatest beneficiaries, with grave damage to the family farmers; and the environment will be gravely damaged.

The bishops of Botswana, South Africa and Swaziland agree: "We do not believe that agro-companies or gene technologies will help our farmers to produce the food that is needed in the 21st century." Roland Lesseps and Peter Henriot, two Jesuits working in Zambia who are experts on agriculture in the developing world, state their opposition on principle: "Nature is not just useful to us as humans, but is valued and loved in itself, for itself, by God in Christ. . . . The right to use other creatures does not give us the right to abuse them."

In a similar but distinct criticism, the executive director of the U.S. National Catholic Rural Conference, David Andrews, C.S.C., feels that "the Pontifical Academy of Sciences has allowed itself to be subordinated to the U.S. government's insistent advocacy of biotechnology and the companies which market it." Sean McDonagh states: "With patents [on genetically engineered food], farmers will never own their own food. . . ." He believes that "corporate greed" is at the heart of the GMO controversy. Biowatch's Elfrieda Pschorn-Strauss agrees: "With GM crops, small-scale farmers will become completely reliant on and controlled by big foreign companies for their food supply."

This position concludes that the use of GMOs is morally irresponsible.

3. *Approach the use of GMOs with caution.* Two years ago Pope John Paul II declared that GMO agriculture could not be judged solely on the basis of "short-term economic interests," but needed to be subject to "a rigorous scientific and ethical process of verification." This cautionary stance has been adopted by the Catholic Bishops Conference of the Philippines in urging its government to postpone authorization of GMO corn until comprehensive studies have been made: "We have to be careful because, once it is there, how can we remedy its consequences?"

In 2003 the Rural Life Committee of the North Dakota Conference of Churches also called for "rigorous examination" to understand fully the outcomes of the use of GMOs. This document endorses the "Precautionary Principle" formulated in 1992 by the United Nations Conference on Environment and Development in order to avoid "potential harm and unforeseen and unintended consequences."

This view mandates restraint and places the fundamental burden on demonstrating safety. The arguments are based on three areas of concern: the impact on the natural environment, the size of the benefit to the small farmer if the owners and distributors are giant companies like Bristol-Myers and Monsanto and the long-term effects of GMOs on human and animal health and nutrition.

This position concludes that the use of GMOs should be approached with caution.

While the "Precautionary Principle" seems prudent, there is simultaneously a strong moral argument that a war on hunger is a grave, universal need. Last year, 10 million people died of starvation. Every 3.6 seconds someone dies from hunger—24,000 people each day. Half of sub-Saharan

Africans are malnourished, and this number is expected to increase to 70 percent by 2010. It was a moral disgrace that in 2002 African governments gave in to GMO opponents and returned to the World Food Program tons of GMO corn simply because it was produced in the U.S. by biotechnology.

The Roman conference gives solid reasons that GMOs are useful, healthful and nonharmful. After all, organisms have been exchanging genetic information for centuries. The tomato, corn and potato would not exist today if human engineering had not transferred genes between species.

The *Catechism of the Catholic Church* teaches that we have a duty to "make accessible to each what is needed to lead a truly human life." The very first example given is food. In *Populorum Progressio* (1967), *Sollicitude Rei Socialis* (1987) and *Centesimus Annus* (1991), Paul VI and later John Paul II forcefully insisted that rich countries have an obligation to help the poor, just as global economic interdependence places us on a moral obligation to be in solidarity with poor nations. Likewise, The Challenge of Faithful Citizenship, published by the U.S. bishops in 2004, argues that the church's preferential option for the poor entails "a moral responsibility to commit ourselves to the common good at all levels."

At the same time, it is critical that farmers in developing countries not become dependent on GMO seeds patented by a small number of companies. Intellectual knowledge must be considered the common patrimony of the entire human family. As the U.S. bishops have stated, "Both public and private entities have an obligation to use their property, including intellectual and scientific property, to promote the good of all people" (*For I Was Hungry and You Gave Me Food, 2003*).

The Catholic Church sees deep sacramental significance in wheat and bread, and insists on the absolute imperative to feed and care for the poor of the world. A vital way to promote and ensure the dignity of every human being is to enable them to have their daily bread.

Sean McDonagh  **NO**

Genetic Engineering
Is Not the Answer

In 1992 the then-chief executive of Monsanto, Robert Shapiro, told the *Harvard Business Review* that genetically modified crops will be necessary to feed a growing world population. He predicted that if population levels were to rise to 10 billion, humanity would face two options: either open up new land for cultivation or increase crop yields. Since the first choice was not feasible, because we were already cultivating marginal land and in the process creating unprecedented levels of soil erosion, we would have to choose genetic engineering. This option, Shapiro argued, was merely a further improvement on the agricultural technologies that gave rise to the Green Revolution that saved Asia from food shortages in the 1960s and 1970s.

Genetically engineered crops might seem an ideal solution. Yet both current data and past examples show problems and provoke doubts as to their necessity.

The Green Revolution

The Green Revolution involved the production of hybrid seeds by crossing two genetically distant parents, which produced an offspring plant that gave increased yield. Critics of genetic engineering question the accepted wisdom that its impact has been entirely positive. Hybrid seeds are expensive and heavily reliant on fertilizers and pesticides. And because they lose their vigor after the first planting, the farmer must purchase new seeds for each successive planting.

In his book *Geopolitics and the Green Revolution*, John H. Perkins describes the environmentally destructive and socially unjust aspects of the Green Revolution. One of its most important negative effects, he says, is that it has contributed to the loss of three-quarters of the genetic diversity of major food crops and that the rate of erosion continues at close to 2 percent per year. The fundamental importance of genetic diversity is illustrated by the fact that when a virulent fungus began to destroy wheat fields in the United States and Canada in 1950, plant breeders staved off disaster by cross-breeding five Mexican wheat varieties with 12 imported ones. In the process they created a new strain that was able to resist so-called "stem rust." The loss of these varieties would have been a catastrophe for wheat production globally.

The Terminator Gene

The development by a Monsanto-owned company of what is benignly called a Technology Protection System—a more apt name is terminator technology—is another reason for asserting that the feed-the-world argument is completely spurious. Genetically engineered seeds that contain the terminator gene self-destruct after the first crop. Once again, this forces farmers to return to the seed companies at the beginning of each planting season. If this technology becomes widely used, it will harm the two billion subsistence farmers who live mainly in the poor countries of the world. Sharing seeds among farmers has been at the very heart of subsistence farming since the domestication of staple food crops 11,000 years ago. The terminator technology will lock farmers into a regime of buying genetically engineered seeds that are herbicide tolerant and insect resistant, tethering them to the chemical treadmill.

On an ethical level, a technology that, according to Professor Richard Lewontin of Harvard University, "introduces a 'killer' transgene that prevents the germ of the harvested grain from developing" must be considered grossly immoral. It is a sin against the poor, against previous generations who freely shared their knowledge of plant life with us, against nature itself and finally against the God of all creativity. To set out deliberately to create seeds that self-destruct is an abomination no civilized society should tolerate. Furthermore, there is danger that the terminator genes could spread to neighboring crops and to the wild and weedy relatives of the plant that has been engineered to commit suicide. This would jeopardize the food security of many poor people.

The current situation promoting genetically modified organisms also means supporting the patenting of living organisms—both crops and animals. I find it difficult to understand the support that Cardinal Renato Martino, prefect of the Pontifical Council for Justice and Peace, seems to be giving to genetically modified organisms, given the Catholic Church's strong pro-life position. In my book *Patenting Life? Stop!* I argue that "patenting life is a fundamental attack on the understanding of life as interconnected, mutually dependent and a gift of God to be shared with everyone. Patenting opts for an atomized, isolated understanding of life." The Indian scientist and activist Dr. Vandana Shiva believes that patented crops will lead to food dictatorship by a handful of northern transnational corporations. This would certainly be a recipe for hunger and starvation—in conflict with Catholic social teaching on food and agriculture.

No Higher Yield, No Reduction in Chemicals

Early in 2003 a researcher at the Institute of Development Studies at Sussex University in England published an analysis of the GMO crops that biotech companies are developing for Africa. Among the plants studied were cotton, maize and the sweet potato. The GMO research on the sweet potato is now approaching its 12th year and has involved the work of 19 scientists; to date it has cost $6 million. Results indicated that yield has increased by 18 percent. On the other hand, conventional sweet potato breeding, working with a small budget, has produced a virus-resistant variety with a 100 percent yield increase.

Claims that GMOs lead to fewer chemicals in agriculture are also being challenged. A comprehensive study using U.S. government data on the use of chemicals on genetically engineered crops was carried out by Charles Benbrook, head of the Northwest Science and Environmental Policy Center in Sandpoint, Idaho. He found that when GMOs were first introduced, they needed 25 percent fewer chemicals for the first three years. But in 2001, 5 percent more chemicals were sprayed compared with conventional crop varieties. Dr. Benbrook stated: "The proponents of biotechnology claim GMO varieties substantially reduce pesticide use. While true in the first few years of widespread planting, it is not the case now. There's now clear evidence that the average pound of herbicide applied per acre planted to herbicide-tolerant varieties have increased compared to the first few years."

Toward a Solution

Hunger and famine around the world have more to do with the absence of land reform, social inequality, bias against women farmers and the scarcity of cheap credit and basic agricultural tools than with lack of agribusiness super-seeds. This fact was recognized by those who attended the World Food Summit in Rome in November 1996. People are hungry because they do not have access to food production processes or the money to buy food. Brazil, for example, is the third largest exporter of food in the world, yet one-fifth of its population, over 30 million people, do not have enough food to eat. Clearly hunger there is not due to lack of food but to the unequal distribution of wealth and the fact that a huge number of people are landless.

Do the proponents of genetically engineered food think that agribusiness companies will distribute such food free to the hungry poor who have no money? There was food in Ireland during the famine in the 1840s, for example, but those who were starving had no access to it or money to buy it.

As a Columban missionary in the Philippines, I saw something similar during the drought caused by El Niño in 1983. There was a severe food shortage among the tribal people in the highlands of Mindanao. The drought destroyed their cereal crops, and they could no longer harvest food in the tropical forest because it had been cleared during the previous decades. Even during the height of the drought, an agribusiness corporation was exporting tropical fruit from the lowlands. There was also sufficient rice and corn in the lowlands, but the tribal people did not have the money to buy it. Had it not been for food aid from nongovernmental organizations, many of the tribal people would have starved.

In 1990 the World Food Program at Brown University calculated that if the global food harvests over the previous few years were distributed equitably among all the people of the world, it could provide a vegetarian diet for over 6 billion people. In contrast, a meat-rich diet, favored by affluent countries and currently available to the global elite, could feed only 2.6 billion people. Human society is going to be faced with the option of getting protein from plants or from animals. If we opt for animal protein, the consequence will be a much less equitable world, with increasing levels of human misery.

Those who wish to banish hunger should address the social and economic inequalities that create poverty and not claim that a magic-bullet technology will solve all the problems.

POSTSCRIPT

Is Genetic Engineering the Answer to Hunger?

Kathleen Hart, in *Eating in the Dark* (Pantheon, 2002), expresses horror at the fact that "Frankenfood" is not labeled and that U.S. consumers are not as alarmed by genetically modified foods as European consumers are. The worries—and the scientific evidence to support them—are summarized by Kathryn Brown, in "Seeds of Concern," and Karen Hopkin, in "The Risks on the Table," *Scientific American* (April 2001). In the same issue, Sasha Nemecek poses the question "Does the World Need GM Foods?" to two prominent figures in the debate: Robert B. Horsch, vice president of the Monsanto Corporation and recipient of the 1998 National Medal of Technology for his work on modifying plant genes, says yes; Margaret Mellon, of the Union of Concerned Scientists, says no, adding that much more work needs to be done with regard to safety. The May 2002 U.S. General Accounting Office Report to Congressional Requesters, "Genetically Modified Foods: Experts View Regimen of Safety Tests as Adequate, but FDA's Evaluation Process Could Be Enhanced," urges more attention to verifying safety testing performed by biotechnology companies. Carl F. Jordan, in "Genetic Engineering, the Farm Crisis, and World Hunger," *Bioscience* (June 2002), says that a major problem is already apparent in the way agricultural biotechnology is widening the gap between the rich and the poor.

Charles Mann, in "Biotech Goes Wild," *Technology Review* (July/August 1999), discusses the continuing "lack of a rigorous regulatory framework to sort out the risks inherent in agricultural biotech." Margaret Kriz, in "Global Food Fight," *National Journal* (March 4, 2000), describes the January 2000 Montreal meeting, in which representatives of 130 countries reached "an agreement that requires biotechnology companies to ask permission before importing genetically altered seeds [and] forces food companies to clearly identify all commodity shipments that may contain genetically altered grain." Also see "Environmental Effects of Transgenic Plants: The Scope and Adequacy of Regulation," a report of the National Research Council's Committee on Environmental Impacts Associated With Commercialization of Transgenic Crops (National Academy Press, 2002).

Gregory Conko and C. S. Prakash, in "The Attack on Plant Biotechnology," in Ronald Bailey, ed., *Global Warming and Other Eco-Myths: How the Environmental Movement Uses False Science to Scare Us to Death* (Prima, 2002), say that genetically engineered crops have successfully increased yields and decreased pesticide usage, have not had notable bad environmental side effects, and will be essential for feeding the world's growing population. Lee Silver, "Why GM Is Good for Us," *Newsweek (Atlantic Edition)* (March 20, 2006), reports that genetically modified foods may actually have fewer allergy-related problems than organic foods. On the other hand, Jeffrey M. Smith, "Frankenstein Peas," *The Ecologist* (April 2006), reports on a study that genetically modified peas fed to mice caused serious immune system problems.

The UN Food and Agriculture Organization's 2004 annual report, *The State of Food and Agriculture 2003-2004*, FAO Agriculture Series No. 35 (Rome, 2004), maintains that the biggest problem with genetically engineered crops is that the technology has focused so far on crops of interest to large commercial firms. GM crops have not spread fast enough to small farmers, although where they have been introduced into developing countries, they have yielded economic gains and reduced the use of toxic chemicals. The report concludes that there have been no adverse health or environmental consequences so far. Continued safety will require more research and governmental regulation and monitoring. Jerry Cayford notes in "Breeding Sanity into the GM Food Debate," *Issues in Science and Technology* (Winter 2004) that the issue is one of social justice as much as it is one of science. Who will control the world's food supply? Which philosophy—democratic competition or technocratic monopoly—will prevail? Meanwhile, researchers are expanding the technology to turn crop plants into "bioreactors," GM crops that produce medically important proteins; see Deborah A. Fitzgerald, "Revving up the Green Express," *The Scientist* (July 14, 2003).

ISSUE 15

Are Marine Reserves Needed to Protect Global Fisheries?

YES: Robert R. Warner, from "Marine Protected Areas," Statement Before the Subcommittee on Fisheries Conservation, Wildlife and Oceans Committee on House Resources, United States House of Representatives (May 23, 2002)

NO: Michel J. Kaiser, from "Are Marine Protected Areas a Red Herring or Fisheries Panacea?" *Canadian Journal of Fisheries and Aquatic Sciences* (May 2005)

ISSUE SUMMARY

YES: Professor of marine ecology Robert R. Warner argues that marine reserves, areas of the ocean completely protected from all extractive activities such as fishing, can be a useful tool for preserving ecosystems and restoring productive fisheries.

NO: Professor Michel J. Kaiser argues that although the use of marine protected areas can be beneficial, limiting fishing effort is a more effective way of achieving sustainable fisheries.

Carl Safina called attention to the poor state of the world's fisheries in "The World's Imperiled Fish," *Scientific American* (November 1995) and "Where Have All the Fishes Gone?" *Issues in Science and Technology* (Spring 1994). Expanding population, improved fishing technology, and growing demand had combined to drive down fish stocks around the world. Fishers going further from shore and deploying larger nets kept the catch growing, but the UN's Food and Agriculture Organization (FAO) had noted that the fisheries situation was already "globally non-sustainable, and major ecological and economic damage [was] already visible."

The UN declared 1998 the International Year of the Ocean. Kieran Mulvaney, in "A Sea of Troubles," *E Magazine* (January/February 1998), reported, "According to the United Nations Food and Agriculture Organization (FAO), an estimated 70 percent of global fish stocks are 'over-exploited,' 'fully exploited,' 'depleted' or recovering from prior over-exploitation. By 1992,

FAO had recorded 16 major fishery species whose global catch had declined by more than 50 percent over the previous three decades—and in half of these, the collapse had begun after 1974." Ocean fishing is not sustainable, concluded Mulvaney.

Daniel Pauly and Reg Watson, in "Counting the Last Fish," *Scientific American* (July 2003), note that desirable fish tend to be top predators, such as tuna and cod. When the numbers of these fish decline due to overfishing, fishers shift their attention to fish lower on the food chain, and consumers see a change in what is available at the market. The cod are smaller, and monkfish and other once less-desirable fish join them on the crushed ice at the market. Such a change is an indicator of trouble in the marine ecosystem.

Responses to the situation have included government buyouts of fishing fleets and closures of fisheries such as the Canadian cod fishery. But the situation has not improved. David Helvarg, in "The Last Fish," *Earth Island Journal* (Spring 2003), concludes that about half of America's commercial seafood species are now overfished. Globally, the figure is still over 70 percent. And the North Atlantic contains only one third as much biomass of commercially valuable fish as it did in the 1950s. Helvarg recommends more buying out of excess fishing capacity; limiting the number of people allowed to enter the fishing industry; creating reserves; and perhaps most importantly, taking fisheries management out of the hands of people with a vested interest in the status quo.

In June 2000, the independent Pew Oceans Commission undertook the first national review of ocean policies in more than 30 years. Its report, *America's Living Oceans: Charting a Course for Sea Change* (Pew Oceans Commission, 2003), available at http://www.pewoceans.org/, notes that many commercially fished species are in decline; North Atlantic cod, haddock, and yellowtail flounder reached historic lows in 1989. The reasons include intense fishing pressure to feed demand for seafood; pollution; coastal development; fishing practices, such as bottom dragging, that destroy habitat; and fragmented ocean policy that makes it difficult to prevent or control the damage. The answers, the commission suggests, must have clear benefits for commercial fishing and include such practices as the no-fishing zones known as marine protected areas, which have been shown to restore habitat and fish populations.

In the following selections, Robert R. Warner argues that marine reserves, areas of the ocean completely protected from all extractive activities such as fishing, lead quickly to increased abundance and size of most species—especially exploited species—even outside the borders of the reserves. He insists that marine reserves can be a useful tool for preserving ecosystems and restoring productive fisheries. Professor Michel J. Kaiser argues that although the use of marine protected areas can be beneficial for some fish species, using them for wide-ranging species such as cod requires very large protected areas and is vulnerable to the same political forces as other management approaches. Limiting fishing effort is a more effective way of achieving sustainable fisheries.

Marine Protected Areas

Statement of Robert R. Warner, Professor of Marine Ecology, University of California, Santa Barbara

Before the Subcommittee on Fisheries Conservation, Wildlife and Oceans Committee on House Resources

United States House of Representatives

May 23, 2002

...We depend on ocean life in many ways, far beyond the 80 million metric tons of food that we draw from the sea each year. The ecologist Stuart Pimm recently estimated that marine ecosystem services have a value of $20 trillion, with most of that being provided from coastal ecosystems. Yet these ecosystems have been altered dramatically over the past decades—in some places, they have essentially collapsed. Many of the fisheries of the world are depleted, and the species we catch are getting smaller and further down the food chain. The problems of habitat alteration, pollution, aquaculture, exotic species, and climate change all converge on the species that make up ecosystems, and effects on one species can severely affect others. For example, in Hawaii, nutrient pollution fuels algal growth, and fishing removes the fishes that eat the algae, and corals die underneath the encroaching seaweeds. In every marine ecosystem that one of the NCEAS [National Center for Ecological Analysis and Synthesis] working groups investigated, there was clear evidence of fundamental change and loss of resources, and these losses are accelerating. Ecosystem health is often measured in terms of productivity and species diversity, and it is precisely these measures that are declining in many coastal habitats.

Entire marine ecosystems are affected by threats at many levels, and evaluating and responding to these threats in an integrated fashion is the challenge we currently face. Let me make this clear: there is a real need to shift our attention to ecosystems—based management of the marine environment, away from the confusing and often conflicting mass of single-species management plans. On the West coast, there are 88 species that generate more than $1 million a year in fisheries revenue. In New England, there are 41 such species, and in both areas invertebrates like urchins, squid, and lobsters are the most valuable resources. Multiple overlapping single-species management plans can

From the United States House of Representatives, May 23, 2002.

become cumbersome and difficult. A complementary approach to this problem is a scheme of ecosystem-based management.

Marine reserves, areas of the ocean completely protected from all extractive activities, can be a useful tool for ecosystem-based management. They cannot solve all of the problems of the coastal ocean, but they can stop habitat alteration and allow the recovery of depleted populations of several species at a time. Reserves are a place- and habitat-based approach to management, distinctly different from single-species management.

Because much of the sea is hidden from our view, and because the ocean is so vast, we have not been as aware of changes in marine ecosystems as we are of terrestrial changes. On land, many of the larger animals went extinct soon after humans arrived on the scene, and commercial hunting disappeared at the turn of the last century. In the sea, many of the large animals are rare but still present, and harvesting of wild animals continues at high levels. There is hope in this fact—it may be possible to restore marine ecosystems in some places to conditions approaching their former glory, because most of the key players are still present. This is a chance to do more than build a small monument to what existed before. We have a much more rewarding goal: rebuilding coastal ecosystems and recharging coastal fisheries. This is one of the few instances where we can combine benefit to both the extractive users and to the conservation community. It can be done.

The simplest question to ask is what happens when reserves are established. That is, can we document the effect of reserves on coastal ecosystems?

Documented Responses of Animals and Plants to Protection Inside Reserves

The overall coastal area currently under full protection in marine reserves is less than a fraction of one per cent. Although reserves are rare in the US, several have been the subject of careful study. The NCEAS working group summarized these studies and scores of other peer-reviewed reports of the responses of animals and plants to reserve protection around the world. The results were striking. Regardless of whether the reserve was in the tropics or in temperate waters, there was strong evidence that reserves function to increase the abundance and size of many species within their borders. On average, population sizes of animals nearly double, and the animals themselves average about 30% larger. This means that the biomass (or capacity for production) of these species showed a dramatic increase, at least doubling regardless of the location of the reserves.

Not surprisingly, it is exploited species that show the strongest positive response to protection, including species thought to be too mobile to benefit from reserve protection. But I want to stress that the changes seen inside reserves are ecosystem-level changes—not just the recovery of exploited species. For example, when reserves were established in New Zealand, the increase in lobsters resulted in a major decrease in sea urchins, the lobster's prey. This, in turn allowed kelp beds to flourish (because urchins eat kelp), and the overall productivity of the area has increased.

When year-round area closures were instituted on the Georges Bank to aid in the recovery of cod and other finfish, it was scallops that responded the most quickly, becoming unbelievably abundant inside the closed areas. Thus many species can be simultaneously affected by any particular closure.

Responses occurred in reserves of all sizes, and they appear rather quickly—reserves only two to four years old showed increased levels of animal abundance and size equivalent to reserves that had been established for decades.

As I mentioned previously, not all species increase inside reserves, but the great majority show a strong positive response. Neither will all species show a rapid response, especially those that are long-lived and slow-growing. However, the overwhelming result from over 20 years of studies is that species recover within reserve borders, becoming more numerous and larger. Although local conditions may affect the exact result in any particular place, the value of reserves in generating broad changes within their boundaries has been demonstrated in scores of well-documented studies in virtually all settings. This is good news for ecosystem-based management.

While reserves cannot stop pollution, prevent catastrophes, or slow the arrival of exotic invaders into marine ecosystems, they can help to withstand these threats simply because they contain larger populations and more species. Many studies have shown that healthy ecosystems are more resilient to chronic or acute threats, and species-rich ecosystems are more resistant to invasion.

Effects Outside of Reserve Borders

While the major role envisioned for reserves is the protection of habitats and ecosystems, there is added benefit if they export some of their population to surrounding areas. This function is particularly important when reserves are viewed as a fishery management tool, because this export could be used to replenish species subject to harvest in non-reserve areas.

The large variety of life histories, movement patterns, and time spent as a planktonic, drifting larva means that spillover will occur differently for different species. There are so few reserves established, and most of them are so small, that there have been relatively few studies done on spillover. Nevertheless, the evidence is compelling that reserves can recharge nearby areas.

Spillover can take two forms. The first is simple movement of adult animals out of reserves. Several studies have shown that numbers and sizes of species are greater in areas near reserve boundaries, and other studies have shown that the catches of fishers near reserves are higher than in other areas. Fishermen may not have read these studies, but they often know where the fish are, and this has led to concentrations of recreational and commercial fishing activity along reserve borders, an activity known as "fishing the line."

The other major potential contribution of marine reserves to fisheries is through larval export. Most marine species produce tiny young that drift in the water for days or weeks. We know that the rate of production of young by animals inside reserves can be tremendous—at the Edmonds Underwater Park in Washington, for example, it is estimated that the large lingcod there produce 20 times as many young than are produced in equivalent areas outside. But do some of these young make their way into the fishery? There has been little documentation of the effects of larval spillover, mostly because reserves are simply too small to have much effect. The Edmonds reserve, for example, is only 25 acres in extent, a tiny fraction of the area over which the larvae produced there could be expected to drift.

In one US example of a marine reserve large enough to have the potential to recharge fisheries through larval export, this apparently has occurred. On the Georges Bank, several large areas were set aside in 1994 to preserve cod and other groundfish, and as I have mentioned the strongest response so far has been in the fast-growing scallops. By 1998 scallops were 14 times more dense in the protected areas than outside, and dense settlement of young was predicted in downcurrent areas near the reserves. These areas are in fact now yielding higher catches than other areas, and overall revenues have increased from $91 million in 1995 to $123 million in 1999.

Reserve Size and Reserve Networks

A common perception is that conservation and fishery objectives for marine reserves are incompatible, and there will be inevitable conflict between these competing interests. That is certainly what appears to be happening at this point, but models of reserve function suggest that this need not be so. It is true that the larger the reserve, the more species will be able to complete their entire life cycles inside reserves. A reserve too small will not be self-sustaining because most larvae produced in it will be transported elsewhere, and thus a small reserve needs to be seeded from a fished area. Very large reserves, on the other hand, leave little area left to in which to fish.

Most single-species fisheries models of reserves suggest that the most substantial impacts on yields occur when between 20 and 50% of the area is set aside. The amount of area required in reserves varies, but few models show significant benefit at levels below 10%. The more depleted the fishery is on the outside, the more substantial the benefit from reserves.

Where does this leave conservation interests? To what extent can setasides at this level work to rebuild ecosystems? Fortunately, the most recent scientific findings have suggested a solution: networks of smaller reserves. While these reserves may individually be too small for self-seeding, they are close enough together so that one reserve can seed another. In addition, networks can provide high amounts of spillover into fished areas because they have extensive borders, and networks can boost regional production of young as long as the aggregate area in reserves is sufficiently large.

Studies also suggest networks of reserves can provide additional protection against catastrophic loss (because we're not putting all of our eggs in one basket), and they may make reserve siting easier and more flexible because there are simply more options available.

Where to Put a Marine Reserve?

Recent scientific work on the criteria for siting marine reserves has emphasized that in any management area, there are many different reserve designs that might fit the biological needs of the protected community. That is, science can suggest a range of options that can then be evaluated for other criteria, like their social, economic, or political impact. This flexibility is good news for the process of establishing marine reserves, because it can include input from many different sectors of the community in forming the final decisions.

The most important criterion for designing reserves is to include representation of all habitat types within an area, preferably adjacent to one another, simply because many species use different habitats over the course of their lives. A common misconception is that reserves should be placed in the areas of best fishing. In fact, reserves should show the best response in areas that were formerly productive but are currently overfished—protection can allow these areas of proven potential to recover.

Conclusions

I realize that much of the regulatory process is constrained by mandated consideration of one species at a time. However, the solution to managing multiple threats to the oceans requires an integrated approach that includes the need to preserve intact marine ecosystems on a regional basis. Single species management is not sufficient for the future, especially since many fisheries already affect many different species through by-catch.

Marine reserves are one of the best tools we have to address management of entire marine ecosystems. While they are not the solution to every problem facing the coastal ocean, they can stem habitat destruction, alleviate the effects of local overfishing, simplify the simultaneous management of multiple species, and restore biodiversity within their borders. The healthier ecosystems inside reserves can be more resistant to threats from the outside, and more resilient in their recovery. A regional network of marine reserves may be the best solution for the broad enhancement of coastal ecosystems, with substantial contributions to biodiversity and recruitment of young both inside and outside their borders. While reserves are ideal tools for habitat protection and ecosystem preservation, they are best used as a complement to traditional fisheries management.

Michel J. Kaiser

 **NO**

Are Marine Protected Areas a Red Herring or Fisheries Panacea?

. . . **M**arine protected areas (MPAs) have been heralded as the saviour of global fisheries by some conservationists, fishers, and managers and are seen as the solution to the perceived failures of current management methods. The potential benefits of excluding fishing activity from parts of the sea is an easy concept for nonspecialists to grasp, making MPAs an alluring alternative to the complex array of current management tools (Roberts et al. 2001; Gell and Roberts 2003). Here, I argue that while MPAs can and have been used with success in certain locations for particular species (Gell and Roberts 2003; Roberts et al. 2005), they are not the cure-all that some purport, and the scientific evidence chosen to support such conclusions on occasion has been drawn from those studies that demonstrate a positive outcome of MPA implementation (Halpern 2003; Zeller and Russ 2004). Needless to say, it may be equally important to understand better those circumstances when the use of MPAs has not been successful, as such instances provide useful metadata. In areas where highly mobile species are the target of the fishery, assessment of the full potential of MPAs as management tools [is] confounded by factors such as previous and concurrent fishing history, changes in size-at-age, community structure, genetic bottlenecks, stock collapse, and Allee effects (Frank et al. 2000; Willis et al. 2003*a*). Furthermore, the successful implementation of MPAs requires a crossdisciplinary approach that is far more complex than the biological conservation goals originally envisaged (Agardy et al. 2003; Ray 2004).

Much of the confusion surrounding the use of MPAs has stemmed from their intended objectives. The purpose of MPAs to conserve habitat and biodiversity of nontarget species is not necessarily consistent with the maintenance of sustainable fish stocks, although it is clear that in some cases these two goals can be met to some degree by the implementation of an MPA. The scientific community is split regarding the efficacy of the unilateral use of MPAs to achieve sustainable exploitation of fish and shellfish stocks (Steele and Hoagland 2003, 2004; Zeller and Russ 2004). The appropriate and rigorous use of fishing effort controls that consider natural, long-term fluctuations in fish populations should achieve sustainable management.

From *Canadian Journal of Fisheries and Aquatic Sciences*, vol. 62, May 2005, pp. 1194-1198, omit references and charts. Copyright © 2005 by National Research Council of Canada. Reprinted by permission.

However, advocates of MPAs counter that the current severe decline in fish stocks has eliminated the potential for population increase even under favourable environmental conditions (Zeller and Russ 2004).

The wider ecosystem effects of fishing are cited as being incompatible with the aim of sustainable harvesting (Gell and Roberts 2003). Harvesting of fish from the seabed and the collateral damage that occurs from this [have] been compared with the clear-cutting of forests to catch deer (Watling and Norse 1998). These viewpoints typically lead to the conclusion that the only way to achieve sustainable use of fisheries is to exclude fishing activities from "large" areas of the ocean (Watling and Norse 1998). However, the implementation of appropriate MPAs is fraught with ecological, economic, and social problems that tend to be overlooked (Agardy et al. 2003). Although some of these problems are shared with current effort control systems (Hilborn et al. 2004), MPAs have a greater potential to exclude access by some fishers according to their geographic location. Worse still, MPAs have the potential to displace current fishing activities and thereby cause wider ecological damage to previously undisturbed and perhaps unknown, critical seabed habitats (i.e., essential fish habitat). However, there are circumstances when the benefits of using spatial effort control measures will offset some of the negative ecological effects of the displacement of fishing activities to alternative areas. Such examples would include the use of temporary closed areas to prevent fishing on spawning aggregations, on nursery areas, or at migration bottlenecks.

Most reviews of the ecological effects of MPAs draw on case studies in which fish biomass increased within areas from which fishing was excluded (Roberts et al. 2001; Gell and Roberts 2003; Halpern 2003). In general, these case studies focused on habitat-specific systems (e.g., coral or rock reefs) that are relatively small in scale (<500 km^2; Halpern 2003). These systems are easier to protect from the effects of fishing, as the target species usually remain in close proximity to a well-delineated habitat (Russ and Alcala 1996; Willis et al. 2003b). Fishers that prosecute these systems usually live within kilometres of the specific habitat and depend upon it for a considerable proportion of their diet or income (Jennings et al. 1996; Blyth et al. 2002). Hence, their motivation to participate in MPAs is much greater compared with those who fish across areas of tens of thousands of square kilometres and who compete against fishers with whom they have little or no social contact.

A recent meta-analysis of MPAs reported general, positive biological effects that were most usually expressed as an increase in biomass within the reserve of both target and nontarget species (Halpern 2003). Interestingly, no relationship was found between the spatial extent of the MPA and the magnitude of the change in biomass or abundance of the protected target species' population (Halpern 2003). While it might be tempting to conclude that an MPA of any size would provide conservation benefits, even if only at a local scale, this holds true for only the majority of studies on which this meta-analysis was based: coral or temperate rock reefs. These conclusions are not valid for many of the commercially important, temperate fish species that are widespread across a variety of habitats, exhibit entirely different behaviours between sea basins, and may move considerable distances within a year (Metcalfe and Arnold 1997; Horwood et al. 1998; Horwood 2000). For these species, the critical and most demanding issue relating to the design of appropriate MPAs is the scale of the area required to achieve effective protection and stock enhancement of highly mobile fish stocks. Recent estimates indicate that a mean area equivalent to approximately 32% (range 10%–65%) of the area available for fishing would require protection (Gell and Roberts 2003). Note that these figures are once again based largely on studies of fish species with strong habitat-specific associations. However, for wide-ranging species such as Atlantic cod (*Gadus morhua*), calculations indicate that excluding fishing from an area as large as 25% of the North Sea would have a negligible effect on their spawning stock biomass (Horwood 2000). Given the multispecies nature of most temperate fisheries, the different MPA requirements of each species would add yet another layer of management complexity.

What would the composite map of species-specific MPAs look like and could they be implemented? These important questions are yet to be addressed; indeed, little if any consideration is given to the likely social and economic consequences for the inevitably reduced number of fishers in the system. With fishing excluded from large areas of the sea, a negotiated placement of MPAs to take into account useable fishing grounds that are accessible to all remaining participants would seem hopeful at best. In regions of the world where total control of exclusive economic zones is exercised out to 200 nautical miles, the placement of MPAs can be imposed through legislation. However, such simple solutions are not likely to be found in more complex systems, such as Europe, where multiple nations share access to the same resources. Even if such systems can be implemented, fishers can expect to wait decades before signs of stock recovery for certain stocks occur (Hutchings 2000; Steele and Beet 2003). Steele and Beet (2003) make the telling observation that the initial stages of MPA implementation would be doubly painful as fishers are deprived access to traditional fishing grounds and forced into less favourable fishing areas. This is further compounded by the added reductions in fishing effort (Zeller and Russ 2004). Furthermore, in the initial stages after MPA implementation, benthic production in the newly protected areas is likely to be low as it enters a phase of recovery once fishing disturbance has been removed, while those areas newly affected by displaced demersal effort will experience a severe decline in production varying according to habitat type (Jennings et al. 2001; Dinmore et al. 2003; J. Hiddink, School of Ocean Sciences, University of Wales-Bangor, Menai Bridge, Anglesey LL59 5AB, UK, unpublished data).

Given the unknown social and variable predicted economic consequences of a switch to management based on MPAs (Sanchiro and Wilen 2001), one has to ask what is supposedly wrong with current fishery management techniques? I would argue that for many temperate species there is nothing fundamentally wrong with either the methodology of stock assessment or the available effort controls (e.g., days at sea) used in conjunction with catch controls, such as total allowable catches and individual transferable quotas. Indeed, when fishing effort control has been implemented effectively, fisheries have been managed successfully (e.g., the Western Australian rock lobster (*Palinurus cygnus*), New Zealand hoki (*Macruronus novaezealandiae*), Alaska salmon (*Oncorhynchus* spp.), and the Thames herring (*Clupea harengus*) fisheries, all of which have achieved Marine Stewardship Council certification as sustainable fisheries).

So why have other fisheries failed so dismally? Current fisheries management depends on annual forecasts of recruitment and catches that incorporate significant uncertainty. The political need to appease a desperate fishing industry has tended to push management decisions towards, and often beyond, the upper confidence limits for future allowable catches (Hutchings and Myers 1995). Such behaviour is not surprising given the typical short life-span of the average government, for whom the shadow of the future does not apply (Hart 1998). Do proponents of MPAs believe that scientific debate and uncertainty over the precise size, position, and configuration of MPAs will not provide the very same pathway to politically compromised management? If we ultimately move to a management system that is underpinned by a system of MPAs, we will still have to set a figure of how much and where. As soon as such figures are set, they will be vulnerable to erosion by a system of political negotiation. Given the effort that will be required to implement these systems, there is likely to be considerable resistance to alter their configuration should we decide at some point in the future that management objectives or large-scale environmental changes dictate their reconfiguration.

Furthermore, it is often stated that we require a portfolio of both MPAs and effort reduction to achieve sustainable use of fish stocks (Zeller and Russ 2004). If the latter is a prerequisite for the successful implementation of the former, the prognosis does not look too promising. If fishers are displaced by the imposition of an MPA from favoured fishing grounds into areas where catch rates of legal-sized target species are lower, the use of quota management is likely to result in a considerable increase in bycatch as fishers work harder (spend more time fishing at sea) to land the same quota. If the use of MPAs is to be successful, we first need to achieve effective fishing effort control. The failure or success of the use of an MPA as a fishery management tool (note the use of an MPA to protect specific habitat is excluded in this context) is inextricably linked to effective fishing effort control in the surrounding waters. As a result, there is a real risk that inadequately designed MPAs instigated without effective parallel effort control will be perceived by the industry and the wider public as failures and their utility discredited.

So far, our focus has been directed almost entirely on the fish and their biology. We have sought to control the activities of a predator, mankind, without devoting an equal effort to understanding the dynamics of a predator that operates in a system with imperfect knowledge about the distribution of the prey and the habitat in which they hunt them. Fishers sample the fish population and record their reward rate in space and time (Gillis and Peterman 1998). Nets are towed over areas of the seabed about which knowledge is known at a coarse scale. Charts with seabed features are not based on metre-accurate remote sensing as is possible on the land, but on sample points located tens or even hundreds of kilometres apart. What lies in between is uncertain. Even using echo-sounders, the instantaneous knowledge relayed to the wheelhouse of a fishing vessel provides information of only a few metres of the seabed directly beneath the hull and tells one nothing about the 20 m of seabed on either side that are about to be swept by the following fishing net. Thus every new exploratory tow using a bottom-towed fishing gear is potentially hazardous.

When fish stocks are healthy, fishers are most likely to repeatedly return to locations that they know from past experience to yield economically rewarding catches with a minimum risk to gear and vessel safety and avoid excessive competition with other fishers (Gillis and Peterman 1998; Holland and Sutinen 2000; Jin et al. 2002). As stocks decline and more fishing effort is required to maintain the same catch, fishers are forced to explore new areas where knowledge of the seabed and catch is uncertain. Evidence for this behaviour comes from vessel-monitoring systems and direct observations that demonstrate that fishing effort is highly aggregated and is only homogeneous at a scale of 1 km^2 (Rijnsdorp et al. 1998). It is clear that large areas of the seabed remain unfished, while other areas receive intensive fishing activity. As fishing effort rises to compensate for falling landings, previously unfished grounds begin to become the focus of attention. However, as effort falls, fishers largely return to their favoured localities. Indeed, the proportion of seabed affected by fishing is highly predictable in line with fluctuations in total effort. Faced with declining stocks and unrestricted fishing effort under a quota management system, fishers have the incentive to explore new grounds, thereby increasing their grounds will be adversely affected in the future.

The aggregative behaviour of fishers has important ecological consequences, since the first few passages of bottomfishing gear across the seabed cause the biggest reduction in seabed community production and [the seabed] thereafter remains in a relatively constant condition of reduced benthic production (Jennings et al. 2001). It is important to appreciate that some fishers repeatedly fish the same tows several (or more) times per year on an annual basis. This aggregative behaviour occurs even over areas that have the same environmental conditions and hence the same potential benthic production in the absence of fishing (J. Hiddink, School of Ocean Sciences, University of Wales-Bangor, Menai Bridge, Anglesey LL59 5AB, UK, unpublished data). Thus restricting bottomfishing activities to confined areas of the seabed minimizes the spatial extent of the negative effects on seabed community production, which is a status that is already achieved as a result of fishers' behaviour and without the imposition of MPAs. The imposition of MPAs without due consideration for fishers' behavioural responses may cause more damage than the status quo as evidenced by the recent "cod box" in the North Sea, which forced fishers to reallocate their effort to areas of the seabed where previously fishing boats had not been recorded (Dinmore et al. 2003).

The use of MPAs clearly has beneficial effects for very habitat-specific fish species associated with coral and temperate rock reefs, and they are the only tool that can effectively protect sensitive habitats such as beds of ancient calcareous algae and deep water corals (Willis et al. 2003*b*). In temperate waters, where the majority of the seabed is characterized by sediments and aggregates, sedentary species such as sea scallops and fish that have restricted movements are the most likely candidates to benefit from the exclusion of fishing activities (Horwood et al. 1998; Murawski et al. 2000). However, the scale of MPAs required to ensure sustainable fisheries of wide-ranging, long-lived species, such as cod and plaice, may be both impractical and equally prone to the same political horse-trading that has neutered many current management systems. New evidence suggests that in the worst case scenario, the ill-considered use of MPAs will displace fishing activity and thereby could result in additional damage to the marine environment before recolonization occurs in the newly protected area (Dinmore et al. 2003), whereas the proper implementation of fishing effort reduction still has the potential to out-perform the use of MPAs in terms of increasing spawning stock biomass (Steele and Beet 2003). As stocks recover and catch efficiency increases, the necessity to fish for prolonged periods and in marginal areas of the seabed should diminish. These beneficial effects of stock recovery can easily be undone by allowing the industry to expand uncontrolled, thereby exacerbating the "ratchet effect." In temperate systems, MPAs are not the singular solution to sustainable fisheries and are perhaps more a symptom of chronic failures to apply scientific advice (Hall 1999). Uncompromised management of fishing effort would simultaneously solve many of the wider negative, ecological, and socio-economic effects of overfishing.

If we are to advocate MPAs as a fundamental tool that can help to achieve sustainable use of marine resources, it is imperative that the uncertainty regarding the outcome of such a management system is made clear to all stakeholders and that it is dependent upon adequate regulation and uncompromised implementation. In effect, any such shift towards the use of MPAs as part of our management tool portfolio should be tested using a rigorously designed, experimental management regime, the performance of which is tested in a scientifically objective manner and reviewed on a regular basis.

References

Agardy, T., Bridgewater, P., Crosby, M.P., Day, J., Dayton, P.K., Kenchington, R., Laffoley, D., McConney, P., Murray, P.A., Parks, J.E., and Peau, L. 2003. Danger-ous targets? Unresolved issues and ideological clashes around marine protected areas. Aquat. Conserv. Mar. Freshw. Ecosyst. **13**: 353–367.

Blyth, R.E., Kaiser, M.J., Edwards-Jones, G., and Hart, P.J.B. 2002. Voluntary man-agement in an inshore fishery has conservation benefits. Environ. Conserv. **29**: 493–508.

Dinmore, T.A., Duplisea, D.E., Rackham, B.D., Maxwell, D.L., and Jennings, S. 2003. Impact of a large-scale area closure on patterns of fishing disturbance and the consequences for benthic communities. ICES J. Mar. Sci. **60**: 371–380.

Frank, K.T., Shackell, N.L., and Simon, J.E. 2000. An evaluation of the Emerald/Western Bank juvenile haddock closed areas. ICES J. Mar. Sci. **57**: 1023–1034.

Gell, F.R., and Roberts, C.M. 2003. Benefits beyond boundaries: the fishery effects of marine reserves. Trends Ecol. Evol. **18**: 448–455.

Gillis, D.M., and Peterman, R.M. 1998. Implications of interference among fishing vessels and the ideal free distribution to the interpretation of CPUE. Can. J. Fish. Aquat. Sci. **55**: 37–46.

Hall, S.J. 1999. The effects of fishing on marine ecosystems and communities. Blackwell Science, Oxford.

Halpern, B.S. 2003. The impact of marine reserves: Do reserves work and does reserve size matter? Ecol. Appl. **13**: S117–S137.

Hart, P.J.B. 1998. Enlarging the shadow of the future: avoiding conflict and conserving fish off south Devon, U.K. *Edited by* T.J. Pitcher, P.J.B. Hart, and D. Pauly. Reinventing Fisheries Management, Kluwer, Dordrecht.

Hilborn, R., Stokes, T.K., Maguire, J.-J., Smith, A., Botsford, L.W., Mangel, M., Orensanz, J., Parma, A., Rice, J., Bell, J.D., Cochrane, K.L., Garcia, S., Hall, S.J., Kirkwood, G.P., Sainsbury, J.C., Stefansson, G., and Walters, C.J. 2004. When can marine reserves improve fisheries management? Ocean Coast. Manag. **47**: 197–205.

Holland, D.S., and Sutinen, J.G. 2000. Location choice in New England trawl fisheries: old habits die hard. Landsc. Ecol. **76**: 133–149.

Horwood, J.W. 2000. No-take zones: a management context. *Edited by* M.J. Kaiser and S.J. de Groot. The effects of fishing on non-target species and habitats: biological, social and economic issues. Blackwell Science, Oxford.

Horwood, J.W., Nichols, J.H., and Milligan, S. 1998. Evaluation of closed areas for fish stock conservation. J. Appl. Ecol. **35**: 893–903.

Hutchings, J.A. 2000. Collapse and recovery of marine fishes. Nature (London), **406**: 882–885.

Hutchings, J.A., and Myers, R.A. 1995. The North Atlantic Fisheries: successes, failures, and challenges. *Edited by* R. Amason and L. Felt. Institute of Island Studies, Charlottetown, P.E.I., Canada. pp. 38–93.

Jennings, S., Marshall, S.S., and Polunin, N.V.C. 1996. Seychelles' marine protected areas: comparative structure and status of reef fish communities. Biol. Conserv. **75**: 201–209.

Jennings, S., Dinmore, T.A., Duplisea, D.E., Warr, K.J., and Lancaster, J.E. 2001. Trawling disturbance can modify benthic production processes. J. Anim. Ecol. **70**: 459–475.

Jin, D., Kite-Powell, H.L., Thunberg, E., Solow, A.R., and Talley, W.K. 2002. A model of fishing vessel accident probability. J. Saf. Res. **33**: 497–510.

Metcalfe, J.D., and Arnold, G.P. 1997. Tracking fish with electronic tags. Nature (London), **387**: 665–666.

Murawski, S.A., Brown, R., Lai, H.-L., Rago, P.J., and Hendrickson, L. 2000. Large-scale closed areas as a fishery-management tool in temperate marine systems: the Georges Bank experiment. Bull. Mar. Sci. **66**: 775–798.

Ray, G.C. 2004. Reconsidering "dangerous targets" for marine protected areas. Aquat. Conserv. Mar. Freshw. Ecosyst. **14**: 211–215.

Rijnsdorp, A.D., Buijs, A.M., Storbeck, F., and Visser, E. 1998. Micro-scale distribution of beam trawl effort in the southern North Sea between 1993 and 1996 in relation to the trawling frequency of the sea bed and the impact on benthic organisms. ICES J. Mar. Sci. **55**: 403–419.

Roberts, C.M., Bohnsack, J.A., Gell, F.R., Hawkins, J., and Goodridge, R. 2001. Marine reserves enhance adjacent fisheries. Science (Washington, D.C.), **294**: 1920–1923.

Roberts, C.M., Hawkins, J.P., and Gell, F.R. 2005. The role of marine reserves in achieving sustainable fisheries. Philos. Trans. R. Soc. Lond. B Biol. Sci., **360**: 123–132

Russ, G., and Alcala, A. 1996. Marine reserves-rates and patterns of recovery and decline of large predatory fish. Ecol. Appl. **6**: 947–961.

Sanchiro, J.N., and Wilen, J. E. 2001. A bioeconomics model of marine reserve creation. J. Environ. Econ. Manag. **42**: 257–276.

Steele, D.H., and Beet, A.R. 2003. Marine protected areas in "nonlinear" ecosystems. Proc. R. Soc. Lond. B Biol. Sci. **270**(Suppl.): S230–S233.

Steele, J.H., and Hoagland, P. 2003. Are fisheries "sustainable"? Fish. Res. **64**: 1–3.

Steele, J.H., and Hoagland, P. 2004. Are fisheries "sustainable"? A counterpoint to Steele and Hoagland: a reply. Fish. Res. **67**: 247–248.

Watling, L., and Norse, E.A. 1998. Disturbance of the seabed by mobile fishing gear: a comparison to forest clearcutting. Conserv. Biol. **12**: 1180–1197.

Willis, T.J., Millar, R.B., Babcock, R.C., and Tolimieri, N. 2003*a*. Burdens of evidence and the benefits of marine reserves: putting Descartes before des horse? Environ. Conserv. **30**: 97–103.

Willis, T.J., Millar, R.B., and Babcock, R.C. 2003*b*. Protection of exploited fishes in temperate regions: high density and biomass of snapper *Pagurus auratus* (Sparidae) in northern New Zealand marine reserves. J. Appl. Ecol. **40**: 214–227.

Zeller, D., and Russ, G.R. 2004. Are fisheries "sustainable"? A counterpoint to Steele and Hoagland. Fish. Res. **67**: 241–245.

POSTSCRIPT

Are Marine Reserves Needed to Protect Global Fisheries?

Oran R. Young, in "Taking Stock: Management Pitfalls in Fisheries Science,"*Environment* (April 2003), notes that despite putting great effort into assessing fish stocks and managing fisheries for sustainable yield, marine fish stocks have continued to decline. This is partly the "result of the inability of managers to resist pressures from interest groups to set total allowable catches too high, even in the face of warnings from scientists about the dangers of triggering stock depletions. The problem also arises, however, from repeated failures on the part of analysts and policy makers to anticipate the collapse of major stocks or to grasp either the current condition or the reproductive dynamics of important stocks." He cautions against putting "blind faith in the validity of scientific assessments" and suggests more use of the *precautionary principle* (see Issue 1) despite the risk that this would set allowed catch levels lower than many would like. Lydia K. Bergen and Mark H. Carr, in "Establishing Marine Reserves," *Environment* (March 2003), favor the development of marine reserves. They review the Channel Islands reserve discussed by Warner, praising its incorporation of scientific data. Lynda D. Rodwell and Callum M. Roberts, "Fishing and the Impact of Marine Reserves in a Variable Environment," *Canadian Journal of Fisheries & Aquatic Sciences* (November 2004), find that in variable marine environments, reserves can increase catches, reduce variability of catches, and make planning more efficient. However, it is worth noting that Martin D. Smith, Junjie Zhang, and Felicia C. Coleman, "Effectiveness of Marine Reserves for Large-Scale Fisheries Management," *Canadian Journal of Fisheries & Aquatic Sciences* (January 2006), find that the effect of at least some marine reserves is not beneficial. Heather M. Leslie, "A Synthesis of Marine Conservation Planning Approaches," *Conservation Biology* (December 2005), casts some light on potential problems with marine reserves when she notes, among other things, that failed reserves tend to be smaller in scale. Results may also vary according to the fish species of interest; see Robert E. Blyth-Skyrme, et al., "Conservation Benefits of Temperate Marine Protected Areas: Variation among Fish Species," *Conservation Biology* (June 2006).

In September 2004, the U.S. Commission on Ocean Policy issued its report, "An Ocean Blueprint for the 21st Century" (http://www.oceancommission.gov/documents/full_color_rpt/welcome.html), calling for improved management systems "to handle mounting pollution, declining fish populations and coral reefs, and promising new industries such as aquaculture." In many ways it agreed with the Pew report, as mentioned in the introduction to this issue. Carl Safina and Sarah Chasis, "Saving the Oceans," *Issues in Science and Technology* (Fall 2004), discuss the two reports and say that it is time for Congress to craft a new approach to ocean policy, including fisheries protection, and put scientists in charge of policy. James N. Sanchirico and Susan S.Hanna, "Sink or Swim Time for U.S. Fishery Policy," *Issues in Science and Technology* (Fall 2004), add that an ecosystem approach is essential. In December 2004, President Bush created the Committee on Ocean Policy in the Council on Environmental Quality. Among its goals, however, were both environmental and economic interests of the U.S. Marine reserves are definitely gaining favor as part of the solution. John Temple Swing, in "What Future for the Oceans?" *Foreign Affairs* (September/October 2003), calls marine protected areas "one of the promising trends of the past two decades." See also "Marine Protected Areas," *Congressional Digest* (September 2003) and Sascha K. Hooker and Leah R. Gerber, "Marine Reserves as a Tool for Ecosystem-Based Management: The Potential Importance of Megafauna," *Bioscience* (January 2004). Jack Sobel's and Craig Dahlgren's *Marine Reserves: A Guide to Science, Design, and Use* (Island Press, Washington, DC, 2004), is a useful treatment of the background of, evidence for, and implementation of marine reserves. But Garry R. Russ and Angela C. Alcala, in "Marine Reserves: Long-Term Protection Is Required for Full Recovery of Predatory Fish Populations," *Oecologia* (March 2004), conclude that reserves cannot be viewed as temporary creations; reserve management systems must be designed to last generations in the face of increasing pressures for food and other resources. However, it is essential to engage the fishing community in the process of designing and establishing reserves. See Mark Helvey, "Seeking Consensus on Designing Marine Protected Areas: Keeping the Fishing Community Engaged," *Coastal Management* (April/June 2004).

It is worth noting that there remains resistance to the idea that overfishing is a problem. See Walter A. Starck, "Fishy Claims of Overfishing on the Great Barrier Reef," *Professional Fisherman* (September 2005). While commercial fishers object to conservation measures such as marine reserves because they see them as affecting their livelihood, sports fishers also object to marine reserves. See Jerry Gibbs, "Freedom to Fish?" *Outdoor Life* (May 2004). Sports fishers, too, may need to be engaged in the process of designing and establishing reserves.

On the Internet . . .

The Pesticide Action Network

The Pesticide Action Network North America (PANNA) challenges the global proliferation of pesticides, defends basic rights to health and environmental quality, and works to insure the transition to a just and viable society.

http://www.panna.org/

Africa Fighting Malaria

Africa Fighting Malaria is an NGO (non-governmental organization) which seeks to educate people about the scourge of malaria and the political economy of malaria control. It believes that global health organizations must be free to employ all available tools to fight malaria.

http://www.fightingmalaria.org/

e.hormone

e.hormone is hosted by the Center for Bioenvironmental Research at Tulane and Xavier Universities in New Orleans. It provides accurate, timely information about environmental hormones and their impacts.

http://e.hormone.tulane.edu/

The Silicon Valley Toxics Coalition

The Silicon Valley Toxics Coalition (SVTC) was formed in 1982 to engage in research, advocacy, and organizing associated with environmental and human problems caused by the rapid growth of the high-tech electronics industry, to advance environmental sustainability and clean production in the industry, and to improve health, promote justice, and ensure democratic decision-making for affected communities and workers in the U.S. and the world.

http://www.svtc.org/

Superfund

The U.S. Environmental Protection Agency provides a great deal of information on the Superfund program, including material on environmental justice.

http://www.epa.gov/superfund/

Yucca Mountain

The EPA also provides information on the proposed Yucca Mountain permanent nuclear waste repository.

http://www.epa.gov/radiation/yucca/

PART 5

Toxic Chemicals

A *great many of today's environmental issues have to do with industrial development, which expanded greatly during the twentieth century. Just since World War II, many thousands of synthetic chemicals—pesticides, plastics, and antibiotics—have flooded the environment. We have become dependent on the production and use of energy, particularly in the form of fossil fuels. We have discovered that industrial processes generate huge amounts of waste, much of it toxic. Air and water pollution have become global problems. And we have discovered that our actions may change the world for generations to come.*

This section deals with two prominent controversies concerning toxic chemicals and two concerning hazardous wastes. There are others in both categories.

- Should DDT Be Banned Worldwide?

- Do Environmental Hormone Mimics Pose a Potentially Serious Health Threat?

- Is the Superfund Program Successfully Protecting the Environment from Hazardous Wastes?

- Should the United States Reprocess Spent Nuclear Fuel?

- Is a Large-Scale Shift to Organic Farming the Best Way to Increase World Food Supply?

- Should the Endangered Species Act Be Strengthened?

ISSUE 16

Should DDT Be Banned Worldwide?

YES: Anne Platt McGinn, from "Malaria, Mosquitoes, and DDT," *World Watch* (May/June 2002)

NO: Donald R. Roberts, from Statement before the U.S. Senate Committee on Environment & Public Works, Hearing on the Role of Science in Environmental Policy-Making (September 28, 2005)

ISSUE SUMMARY

YES: Anne Platt McGinn, a senior researcher at the Worldwatch Institute, argues that although DDT is still used to fight malaria, there are other, more effective and less environmentally harmful methods. She maintains that DDT should be banned or reserved for emergency use.

NO: Donald R. Roberts argues that the scientific evidence regarding the environmental hazards of DDT has been seriously misrepresented by anti-pesticide activists. The hazards of malaria are much greater and, properly used, DDT can prevent them and save lives.

DDT is a crucial element in the story of environmentalism. The chemical was first synthesized in 1874. Swiss entomologist Paul Mueller was the first to notice that DDT has insecticidal properties, which, it was quickly realized, implied that the chemical could save human lives. It had long been known that more soldiers died during wars because of disease than because of enemy fire. During World War I, for example, some 5 million lives were lost to typhus, a disease carried by body lice. DDT was first deployed during World War II to halt a typhus epidemic in Naples, Italy. It was a dramatic success, and DDT was soon used routinely as a dust for soldiers and civilians. During and after the war, DDT was also deployed successfully against the mosquitoes that carry malaria and other diseases. In the United States cases of malaria fell from 120,000 in 1934 to 72 in 1960, and cases of yellow fever dropped from 100,000 in 1878 to none. In 1948 Mueller received the Nobel Prize for medicine and physiology because DDT had saved so many civilian lives.

DDT was by no means the first pesticide. But its predecessors—arsenic, strychnine, cyanide, copper sulfate, and nicotine—were all markedly toxic to humans. DDT was not only more effective as an insecticide, it was also less hazardous to users. It is therefore not surprising that DDT was seen as a beneficial substance. It was soon applied routinely to agricultural crops and used to control mosquito populations in American suburbs. However, insects quickly became resistant to the insecticide. (In any population of insects, some will be more resistant than others; when the insecticide kills the more vulnerable members of the population, the resistant ones are left to breed and multiply. This is an example of natural selection.) In *Silent Spring* (Houghton Mifflin, 1962), marine scientist Rachel Carson demonstrated that DDT was concentrated in the food chain and affected the reproduction of predators such as hawks and eagles. In 1972 the U.S. Environmental Protection Agency banned almost all uses of DDT (it could still be used to protect public health). Other developed countries soon banned it as well, but developing nations, especially those in the tropics, saw it as an essential tool for fighting diseases such as malaria. Roger Bate, director of Africa Fighting Malaria, argues in "A Case of the DDTs," *National Review* (May 14, 2001) that DDT remains the cheapest and most effective way to combat malaria and that it should remain available for use.

It soon became apparent that DDT is by no means the only pesticide or organic toxin with environmental effects. As a result, on May 24, 2001, the United States joined 90 other nations in signing the Stockholm Convention on Persistent Organic Pollutants (POPs). This treaty aims to eliminate from use the entire class of chemicals to which DDT belongs, beginning with the "dirty dozen," pesticides DDT, aldrin, dieldrin, endrin, chlordane, heptachlor, mirex, and toxaphene, and the industrial chemicals polychlorinated biphenyls (PCBs), hexachlorobenzene (HCB), dioxins, and furans. Since then, 59 countries, not including the United States and the European Union (EU), have formally ratified the treaty. It took effect in May 2004. Fiona Proffitt, in "U.N. Convention Targets Dirty Dozen Chemicals," *Science* (May 21, 2004), notes, "About 25 countries will be allowed to continue using DDT against malaria-spreading mosquitoes until a viable alternative is found."

In the following selection, Anne Platt McGinn, granting that malaria remains a serious problem in the developing nations of the tropics, especially Africa, contends that although DDT is still used to fight malaria in these nations, it is far less effective than it used to be. She argues that the environmental effects are also serious concerns and that DDT should be banned or reserved for emergency use. In the second selection, Professor Donald R. Roberts argues that the scientific evidence regarding the environmental hazards of DDT has been seriously misrepresented by anti-pesticide activists. The hazards of malaria are much greater and, properly used, DDT can prevent them and save lives. Efforts to prevent the use of DDT have produced a "global humanitarian disaster."

YES

Anne Platt McGinn

Malaria, Mosquitoes, and DDT

This year, like every other year within the past couple of decades, uncountable trillions of mosquitoes will inject malaria parasites into human blood streams billions of times. Some 300 to 500 million full-blown cases of malaria will result, and between 1 and 3 million people will die, most of them pregnant women and children. That's the official figure, anyway, but it's likely to be a substantial underestimate, since most malaria deaths are not formally registered, and many are likely to have escaped the estimators. Very roughly, the malaria death toll rivals that of AIDS, which now kills about 3 million people annually.

But unlike AIDS, malaria is a low-priority killer. Despite the deaths, and the fact that roughly 2.5 billion people (40 percent of the world's population) are at risk of contracting the disease, malaria is a relatively low public health priority on the international scene. Malaria rarely makes the news. And international funding for malaria research currently comes to a mere $150 million annually. Just by way of comparison, that's only about 5 percent of the $2.8 billion that the U.S. government alone is considering for AIDS research in fiscal year 2003.

The low priority assigned to malaria would be at least easier to understand, though no less mistaken, if the threat were static. Unfortunately it is not. It is true that the geographic range of the disease has contracted substantially since the mid-20th century, but over the past couple of decades, malaria has been gathering strength. Virtually all areas where the disease is endemic have seen drug-resistant strains of the parasites emerge—a development that is almost certainly boosting death rates. In countries as various as Armenia, Afghanistan, and Sierra Leone, the lack or deterioration of basic infrastructure has created a wealth of new breeding sites for the mosquitoes that spread the disease. The rapidly expanding slums of many tropical cities also lack such infrastructure; poor sanitation and crowding have primed these places as well for outbreaks—even though malaria has up to now been regarded as predominantly a rural disease.

What has current policy to offer in the face of these threats? The medical arsenal is limited; there are only about a dozen antimalarial drugs commonly in use, and there is significant malaria resistance to most of them. In the absence of a reliable way to kill the parasites, policy has tended to focus on killing the mosquitoes that bear them. And that has led to an abundant use of synthetic pesticides, including one of the oldest and most dangerous: dichlorodiphenyl trichloroethane, or DDT.

DDT is no longer used or manufactured in most of the world, but because it does not break down readily, it is still one of the most commonly detected pesticides in the milk of nursing mothers. DDT is also one of the "dirty dozen" chemicals included in the 2001 Stockholm Convention on Persistent Organic Pollutants [POPs]. The signatories to the "POPs Treaty" essentially agreed to ban all uses of DDT except as a last resort against disease-bearing mosquitoes. Unfortunately, however, DDT is still a routine option in 19 countries, most of them in Africa. (Only 11 of these countries have thus far signed the treaty.) Among the signatory countries, 31—slightly fewer than one-third—have given notice that they are reserving the right to use DDT against malaria. On the face of it, such use may seem unavoidable, but there are good reasons for thinking that progress against the disease is compatible with *reductions* in DDT use.

⚜️

Malaria is caused by four protozoan parasite species in the genus *Plasmodium.* These parasites are spread exclusively by certain mosquitoes in the genus *Anopheles.* An infection begins when a parasite-laden female mosquito settles onto someone's skin and pierces a capillary to take her blood meal. The parasite, in a form called the *sporozoite,* moves with the mosquito's saliva into the human bloodstream. About 10 percent of the mosquito's lode of sporozoites is likely to be injected during a meal, leaving plenty for the next bite. Unless the victim has some immunity to malaria—normally as a result of previous exposure—most sporozoites are likely to evade the body's immune system and make their way to the liver, a process that takes less than an hour. There they invade the liver cells and multiply asexually for about two weeks. By this time, the original several dozen sporozoites have become millions of *merozoites*—the form the parasite takes when it emerges from the liver and moves back into the blood to invade the body's red blood cells. Within the red blood cells, the merozoites go through another cycle of asexual reproduction, after which the cells burst and release millions of additional merozoites, which invade yet more red blood cells. The high fever and chills associated with malaria are the result of this stage, which tends to occur in pulses. If enough red blood cells are destroyed in one of these pulses, the result is convulsions, difficulty in breathing, coma, and death.

As the parasite multiplies inside the red blood cells, it produces not just more merozoites, but also *gametocytes,* which are capable of sexual reproduction. This occurs when the parasite moves back into the mosquitoes; even as they inject sporozoites, biting mosquitoes may ingest gametocytes if they are feeding on a person who is already infected. The gametocytes reproduce in the insect's gut and the resulting eggs move into the gut cells. Eventually, more sporozoites emerge from the gut and penetrate the mosquito's salivary glands, where they await a chance to enter another human bloodstream, to begin the cycle again.

Of the roughly 380 mosquito species in the genus *Anopheles,* about 60 are able to transmit malaria to people. These malaria vectors are widespread throughout the tropics and warm temperate zones, and they are very efficient at spreading the disease. Malaria is highly contagious, as is apparent from a measurement that epidemiologists call the "basic reproduction number," or BRN. The BRN indicates, on average, how many new cases a single infected person is likely to cause. For example, among the nonvectored diseases (those in which the pathogen travels directly from person to person without an intermediary like a mosquito), measles is one of the most contagious. The BRN for measles is 12 to 14, meaning that someone with measles is likely to infect 12 to 14 other people. (Luckily, there's an inherent limit in this process: as a pathogen spreads through any particular area, it will encounter fewer and fewer susceptible people who aren't already sick, and the outbreak will eventually subside.) HIV/AIDS is on the other end of the scale: it's deadly, but it burns through a population slowly. Its BRN is just above 1, the minimum necessary for the pathogen's survival. With malaria, the BRN varies considerably, depending on such factors as which mosquito species are present in an area and what the temperatures are. (Warmer is worse, since the parasites mature more quickly.) But malaria can have a BRN in excess of 100: over an adult life that may last about a week, a single, malaria-laden mosquito could conceivably infect more than 100 people.

Seven Years, Seven Months

"Malaria" comes from the Italian "mal'aria." For centuries, European physicians had attributed the disease to "bad air." Apart from a tradition of associating bad air with swamps—a useful prejudice, given the amount of mosquito habitat in swamps—early medicine was largely ineffective against the disease. It wasn't until 1897 that the British physician Ronald Ross proved that mosquitoes carry malaria.

The practical implications of Ross's discovery did not go unnoticed. For example, the U.S. administration of Theodore Roosevelt recognized malaria and yellow fever (another mosquito-vectored disease) as perhaps the most serious obstacles to the construction of the Panama Canal. This was hardly a surprising conclusion, since the earlier and unsuccessful French attempt to build the canal—an effort that predated Ross's discovery—is thought to have lost between 10,000 and 20,000 workers to disease. So the American workers draped their water supplies and living quarters with mosquito netting, attempted to fill in or drain swamps, installed sewers, poured oil into standing water, and conducted mosquito-swatting campaigns. And it worked: the incidence of malaria declined. In 1906, 80 percent of the workers had the disease; by 1913, a year before the Canal was completed, only 7 percent did. Malaria could be suppressed, it seemed, with a great deal of mosquito netting, and by eliminating as much mosquito habitat as possible. But the labor involved in that effort could be enormous.

That is why DDT proved so appealing. In 1939, the Swiss chemist Paul Müller discovered that this chemical was a potent pesticide. DDT was first used during World War II, as a delousing agent. Later on, areas in southern Europe, North Africa, and Asia were fogged with DDT, to clear malaria-laden mosquitoes from the paths of invading Allied troops. DDT was cheap and it seemed to be harmless to anything other than insects. It was also long-lasting: most other insecticides lost their potency in a few days, but in the early years of its use, the effects of a single dose of DDT could last for up to six months. In 1948, Müller won a Nobel Prize for his work and DDT was hailed as a chemical miracle.

A decade later, DDT had inspired another kind of war—a general assault on malaria. The "Global Malaria Eradication Program," launched in 1955, became one of the first major undertakings of the newly created World Health Organization [WHO]. Some 65 nations enlisted in the cause. Funding for DDT factories was donated to poor countries and production of the insecticide climbed.

The malaria eradication strategy was not to kill every single mosquito, but to suppress their populations and shorten the lifespans of any survivors, so that the parasite would not have time to develop within them. If the mosquitoes could be kept down long enough, the parasites would eventually disappear from the human population. In any particular area, the process was expected to take three years—time enough for all infected people either to recover or die. After that, a resurgence of mosquitoes would be merely an annoyance, rather than a threat. And initially, the strategy seemed to be working. It proved especially effective on islands—relatively small areas insulated from reinfestation. Taiwan, Jamaica, and Sardinia were soon declared malaria-free and have remained so to this day. By 1961, arguably the year at which the program had peak momentum, malaria had been eliminated or dramatically reduced in 37 countries.

One year later, Rachel Carson published *Silent Spring,* her landmark study of the ecological damage caused by the widespread use of DDT and other pesticides. Like other organochlorine pesticides, DDT bioaccumulates. It's fat soluble, so when an animal ingests it—by browsing contaminated vegetation, for example—the chemical tends to concentrate in its fat, instead of being excreted. When another animal eats that animal, it is likely to absorb the prey's burden of DDT. This process leads to an increasing concentration of DDT in the higher links of the food chain. And since DDT has a high chronic toxicity—that is, long-term exposure is likely to cause various physiological abnormalities—this bioaccumulation has profound implications for both ecological and human health.

With the miseries of malaria in full view, the managers of the eradication campaign didn't worry much about the toxicity of DDT, but they were greatly concerned about another aspect of the pesticide's effects: resistance. Continual exposure to an insecticide tends to "breed" insect populations that are at least partially immune to the poison. Resistance to DDT had been reported as early as 1946. The campaign managers knew that in mosquitoes, regular exposure to DDT tended to produce widespread resistance in four to seven years. Since it took three years to clear malaria from a human population, that didn't leave a lot of leeway for the eradication effort. As it turned out, the logistics simply couldn't be made to work in large, heavily infested areas with high human populations, poor housing and roads, and generally minimal infrastructure. In 1969, the campaign was abandoned. Today, DDT resistance is widespread in *Anopheles,* as is resistance to many more recent pesticides.

Undoubtedly, the campaign saved millions of lives, and it did clear malaria from some areas. But its broadest legacy has been of much more dubious value. It engendered the idea of DDT as a first resort against mosquitoes and it established the unstable dynamic of DDT resistance in *Anopheles* populations. In mosquitoes, the genetic mechanism that confers resistance to DDT does not usually come at any great competitive "cost"—that is, when no DDT is being sprayed, the resistant mosquitoes may do just about as well as nonresistant mosquitoes. So once a population acquires resistance, the trait is not likely to disappear even if DDT isn't used for years. If DDT is reapplied to such a population, widespread resistance will reappear very rapidly. The rule of thumb among entomologists is that you may get seven years of resistance-free use the first time around, but you only get about seven months the second time. Even that limited respite, however, is enough to make the chemical an attractive option as an emergency measure—or to keep it in the arsenals of bureaucracies committed to its use.

Malaria Taxes

In December 2000, the POPs Treaty negotiators convened in Johannesburg, South Africa, even though, by an unfortunate coincidence, South Africa had suffered a potentially embarrassing setback earlier that year in its own POPs policies. In 1996, South Africa had switched its mosquito control programs from DDT to a less persistent group of pesticides known as pyrethroids. The move seemed solid and supportable at the time, since years of DDT use had greatly reduced *Anopheles* populations and largely eliminated one of the most troublesome local vectors, the appropriately named *A. funestus* ("funestus" means deadly). South Africa seemed to have beaten the DDT habit: the chemical had been used to achieve a worthwhile objective; it had then been discarded. And the plan worked—until a year before the POPs summit, when malaria infections rose to 61,000 cases, a level not seen in decades. *A. funestus* reappeared as well, in KwaZulu-Natal, and in a form resistant to pyrethroids. In early 2000, DDT was reintroduced, in an indoor spraying program. (This is now a standard way of using DDT for mosquito control; the pesticide is usually applied only to walls, where mosquitoes alight to rest.) By the middle of the year, the number of infections had dropped by half.

Initially, the spraying program was criticized, but what reasonable alternative was there? This is said to be the African predicament, and yet the South African situation is hardly representative of sub-Saharan Africa as a whole.

Malaria is considered endemic in 105 countries throughout the tropics and warm temperate zones, but by far the worst region for the disease is sub-Saharan Africa. The deadliest of the four parasite species, *Plasmodium falciparum*, is widespread throughout this region, as is one of the world's most effective malaria vectors, *Anopheles gambiae*. Nearly half the population of sub-Saharan Africa is at risk of infection, and in much of eastern and central Africa, and pockets of west Africa, it would be difficult to find anyone who has not been exposed to the parasites. Some 90 percent of the world's malaria infections and deaths occur in sub-Saharan Africa, and the disease now accounts for 30 percent of African childhood mortality. It is true that malaria is a grave problem in many parts of the world, but the African experience is misery on a very different order of magnitude. The average Tanzanian suffers more infective bites each *night* than the average Thai or Vietnamese does in a year.

As a broad social burden, malaria is thought to cost Africa between $3 billion and $12 billion annually. According to one economic analysis, if the disease had been eradicated in 1965, Africa's GDP would now be 35 percent higher than it currently is. Africa was also the gaping hole in the global eradication program: the WHO planners thought there was little they could do on the continent and limited efforts to Ethiopia, Zimbabwe, and South Africa, where eradication was thought to be feasible.

But even though the campaign largely passed Africa by, DDT has not. Many African countries have used DDT for mosquito control in indoor spraying programs, but the primary use of DDT on the continent has been as an agricultural insecticide. Consequently, in parts of west Africa especially, DDT resistance is now widespread in *A. gambiae*. But even if *A. gambiae* were not resistant, a full-bore campaign to suppress it would probably accomplish little, because this mosquito is so efficient at transmitting malaria. Unlike most *Anopheles* species, *A. gambiae* specializes in human blood, so even a small population would keep the disease in circulation. One way to get a sense for this problem is to consider the "transmission index"—the threshold number of mosquito bites necessary to perpetuate the disease. In Africa, the index overall is 1 bite per person per month. That's all that's necessary to keep malaria in circulation. In India, by comparison, the TI is 10 bites per person per month.

And yet Africa is not a lost cause—it's simply that the key to progress does not lie in the general suppression of mosquito populations. Instead of spraying, the most promising African programs rely primarily on "bednets"— mosquito netting that is treated with an insecticide, usually a pyrethroid, and that is suspended over a person's bed. Bednets can't eliminate malaria, but they can "deflect" much of the burden. Because *Anopheles* species generally feed in the evening and at night, a bednet can radically reduce the number of infective bites a person receives. Such a person would probably still be infected from time to time, but would usually be able to lead a normal life.

In effect, therefore, bednets can substantially reduce the disease. Trials in the use of bednets for children have shown a decline in malaria-induced mortality by 25 to 40 percent. Infection levels and the incidence of severe anemia also declined. In Kenya, a recent study has shown that pregnant women who use bednets tend to give birth to healthier babies. In parts of Chad, Mali, Burkina Faso, and Senegal, bednets are becoming standard household items. In the tiny west African nation of The Gambia, somewhere between 50 and 80 percent of the population has bednets.

Bednets are hardly a panacea. They have to be used properly and retreated with insecticide occasionally. And there is still the problem of insecticide resistance, although the nets themselves are hardly likely to be the main cause of it. (Pyrethroids are used extensively in agriculture as well.) Nevertheless, bednets can help transform malaria from a chronic disaster to a manageable public health problem—something a healthcare system can cope with.

So it's unfortunate that in much of central and southern Africa, the nets are a rarity. It's even more unfortunate that, in 28 African countries, they're taxed or subject to import tariffs. Most of the people in these countries would have trouble paying for a net even without the tax. This problem was addressed in the May 2000 "Abuja Declaration," a summit agreement on infectious diseases signed by 44 African countries. The Declaration included a pledge to do away with "malaria taxes." At last count, 13 countries have actually acted on the pledge, although in some cases only by reducing rather than eliminating the taxes. Since the Declaration was signed, an estimated 2 to 5 million Africans have died from malaria.

This failure to follow through with the Abuja Declaration casts the interest in DDT in a rather poor light. Of the 31 POPs treaty signatories that have reserved the right to use DDT, 21 are in Africa. Of those 21, 10 are apparently still taxing or imposing tariffs on bednets. (Among the African countries that have *not* signed the POPs treaty, some are almost certainly both using DDT and taxing bednets, but the exact number is difficult to ascertain because the status of DDT use is not always clear.) It is true that a case can be made for the use of DDT in situations like the one in South Africa in 1999—an infrequent flare-up in a context that lends itself to control. But the routine use of DDT against malaria is an exercise in toxic futility, especially when it's pursued at the expense of a superior and far more benign technology.

Learning to Live with the Mosquitoes

A group of French researchers recently announced some very encouraging results for a new anti-malarial drug known as G25. The drug was given to infected aotus monkeys, and it appears to have cleared the parasites from their systems. Although extensive testing will be necessary before it is known whether the drug can be safely given to people, these results have raised the hope of a cure for the disease.

Of course, it would be wonderful if G25, or some other new drug, lives up to that promise. But even in the absence of a cure, there are opportunities for progress that may one day make the current incidence of malaria look like some dark age horror. Many of these opportunities have been incorporated into an initiative that began in 1998, called the Roll Back Malaria (RBM) campaign, a collaborative effort between WHO, the World Bank, UNICEF, and the UNDP [United Nations Development Programme]. In contrast to the earlier WHO eradication program, RBM grew out of joint efforts between WHO and various African governments specifically to address African malaria. RBM focuses on household- and community-level intervention and it emphasizes apparently modest changes that could yield major progress. Below are four "operating principles" that are, in one way or another, implicit in RBM or likely to reinforce its progress.

1. Do away with all taxes and tariffs on bednets, on pesticides intended for treating bednets, and on antimalarial drugs. Failure to act on this front certainly undercuts claims for the necessity of DDT; it may also undercut claims for antimalaria foreign aid.

2. Emphasize appropriate technologies. Where, for example, the need for mud to replaster walls is creating lots of pothole sized cavities near houses—cavities that fill with water and then with mosquito larvae—it makes more sense to help people improve their housing maintenance than it does to set up a program for squirting pesticide into every pothole. To be "appropriate," a technology has to be both affordable and culturally acceptable. Improving home maintenance should pass this test; so should bednets. And of course there are many other possibilities. In Kenya, for example, a research institution called the International Center for Insect Physiology and Ecology has identified at least a dozen native east African plants that repel *Anopheles gambiae* in lab tests. Some of these plants could be important additions to household gardens.

3. Use existing networks whenever possible, instead of building new ones. In Tanzania, for example, an established healthcare program (UNICEF's Integrated Management of Childhood Illness Program) now dispenses antimalarial drugs—and instruction on how to use them. The UNICEF program was already operating, so it was simple and cheap to add the malaria component. Reported instances of severe malaria and anemia in infants have declined, apparently as a result. In Zambia, the government is planning to use health and prenatal clinics as the network for a coupon system that subsidizes bednets for the poor. Qualifying patients would pick up coupons at the clinics and redeem them at stores for the nets.

4. Assume that sound policy will involve action on many fronts. Malaria is not just a health problem—it's a social problem, an economic problem, an environmental problem, an agricultural problem, an urban planning problem. Health officials alone cannot possibly just make it go away. When the disease flares up, there is a strong and understandable temptation to strap on the spray equipment and douse the mosquitoes. But if this approach actually worked, we wouldn't be in this situation today. Arguably the biggest opportunity for progress against the disease lies, not in our capacity for chemical innovation, but in our capacity for *organizational innovation*—in our ability to build an awareness of the threat across a broad range of policy activities. For example, when government officials are considering loans to irrigation projects, they should be asking: has the potential for malaria been addressed? When foreign donors are designing antipoverty programs, they should be asking: do people need bednets? Routine inquiries of this sort could go a vast distance to reducing the disease.

Where is the DDT in all of this? There isn't any, and that's the point. We now have half a century of evidence that routine use of DDT simply will not prevail against the mosquitoes. Most countries have already absorbed this lesson, and banned the chemical or relegated it to emergency only status. Now the RBM campaign and associated efforts are showing that the frequency and intensity of those emergencies can be reduced through systematic attention to the chronic aspects of the disease. There is less and less justification for DDT, and the futility of using it as a matter of routine is becoming increasingly apparent: in order to control a disease, why should we poison our soils, our waters, and ourselves?

Donald R. Roberts

 **NO**

Statement before the U.S. Senate Committee on Environment & Public Works, Hearing on the Role of Science in Environmental Policy-Making

Thank you, Chairman Inhofe, and distinguished members of the Committee on Environment and Public Works, for the opportunity to present my views on the misuse of science in public policy. My testimony focuses on misrepresentations of science during decades of environmental campaigning against DDT.

Before discussing how and why DDT science has been misrepresented, you first must understand why this misrepresentation has not helped, but rather harmed, millions of people every year all over the world. Specifically you need to understand why the misrepresentation of DDT science has been and continues to be deadly. By way of explanation, I will tell you something of my experience.

I conducted malaria research in the Amazon Basin in the 1970s. My Brazilian colleague—who is now the Secretary of Health for Amazonas State—and I worked out of Manaus, the capitol of Amazonas State. From Manaus we traveled two days to a study site where we had sufficient numbers of cases for epidemiological studies. There were no cases in Manaus, or anywhere near Manaus. For years before my time there and for years thereafter, there were essentially no cases of malaria in Manaus. However, in the late 1980s, environmentalists and international guidelines forced Brazilians to reduce and then stop spraying small amounts of DDT inside houses for malaria control. As a result, in 2002 and 2003 there were over 100,000 malaria cases in Manaus alone.

Brazil does not stand as the single example of this phenomenon. A similar pattern of declining use of DDT and reemerging malaria occurs in other countries as well, Peru for example. Similar resurgences of malaria have occurred in rural communities, villages, towns, cities, and countries around the world. As illustrated by the return of malaria in Russia, South Korea, urban areas of the Amazon Basin, and increasing frequencies of outbreaks in the United States, our malaria problems are growing worse. Today there are 1 to 2 million malaria deaths each year and hundreds of millions of cases. The

U.S. Senate Committee on Environment & Public Works Hearing on the Role of Science in Environmental Policy-Making, September 28, 2005.

poorest of the world's people are at greatest risk. Of these, children and pregnant women are the ones most likely to die.

We have long known about DDT's effectiveness in curbing insect-borne disease. Othmar Zeidler, a German chemistry student, first synthesized DDT in 1874. Over sixty years later in Switzerland, Paul Müller discovered the insecticidal property of DDT. Allied forces used DDT during WWII, and the new insecticide gained fame in 1943 by successfully stopping an epidemic of typhus in Naples, an unprecedented achievement. By the end of the war, British, Italian, and American scientists had also demonstrated the effectiveness of DDT in controlling malaria-carrying mosquitoes. DDT's proven efficacy against insect-borne diseases, diseases that had long reigned unchecked throughout the world, won Müller the Nobel Prize for Medicine in 1948. After WWII, the United States conducted a National Malaria Eradication Program, commencing operations on July 1, 1947. The spraying of DDT on internal walls of rural homes in malaria endemic counties was a key component of the program. By the end of 1949, the program had sprayed over 4,650,000 houses. This spraying broke the cycle of malaria transmission, and in 1949 the United States was declared free of malaria as a significant public health problem. Other countries had already adopted DDT to eradicate or control malaria, because wherever malaria control programs sprayed DDT on house walls, the malaria rates dropped precipitously. The effectiveness of DDT stimulated some countries to create, for the first time, a national malaria control program. Countries with pre-existing programs expanded them to accommodate the spraying of houses in rural areas with DDT. Those program expansions highlight what DDT offered then, and still offers now, to the malaria endemic countries. As a 1945 U.S. Public Health Service manual explained about the control of malaria: "Drainage and larviciding are the methods of choice in towns of 2,500 or more people. But malaria is a rural disease. Heretofore there has been no economically feasible method of carrying malaria control to the individual tenant farmer or sharecropper. Now, for the first time, a method is available—the application of DDT residual spray to walls and ceilings of homes." Health workers in the United States were not the only ones to recognize the particular value of DDT. The head of malaria control in Brazil characterized the changes that DDT offered in the following statement: "Until 1945–1946, preventive methods employed against malaria in Brazil, as in the rest of the world, were generally directed against the aquatic phases of the vectors (draining, larvicides, destruction of bromeliads, etc. . . .). These methods, however, were only applied in the principal cities of each state and the only measure available for rural populations exposed to malaria was free distribution of specific drugs."

DDT was a new, effective, and exciting weapon in the battle against malaria. It was cheap, easy to apply, long-lasting once sprayed on house walls, and safe for humans. Wherever and whenever malaria control programs sprayed it on house walls, they achieved rapid and large reductions in malaria rates. Just as there was a rush to quickly make use of DDT to control disease, there was also a rush to judge how DDT actually functioned to control malaria. That rush to judgment turned out to be a disaster. At the heart of the debate—to the extent there was a debate—was a broadly accepted model that established a mathematical framework for using DDT to kill mosquitoes and eradicate malaria. Instead of studying real data to see how DDT actually worked in controlling malaria, some scientists settled upon what they thought was a logical conclusion: DDT worked solely by killing mosquitoes. This conclusion was based on their belief in the model. Scientists who showed that DDT did not function by killing mosquitoes were ignored. Broad acceptance of the mathematical model led to strong convictions about DDT's toxic actions. Since they were convinced that DDT worked only by killing mosquitoes, malaria control specialists became very alarmed when a mosquito was reported to be resistant to DDT's toxic actions. As a result of concern about DDT resistance, officials decided to make rapid use of DDT before problems of resistance could eliminate their option to use DDT to eradicate malaria. This decision led to creation of the global malaria eradication program. The active years of the global malaria eradication program were from 1959 to 1969. Before, during, and after the many years of this program, malaria workers and researchers carried out their responsibilities to conduct studies and report their research. Through those studies, they commonly found that DDT was functioning in ways other than by killing mosquitoes. In essence, they found that DDT was functioning through mechanisms of repellency and irritancy. Eventually, as people forgot early observations of DDT's repellent actions, some erroneously interpreted new findings of repellent actions as the mosquitoes' adaptation to avoid DDT toxicity, even coining a term, "behavioral resistance," to explain what they saw. This new term accommodated their view that toxicity was DDT's primary mode of action and categorized behavioral responses of mosquitoes as mere adaptations to toxic affects. However this interpretation depended upon a highly selective use of scientific data. The truth is that toxicity is not DDT's primary mode of action when sprayed on house walls. Throughout the history of DDT use in malaria control programs there has always been clear and persuasive data that DDT functioned primarily as a spatial repellent. Today we know that there is no insecticide recommended for malaria control that rivals, much less equals, DDT's spatial repellent actions, or that is as long-acting, as cheap, as easy to apply, as safe for human exposure, or as efficacious in the control of malaria as DDT. . . . The 30 years of data from control programs of the Americas plotted . . . illustrate just how effective DDT is in malaria control. The period 1960s through 1979 displays a pattern of malaria controlled through house spraying. In 1979 the World Health Organization (WHO) changed its strategy for malaria control, switching emphasis from spraying houses to case detection and treatment. In other words, the WHO changed emphasis from malaria prevention to malaria treatment. Countries complied with WHO guidelines and started to dismantle their spray programs over the next several years. . . .

I find it amazing that many who oppose the use of DDT describe its earlier use as a failure. Our own citizens who suffered under the burden of malaria, especially in the rural south, would hardly describe it thus.

Malaria was a serious problem in the United States and for some localities, such as Dunklin County, Missouri, it was a very serious problem indeed. For four counties in Missouri, the average malaria mortality from 1910 to 1914 was 168.8 per 100,000 population. For Dunklin County, it was 296.7 per 100,000, a rate almost equal to malaria deaths in Venezuela and actually greater than the mortality rate for Freetown, Sierra Leone. Other localities in other states were equally as malarious. Growing wealth and improved living conditions were gradually reducing malaria rates, but cases resurged during WWII. The advent of DDT, however, quickly eradicated malaria from the United States.

DDT routed malaria from many other countries as well. The Europeans who were freed of malaria would hardly describe its use as a failure. After DDT was introduced to malaria control in Sri Lanka (then Ceylon), the number of malaria cases fell from 2.8 million in 1946 to just 110 in 1961. Similar spectacular decreases in malaria cases and deaths were seen in all the regions that began to use DDT. The newly formed Republic of China (Taiwan) adopted DDT use in malaria control shortly after World War II. In 1945 there were over 1 million cases of malaria on the island. By 1969 there were only 9 cases and shortly thereafter the disease was eradicated from the island and remains so to this day. Some countries were less fortunate. South Korea used DDT to eradicate malaria, but without house spray programs, malaria has returned across the demilitarized zone with North Korea. As DDT was eliminated and control programs reduced, malaria has returned to other countries such as Russia and Argentina. Small outbreaks of malaria are even beginning to appear more frequently in the United States.

These observations have been offered in testimony to document first that there were fundamental misunderstandings about how DDT functioned to exert control over malaria. Second, that regardless of systematic misunderstandings on the part of those who had influence over malaria control strategies and policies, there was an enduring understanding that DDT was the most cost-effective compound yet discovered for protecting poor rural populations from insect-borne diseases like malaria, dengue, yellow fever, and leishmaniasis. I want to emphasize that misunderstanding the mode of DDT action did not lead to the wholesale abandonment of DDT. It took an entirely new dimension in the misuse of science to bring us to the current humanitarian disaster represented by DDT elimination.

The misuse of science to which I refer has found fullest expression in the collection of movements within the environmental movement that seek to stop production and use of specific man-made chemicals. Operatives within these movements employ particular strategies to achieve their objectives. By characterizing and understanding the strategies these operatives use, we can identify their impact in the scientific literature or in the popular press.

The first strategy is to develop and then distribute as widely as possible a broad list of claims of chemical harm. This is a sound strategy because individual scientists can seldom rebut the scientific foundations of multiple and diverse claims. Scientists generally develop expertise in a single, narrow field and are disinclined to engage issues beyond their area of expertise. Even if an authoritative rebuttal of one claim occurs, the other claims still progress. A broad list of claims also allows operatives to tailor platforms for constituencies, advancing one set of claims with one constituency and a different combination for another. Clever though this technique is, a list of multiple claims of harm is hardly sufficient to achieve the objective of a ban. The second strategy then is to mount an argument that the chemical is not needed and propose that alternative chemicals or methods can be used instead. The third strategy is to predict that grave harm will occur if the chemical continues to be used.

The success of Rachel Carson's *Silent Spring* serves as a model for this tricky triad. In *Silent Spring*, Rachel Carson used all three strategies on her primary target, DDT. She described a very large list of potential adverse effects of insecticides, DDT in particular. She argued that insecticides were not really needed and that the use of insecticides produces insects that are insecticide resistant, which only exacerbates the insect control problems. She predicted scary scenarios of severe harm with continued use of DDT and other insecticides. Many have written rebuttals to Rachel Carson and others who have, without scientific justification, broadcast long lists of potential harms of insecticides. . . .

[T]ime and science have discredited most of Carson's claims. Rachel Carson's descriptions of inappropriate uses of insecticides that harmed wildlife are more plausible. However, harm from an inappropriate use does not meet the requirements of anti-pesticide activists. They can hardly lobby for eliminating a chemical because someone used it wrongly. No, success requires that even the proper use of an insecticide will cause a large and systematic adverse effect. However, the proper uses of DDT yield no large and systematic adverse effects. Absent such adverse actions, the activists must then rely on claims about insidious effects, particularly insidious effects that scientists will find difficult to prove one way or the other and that activists can use to predict a future catastrophe.

Rachel Carson relied heavily on possible insidious chemical actions to alarm and frighten the public. Many of those who joined her campaign to ban DDT and other insecticides made extensive use of claims of insidious effects. These claims were amplified by the popular press and became part of the public perception about modern uses of chemicals. For example, four well-publicized claims about DDT were:

1. DDT will cause the obliteration of higher trophic levels. If not obliterated, populations will undergo reproductive failure. Authors of this claim speculated that, even if the use of DDT were stopped, systematic and ongoing obliterations would still occur.
2. DDT causes the death of algae. This report led to speculations that use of DDT could result in global depletion of oxygen.

3. DDT pushed the Bermuda petrel to the verge of extinction and that full extinction might happen by 1978.
4. DDT was a cause of premature births in California sea lions.

Science magazine, the most prestigious science journal in the United States, published these and other phantasmagorical allegations and/or predictions of DDT harm. Nonetheless, history has shown that each and every one of these claims and predictions were false.

1. The obliteration of higher trophic levels did not occur; no species became extinct; and levels of DDT in all living organisms declined precipitously after DDT was de-listed for use in agriculture. How could the prediction have been so wrong? Perhaps it was so wrong because the paper touting this view used a predictive model based on an assumption of no DDT degradation. This was a startling assertion even at the time as *Science* and other journals had previously published papers that showed DDT was ubiquitously degraded in the environment and in living creatures. It was even more startling that *Science* published a paper that flew so comprehensively in the face of previous data and analysis.
2. DDT's action against algae reportedly occurred at concentrations of 500 parts per billion. But DDT cannot reach concentrations in water higher than about 1.2 parts per billion, the saturation point of DDT in water.
3. Data on the Bermuda petrel did not show a cause-and-effect relationship between low numbers of birds and DDT concentrations. DDT had no affect on population numbers, for populations increased before DDT was de-listed for use in agriculture and after DDT was delisted as well.
4. Data gathered in subsequent years showed that "despite relatively high concentrations [of DDT], no evidence that population growth or the health of individual California sea lions have been compromised. The population has increased throughout the century, including the period when DDT was being manufactured, used, and its wastes discharged off southern California."

If time and science have refuted all these catastrophic predictions, why do many scientists and the public not know these predictions were false? In part, we do not know the predictions were false because the refutations of such claims rarely appear in the literature.

When scientists hear the kinds of claims described above, they initiate research to confirm or refute the claims. After Charles Wurster published his claim that DDT kills algae and impacts photosynthesis, I initiated research on planktonic algae to quantify DDT's effects. From 1968–1969, I spent a year of honest and demanding research effort to discover that not enough DDT would even go into solution for a measurable adverse effect on planktonic algae. In essence, I conducted a confirmatory study that failed to confirm an expected result. I had negative data, and journals rarely accept negative data for publication. My year was practically wasted. Without a doubt, hundreds of other scientists around the world have conducted similar studies and obtained negative results, and they too were unable to publish their experimental findings. Much in the environmental science literature during the last 20–30 years indicates that an enormous research effort went into proving specific insidious effects of DDT and other insecticides. Sadly, the true magnitude of such efforts will never be known because while the positive results of research find their way into the scientific literature, the negative results rarely do. Research on insidious actions that produce negative results all too often ends up only in laboratory and field notebooks and is forgotten. For this reason, I place considerable weight on a published confirmatory study that fails to confirm an expected result.

The use of the tricky triad continues. A . . . recent paper . . . published in *The Lancet* illustrates the triad's modern application. Two scientists at the National Institute of Environmental Health Sciences, Walter Rogan and Aimin Chen, wrote this paper, entitled "Health risks and benefits of bis(4-chlorophenyl)-1,1,1-trichloroethane (DDT)." It is interesting to see how this single paper spins all three strategies that gained prominence in Rachel Carson's *Silent Spring*.

The journal *Emerging Infectious Diseases* had already published a slim version of this paper, which international colleagues and I promptly rebutted. The authors then filled in some parts, added to the claims of harm, and republished the paper in the British journal, *The Lancet*. To get the paper accepted by editors, the authors described studies that support (positive results) as well as studies that do not support (negative results) each claim. Complying with strategy number 1 of the triad, Rogan and Chen produce a long list of possible harms, including the charge that DDT causes cancer in nonhuman primates. The literature reference for Rogan and Chen's claim that DDT causes cancer in nonhuman primates was a paper by Takayama et al. Takayama and coauthors actually concluded from their research on the carcinogenic effect of DDT in nonhuman primates that "the two cases involving malignant tumors of different types are inconclusive with respect to a carcinogenic effect of DDT in nonhuman primates." Clearly, the people who made the link of DDT with cancer were not the scientists who actually conducted the research.

The authors enacted strategy number two of the triad by conducting a superficial review of the role of DDT in malaria control with the goal of discrediting DDT's value in modern malaria control programs. The authors admitted that DDT had been very effective in the past, but then argued that malaria control programs no longer needed it and should use alternative methods of control. Their use of the second strategy reveals, in my opinion, the greatest danger of granting authority to anti-pesticide activists and their writings. As *The Lancet* paper reveals, the NIEHS scientists assert great authority over the topic of DDT, yet they assume no responsibility for the harm that might result from their erroneous conclusions. After many malaria control specialists have expressed the necessity for DDT in malaria control, it is possible for Rogan and Chen to conclude that DDT is not necessary in malaria control only if they have no sense of responsibility for levels of disease and death that will occur if DDT is not used.

Rogan and Chen also employ the third strategy of environmentalism. Their list of potential harms caused by DDT includes toxic effects, neurobehavior effects, cancers, decrements in various facets of reproductive health, decrements in infant and child development, and immunology and DNA damage. After providing balanced coverage of diverse claims of harm, the authors had no option but to conclude they could not prove that DDT caused harm. However, they then promptly negated this honest conclusion by asserting that if DDT is used for malaria control, then great harm might occur. So, in an amazing turn, they conclude they cannot prove DDT causes harm, but still predict severe harm if it is used.

Rogan and Chen end their paper with a call for more research. One could conclude that the intent of the whole paper is merely to lobby for research to better define DDT harm, and what's the harm in that? Surely increasing knowledge is a fine goal. However, if you look at the specific issue of the relative need for research, you will see that the harm of this technique is great. Millions of children and pregnant women die from malaria every year, and the disease sickens hundreds of millions more. This is an indisputable fact: impoverished people engage in real life and death struggles every day with malaria. This also is a fact: not one death or illness can be attributed to an environmental exposure to DDT. Yet, a National Library of Medicine literature search on DDT reveals over 1,300 published papers from the year 2000 to the present, almost all in the environmental literature and many on potential adverse effects of DDT. A search on malaria and DDT reveals only 159 papers. DDT is a spatial repellent and hardly an insecticide at all, but a search on DDT and repellents will reveal only 7 papers. Is this not an egregiously disproportionate research emphasis on non-sources of harm compared to the enormous harm of malaria? Does not this inequity contribute to the continued suffering of those who struggle with malaria? Is it possibly even more than an inequity? Is it not an active wrong?

Public health officials and scientists should not be silent about enormous investments into the research of theoretical risks while millions die of preventable diseases. We should seriously consider our motivations in apportioning research money as we do. For consider this: the U.S. used DDT to eradicate malaria. After malaria disappeared as an endemic disease in the United States, we became richer. We built better and more enclosed houses. We screened our windows and doors. We air conditioned our homes. We also developed an immense arsenal of mosquito control tools and chemicals. Today, when we have a risk of mosquito borne disease, we can bring this arsenal to bear and quickly eliminate risks. And, as illustrated by aerial spray missions in the aftermath of hurricane Katrina, we can afford to do so. Yet, our modern and very expensive chemicals are not what protect us from introductions of the old diseases. Our arsenal responds to the threat; it does not prevent the appearance of old diseases in our midst. What protects us is our enclosed, screened, air-conditioned housing, the physical representation of our wealth. Our wealth is the factor that stops dengue at the border with

Mexico, not our arsenal of new chemicals. Stopping mosquitoes from entering and biting us inside our homes is critical in the prevention of malaria and many other insect-borne diseases. This is what DDT does for poor people in poor countries. It stops large proportions of mosquitoes from entering houses. It is, in fact, a form of chemical screening, and until these people can afford physical screening or it is provided for them, this is the only kind of screening they have.

DDT is a protective tool that has been taken away from countries around the world, mostly due to governments acceding to the whims of the anti-pesticide wing of environmentalism, but it is not only the anti-pesticide wing that lobbies against DDT. The activists have a sympathetic lobbying ally in the pesticide industry. As evidence of insecticide industry working to stop countries from using DDT, I am attaching an email message dated 23rd September and authored by a Bayer official. . . . The Bayer official states

> "[I speak] Not only as the responsible manager for the vector control business in Bayer, being the market leader in vector control and pointing out by that we know what we are talking about and have decades of experiences in the evolution of this very particular market. [but] Also as one of the private sector representatives in the RBM Partnership Board and being confronted with that discussion about DDT in the various WHO, RBM et al circles. So you can take it as a view from the field, from the operational commercial level—but our companies [sic] point of view. I know that all of my colleagues from other primary manufacturers and internationally operating companies are sharing my view."

The official goes on to say that

> "DDT use is for us a commercial threat (which is clear, but it is not that dramatical because of limited use), it is mainly a public image threat."

However the most damning part of this message was the statement that

> "we fully support EU to ban imports of agricultural products coming from countries using DDT"

[There is] . . . clear evidence of international and developed country pressures to stop poor countries from using DDT to control malaria. This message also shows the complicity of the insecticide industry in those internationally orchestrated efforts.

Pressures to eliminate spray programs, and DDT in particular, are wrong. I say this not based on some projection of what might theoretically happen in the future according to some model, or some projection of theoretical harms, I say this based firmly on what has already occurred. The track record of the anti-pesticide lobby is well documented, the pressures on developing countries to abandon their spray programs are well documented, and the struggles of developing countries to maintain their programs or restart their uses of DDT for malaria control are well documented. The tragic results of pressures against the use of DDT, in terms of increasing disease and death, are quantified and well documented. How long will scientists, public health officials, the voting public, and the politicians who lead us continue policies, regulations and funding that have led us to the current state of a global humanitarian disaster? How long will support continue for policies and programs that favor phantoms over facts?

POSTSCRIPT

Should DDT Be Banned Worldwide?

Professor Roberts is not alone in his disapproval of the efforts to halt the use of DDT. Alexander Gourevitch, "Better Living Through Chemistry," *Washington Monthly* (March 2003), comes close to accusing environmentalists of condemning DDT on the basis of politics or ideology rather than of science. Angela Logomasini comes even closer in "Chemical Warfare: Ideological Environmentalism's Quixotic Campaign Against Synthetic Chemicals," in Ronald Bailey, ed., *Global Warming and Other Eco-Myths: How the Environmental Movement Uses False Science to Scare Us to Death* (Prima, 2002). Her admission that public health demands have softened some environmentalists' resistance to the use of DDT points to a basic truth about environmental debates: over and over again, they come down to what we should do first: Should we meet human needs regardless of whether or not species die and air and water are contaminated? Or should we protect species, air, water, and other aspects of the environment even if some human needs must go unmet? What if this means endangering the lives of children? In the debate over DDT, the human needs are clear, for insect-borne diseases have killed and continue to kill a great many people. Yet the environmental needs are also clear. The question is one of choosing priorities and balancing risks. See John Danley, "Balancing Risks: Mosquitoes, Malaria, Morality, and DDT," *Business and Society Review* (Spring 2002). It is worth noting that John Beard, "DDT and Human Health," *Science of the Total Environment* (February 2006), finds the evidence for the ill effects of DDT more convincing and says that it is still too early to say it does not contribute to human disease.

Mosquitoes can be controlled in various ways: Swamps can be drained (which carries its own environmental price), and other breeding opportunities can be eliminated. Fish can be introduced to eat mosquito larvae. And mosquito nets can be used to keep the insects away from people. But these (and other) alternatives do not mean that there does not remain a place for chemical pesticides. In "Pesticides and Public Health: Integrated Methods of Mosquito Management," *Emerging Infectious Diseases* (January–February 2001), Robert I. Rose, an arthropod biotechnologist with the Animal and Plant Health Inspection Service of the U.S. Department of Agriculture, says, "Pesticides have a role in public health as part of sustainable integrated mosquito management. Other components of such management include surveillance, source reduction or prevention, biological control, repellents, traps, and pesticide-resistance management." "The most effective programs today rely on a range of tools," says McGinn in "Combating Malaria," *State of the World 2003* (W. W. Norton, 2003). Still, some countries see DDT as essential. See Charles Wendo, "Uganda Considers DDT to Protect Homes From

Malaria," *The Lancet* (April 24, 2004). Indonesia makes its case in "Bring Back DDT," *Far Eastern Economic Review* (March 4, 2004). Tina Rosenberg attempts to speak for all in "What the World Needs Now Is DDT," *The New York Times Magazine* (April 11, 2004). Kirsten Weir, "The Exterminator," *Current Science* (November 5, 2004), argues that malaria is not going away and DDT is an essential weapon against it.

It has proven difficult to find effective, affordable drugs against malaria; see Ann M. Thayer, "Fighting Malaria," *Chemical and Engineering News* (October 24, 2005), and Claire Panosian Dunavan, "Tackling Malaria," *Scientific American* (December 2005). A great deal of effort has gone into developing vaccines against malaria, but the parasite has demonstrated a persistent talent for evading all attempts to arm the immune system against it. See Z. H. Reed, M. Friede, and M. P. Kieny, "Malaria Vaccine Development: Progress and Challenges," *Current Molecular Medicine* (March 2006). A newer approach is to develop genetically engineered (transgenic) mosquitoes that either cannot support the malaria parasite or cannot infect humans with it; see Louis Lambrechts, Jean-Marc Chavatte, Georges Snounou, and Jacob C. Koella, "Environmental Influence on the Genetic Basis of Mosquito Resistance to Malaria Parasites," *Proceedings: Biological Sciences* (June 2006), and Jane Bradbury, "Transgenic Mosquitoes Bring Malarial Control Closer," *The Lancet* (May 25, 2002).

It is worth stressing that malaria is only one of several mosquito-borne diseases that pose threats to public health. Two others are yellow fever and dengue. A recent arrival to the United States is West Nile virus, which mosquitoes can transfer from birds to humans. However, West Nile virus is far less fatal than malaria, yellow fever, or dengue, and a vaccine is in development. See Dwight G. Smith, "A New Disease in the New World," *The World & I* (February 2002) and Michelle Mueller, "The Buzz on West Nile Virus," *Current Health 2* (April/May 2002). But "West Nile Virus Still a Threat," says *Clinical Infectious Diseases* (April 15, 2006).

It is also worth stressing that global warming means climate changes that may increase the geographic range of disease-carrying mosquitoes. Many climate researchers are concerned that malaria, yellow fever, and other now mostly tropical and subtropical diseases may return to temperate-zone nations and even spread into areas where they have never been known.

ISSUE 17

Do Environmental Hormone Mimics Pose a Potentially Serious Health Threat?

YES: **Michele L. Trankina**, from "The Hazards of Environmental Estrogens," *The World & I* (October 2001)

NO: **Michael Gough**, from "Endocrine Disrupters, Politics, Pesticides, the Cost of Food and Health," *Cato Institute* (December 15, 1997)

ISSUE SUMMARY

YES: Professor of biological sciences Michele L. Trankina argues that a great many synthetic chemicals behave like estrogen, alter the reproductive functioning of wildlife, and may have serious health effects—including cancer—on humans.

NO: Michael Gough, a biologist and expert on risk assessment and environmental policy, argues that only "junk science" supports the hazards of environmental estrogens.

Following World War II there was an exponential growth in the industrial use and marketing of synthetic chemicals. These chemicals, known as "xeno-biotics," were used in numerous products, including solvents, pesticides, refrigerants, coolants, and raw materials for plastics. This resulted in increasing environmental contamination. Many of these chemicals, such as DDT, PCBs, and dioxins, proved to be highly resistant to degradation in the environment; they accumulated in wildlife and were serious contaminants of lakes and estuaries. Carried by winds and ocean currents, these chemicals were soon detected in samples taken from the most remote regions of the planet, far from their points of introduction into the ecosphere.

Until very recently most efforts to assess the potential toxicity of synthetic chemicals to bio-organisms, including human beings, focused almost exclusively on their possible role as carcinogens. This was because of legitimate public concern about rising cancer rates and the belief that cancer causation was the most likely outcome of exposure to low levels of synthetic chemicals.

Some environmental scientists urged public health officials to give serious consideration to other possible health effects of xenobiotics. They were generally ignored because of limited funding and the common belief that toxic effects other than cancer required larger exposures than usually resulted from environmental contamination.

In the late 1980s Theo Colborn, a research scientist for the World Wildlife Fund who was then working on a study of pollution in the Great Lakes, began linking together the results of a growing series of isolated studies. Researchers in the Great Lakes region, as well as in Florida, the West Coast, and Northern Europe, had observed widespread evidence of serious and frequently lethal physiological problems involving abnormal reproductive development, unusual sexual behavior, and neurological problems exhibited by a diverse group of animal species, including fish, reptiles, amphibians, birds, and marine mammals. Through Colborn's insights, communications among these researchers, and further studies, a hypothesis was developed that all of these wildlife problems were manifestations of abnormal estrogenic activity. The causative agents were identified as more than 50 synthetic chemical compounds that have been shown in laboratory studies to either mimic the action or disrupt the normal function of the powerful estrogenic hormones responsible for female sexual development and many other biological functions.

Concern that human exposure to these ubiquitous environmental contaminants may have serious health repercussions was heightened by a widely publicized European research study, which concluded that male sperm counts had decreased by 50 percent over the past several decades (a result that is disputed by other researchers) and that testicular cancer rates have tripled. Some scientists have also proposed a link between breast cancer and estrogen disrupters.

In response to the mounting scientific evidence that environmental estrogens may be a serious health threat, the U.S. Congress passed legislation requiring that all pesticides be screened for estrogenic activity and that the Environmental Protection Agency (EPA) develop procedures for detecting environmental estrogenic contaminants in drinking water supplies; see the EPA's Endocrine Disruptor Screening Program Web site at http://www.epa.gov/scipoly/oscpendo/index.htm. Government-sponsored studies of synthetic endocrine disrupters and other hormone mimics are also under way in the United Kingdom and in Germany.

In the following selections, Michele L. Trankina argues that a great many synthetic chemicals behave like estrogen, alter the reproductive functioning of wildlife, and may have serious health effects—including cancer—on humans. She insists that regulatory agencies must minimize public exposure. Michael Gough argues that only "junk science" supports the hazards of environmental estrogens. Expensive testing and regulatory programs can only drive up the cost of food, he says, which will make it harder for the poor to afford fresh fruits and vegetables. Furthermore, health protection will not be increased.

YES

Michele L. Trankina

The Hazards of Environmental Estrogens

What do Barbie dolls, food wrap, and spermicides have in common? And what do they have to do with low sperm counts, precocious puberty, and breast cancer? "Everything," say those who support the notion that hormone mimics are disrupting everything from fish gender to human fertility. "Nothing," counter others who regard the connection as trumped up, alarmist chemophobia. The controversy swirls around the significance of a number of substances that behave like estrogens and appear to be practically everywhere—from plastic toys to topical sunscreens.

Estrogens are a group of hormones produced in both the female ovaries and male testes, with larger amounts made in females than in males. They are particularly influential during puberty, menstruation, and pregnancy, but they also help regulate the growth of bones, skin, and other organs and tissues.

Over the past 10 years, many synthetic compounds and plant products present in the environment have been found to affect hormonal functions in various ways. Those that have estrogenic activity have been labeled as environmental estrogens, ecoestrogens, estrogen mimics, or xenoestrogens (*xenos* means foreign). Some arise as artifacts during the manufacture of plastics and other synthetic materials. Others are metabolites (breakdown products) generated from pesticides or steroid hormones used to stimulate growth in livestock. Ecoestrogens that are produced naturally by plants are called phytoestrogens (*phyton* means plant).

Many of these estrogen mimics bind to estrogen receptors (within specialized cells) with roughly the same affinity as estrogen itself, setting up the potential to wreak havoc on reproductive anatomy and physiology. They have therefore been labeled as disruptors of endocrine function.

Bizarre Changes in Reproductive Systems

Heightened concern about estrogen-mimicking substances arose when several nonmammalian vertebrates began to exhibit bizarre changes in reproductive anatomy and fertility. Evidence that something was amiss came serendipitously in 1994, from observations by reproductive physiologist Louis Guillette of the University of Florida. In the process of studying the decline of alligator populations at Lake Apopka, Florida, Guillette and coworkers noticed that many male alligators had smaller penises than normal. In addition, females superovulated, with multiple nuclei in some of the surplus ova. Closer scrutiny linked these findings to a massive spill of DDT (dichloro-diphenyl-trichloroethane) into Lake Apopka in 1980. Guillette concluded that the declining alligator population was related to the effects of DDT exposure on the animals' reproductive systems.

Although DDT was banned for use in the United States in the early 1970s, it continues to be manufactured in this country and marketed abroad, where it is sprayed on produce that is then sold in U.S. stores. The principal metabolite derived from DDT is called DDE, a xenoestrogen that lingers in fat deposits in the human body for decades. Historically, there have been reports of estrogen-mimicking effects in various fish species, especially in the Great Lakes, where residual concentrations of DDT and PCBs (polychlorinated biphenyls) are high. These effects include feminization and hermaphroditism in males. In fact, fish serve as barometers of the effects of xenoestrogen contamination in bodies of water. An index of exposure is the presence of vitellogenin, a protein specific to egg yolk, in the blood of male fish. Normally, only females produce vitellogenin in their livers, upon stimulation by estrogen from the ovaries.

Ecoestrogens are further concentrated in animals higher up in the food chain. In the Great Lakes region, birds including male herring gulls, terns, and bald eagles exhibit hermaphroditic changes after feeding on contaminated fish. Increased embryo mortality among these avians has also been observed. In addition, evidence from Florida links sterility in male and female panthers to their predation on animals exposed to pesticides with estrogenic activity.

The harmful effects of DDT have also been observed in rodents. Female rodents treated with high concentrations of DDT become predisposed to mammary tumors, while males tend to develop testicular cancer. These observations raise the question of whether pharmacological (that is, low) doses of substances with estrogenic activity translate into physiological effects. Usually they do not, but chronic exposure may be enough to trigger such effects.

Dangers to Humans

If xenoestrogens cause such dramatic reproductive effects in vertebrates, including mammals, what might be the consequences for humans? Nearly a decade ago, Frederick vom Saal, a developmental biologist at the University of Missouri Columbia, cautioned that mammalian reproductive mechanisms are similar enough to warrant concern about the effects of hormone disruptors on humans.

A 1993 article in *Lancet* looked at decreasing sperm counts in men in the United States and 20 other countries and correlated these decreases with the growing concentration of environmental estrogens. The authors—Niels Skakkenbaek, a Danish reproductive endocrinologist, and Richard Sharpe, of the British Medical Research Council Reproductive Biology Unit in Scotland—performed a meta-analysis of 61 sperm-count studies published between 1938 and 1990 to make their connection.

"Nonbelievers" state that this interpretation is contrived. Others suggest alternative explanations. For instance, the negative effects on sperm counts could have resulted from simultaneous increases in the incidence of venereal diseases. Besides, there are known differences in steroid hormone metabolism between lower vertebrates (including nonprimate mammals) and primates, so one cannot always extrapolate from the former group to the latter.

Even so, the incidence of testicular cancer, which typically affects young men in their 20s and 30s, has increased worldwide. Between 1979 and 1991, over 1,100 new cases were reported in England and Wales—a 55 percent increase over previous rates. In Denmark, the rate of testicular cancer increased by 300 percent from 1945 to 1990. Intrauterine exposure to xenoestrogens during testicular development is thought to be the cause.

Supporting evidence comes from Michigan, where accidental contamination of cattle feed with PCBs in 1973 resulted in high concentrations in the breast milk of women who consumed tainted beef. Their sons exhibited defective genitalia. Furthermore, observers in England have noted increased incidences of cryptorchidism (undescended testes), which results in permanent sterility if left untreated, and hypospadias (urethral orifice on the underside of the penis instead of at the tip). The one area of agreement between those who attribute these effects to ecoestrogens and those who deny such a connection is that more research is needed.

Perhaps one of the most disturbing current trends is the alarming increase in breast cancer incidence. Fifty years ago, the risk rate was 1 woman in 20; today it is 1 in 8. Numerous studies have implicated xenoestrogens as the responsible agents. For instance, high concentrations of pesticides, especially DDT, have been found in the breast tissue of breast cancer patients on Long Island. In addition, it has long been known that under certain conditions, estrogen from any source can be tumor promoting, and that most breast cancer cell types have estrogen receptors.

Precocious Puberty

If that were not enough, the growing number of estrogen mimics in the environment has been linked to early puberty in girls. The normal, average age of onset is between 12 and 13. A recent study of 17,000 girls in the United States indicated that 7 percent of white and 27 percent of black girls exhibited physical signs of puberty by age seven. For 10-year-old girls, the percentages increased to 68 and 95, respectively. Studies from the United Kingdom, Canada, and New Zealand have shown similar changes in the age of puberty onset.

It is difficult, however, to elucidate the exact mechanisms that underlie these trends toward precocious puberty. One explanation, which applies especially to cases in the United States, points to the increasing number of children who are overweight or obese as a result of high-calorie diets and lack of regular exercise. Physiologically, an enhanced amount of body fat implies reproductive readiness and signals the onset of puberty in both boys and girls.

For girls, more body fat ensures that there is enough stored energy to support pregnancy and lactation. Young women with low percentages of body fat caused by heavy exercise, sports, or eating disorders usually do not experience the onset of menses until their body composition reflects adequate fat mass.

Many who study the phenomenon of premature puberty attribute it to environmental estrogens in plastics and secondhand exposure through the meat and milk of animals treated with steroid hormones. An alarming increase in the numbers of girls experiencing precocious puberty occurred in the 1970s and '80s in Puerto Rico. Among other effects, breast development occurred in girls as young as one year. Premature puberty was traced to consumption of beef, pork, and dairy products containing high concentrations of estrogen.

Another study from Puerto Rico revealed higher concentrations of phthalate—a xenoestrogen present in certain plastics—in girls who showed signs of early puberty, compared with controls.

It may be that excess body fat and exposure to estrogenic substances operate in concert to hasten puberty. Body fat is one site of endogenous estrogen synthesis. Exposure to environmental estrogens may add just enough exogenous hormone to exert the synergistic effect necessary to bring on puberty, much like the last drop of water that causes the bucket to overflow.

Although most xenoestrogens produce detrimental effects, at least one subgroup—the phytoestrogens—includes substances that can have beneficial effects. Phytoestrogens are generally weaker than natural estrogens. They are found in various foods, such as flax seeds, soybeans and other legumes, some herbs, and many fruits and vegetables. Some studies suggest that soy products may offer protection against certain cancers, including breast, prostate, uterus, and colon cancers. On the other hand, high doses of certain phytoestrogens— such as coumestrol (in sunflower seeds and alfalfa sprouts)—have been found to adversely affect the fertility and reproductive cycles of animals.

Unlike artificially produced xenoestrogens, phytoestrogens are generally not stored in the body but are readily metabolized and excreted. Their health effects should be evaluated on a case-by-case basis, considering such factors as the individual's age, medical and family history, and potential interactions with medications or supplements.

Plastics, Plastics Everywhere

It has been estimated that perhaps 100,000 synthetic chemicals are registered for commercial use in the world today and 1,000 new ones are formulated every year. While many are toxic and carcinogenic, little is known about the chronic effects of the majority of them. And there is growing concern about their potential hormone-disrupting effects. The problem of exposure is complicated by numerous carrier routes, including air, food, water, and consumer products.

Consider certain synthetics that turn up in familiar places, including food and consumer goods. Such seemingly inert products as plastic soda and water bottles, baby bottles, food wrap, Styrofoam, many toys, cosmetics, sunscreens, and even spermicides either contain or break down to yield xenoestrogens. In addition, environmental estrogens are among the byproducts created by such processes as the incineration of biological materials or industrial waste and chlorine bleaching of paper products.

In April 1999, Consumers Union confirmed information previously reported by the Food and Drug Administration regarding 95 percent of baby bottles sold in the United States. The bottles, made of a hard plastic known as polycarbonate, leach out the synthetic estrogen named bisphenol-A, especially when heated or scratched. Studies verifying the estrogenic activity of bisphenol-A were published in *Nature* in the 1930s, but it did not arouse much concern then. A 1993 report published in *Endocrinology* showed that bisphenol-A produced estrogenic effects in a culture of human breast cancer cells.

Additional studies published by vom Saal in 1997 and '98 have shown that bisphenol-A stimulates precocious puberty and obesity in mice. Others have detected leaching of bisphenol-A from polycarbonate products such as plastic tableware, water cooler jugs, and the inside coatings of certain cans (used for some canned foods) and bottle tops. Autoclaving in the canning process causes bisphenol-A to migrate into the liquid in cans.

Spokespersons from polycarbonate manufacturers have stated that they cannot replicate vom Saal's results, but he counters that industry researchers have not done the experiments correctly.

DEHA (di-[2-ethylhexyl] adipate) is a liquid plasticizer added to some plastic food wraps made of polyvinyl chloride (PVC). Scientific studies have shown that the fat-soluble DEHA can migrate into foods, especially luncheon meats, cheese, and other products with a high fat content. For a 45-pound child eating cheese wrapped in such plastic, the limit of safe intake is 1.5 ounces by European standards or 2.5 ounces by Environmental Protection Agency criteria. Studies conducted by Consumers Union indicate that DEHA leaching from commercial plastic wraps is eight times higher than European directives allow. Fortunately, alternative wraps—such as Handi-Wrap™ and Glad Cling Wrap—are made of polyethylene; chemicals in them do not appear to leach into foods.

Barbie dolls manufactured in the 1950s and '60s are made of PVC containing a stabilizer that degrades to a sticky, estrogen-mimicking residue and accumulates on the dolls' bodies. This phenomenon was noted by Danish museum officials in August 2000. Yvonne Shashoua, an expert in materials preservation at the National Museum of Denmark, warns that young children who play with older Barbie and Ken dolls expose themselves to this estrogenic chemical, and she suggests enclosing the dolls with xenoestrogen-free plastic wrap. Storing the dolls in a cool, dark place also helps prevent the harmful stabilizer from oozing out. Who would have thought that these models of glamour would become health hazards?

Another consumer nightmare is a group of chemical plasticizers known as phthalates. They have been associated with various problems, including testicular depletion of zinc, a necessary nutrient for spermatogenesis. Zinc deficiency results in sperm death and consequent infertility. Many products that previously contained phthalates have been reformulated to eliminate them, but phthalates continue to be present in vinyl flooring, medical tubing and bags, adhesives, infants' toys, and ink used to print on food wrap made of plastic and cardboard. They have been detected in fat-soluble foods such as infant formula, cheese, margarine, and chips.

Some environmental estrogens can be found in unusual places, such as contraceptive products containing the spermicide nonoxynol-9. This chemical degrades to nonylphenol, a xenoestrogen shown to stimulate breast cancer cells. Nonylphenol and other alkylphenols have been detected in human umbilical cords, plastic test tubes, and industrial detergents.

Various substances added to lotions (including sunscreens) and cosmetics serve as preservatives. Some of them, members of a chemical family called parabens, were shown to be estrogen mimics by a study at Brunel University in the United Kingdom. The researchers warned that "the safety of these chemicals should be reassessed with particular attention being paid to .... levels of systemic exposure to humans." Officials of the European Cosmetic Toiletry and Perfumery Association dismissed the Brunel study as "irrelevant," on the grounds that parabens do not enter the systemic circulation. But they ignored the possibility of transdermal introduction.

Additional questions have been raised about the safety—in terms of xenoestrogen content—of recycled materials, especially plastics and paper. Because it is unlikely that a moratorium on chemical synthesis will occur anytime soon, such questions will continue to surface until the public is satisfied that regulatory agencies are doing all they can to minimize exposure. Fortunately, organizations such as the National Institutes of Health, the National Academy of Sciences, the Environmental Protection Agency, the Centers for Disease Control, many universities, and other institutions are involved in efforts to monitor and minimize the effects of environmental estrogens on wildlife and humans.

Michael Gough  **NO**

Endocrine Disrupters, Politics, Pesticides, the Cost of Food and Health

Environmentalists and politicians and federal regulators have added environmental estrogens or endocrine disrupters to the "concerns" or scares that dictate "environmental health policy." That policy, from its beginning, has been based on ideology, not on science. To provide some veneer to the ideology, its proponents have spawned bad science and junk science that claims chemicals in the environment are a major cause of human illness. There is no substance to the claims, but the current policies threaten to cost billions of dollars in wasted estrogen testing programs and to drive some substantial proportion of pesticides from the market.

Rachel Carson, conjuring up a cancer-free, pre-industrial Garden of Eden launched the biggest environmental scare of all in the 1960s. She charged that modern industrial chemicals in the environment caused human cancers. It mattered not at all to her or to her readers that cancers are found in every society, pre-industrial and modern. What mattered were opinions of people such as Umberto Safiotti of the National Cancer Institute, who wrote:

> I consider cancer as a social disease, largely caused by external agents which are derived from our technology, conditioned by our societal lifestyle and whose control is dependent on societal actions and policies.

When Saffioti said "societal actions and policies," he meant government regulations.

By 1968, environmental groups and individuals—including some scientists—appeared on TV and on the floors of the House and Senate to say, over and over again, "The environment causes 90 percent of cancers." They didn't have to say "environment" meant pollution from modern industry and chemicals—especially pesticides—everyone already knew that. Saffioti and others had told them.

In the 1970s, the National Cancer Institute [NCI] released reports that blamed elevated rates for all kinds of cancers on chemicals in the workplace or in the environment. The institute did not have evidence to link those exposures to cancer. It didn't exist then, and it doesn't, except for a limited number of high exposures in the workplace, exist now. So what? The reports were gobbled up by the press, politicians, and the public.

In our ignorance of what causes most cancers, the "90 percent" misstatement provided great hope. If the carcinogenic agents in the environment could be identified and eliminated, cancer rates should drop. NCI scientists said so, and they said success was just around the corner if animal tests were used to identify carcinogens. Congress responded. It created the Environmental Protection Agency [EPA] and the Occupational Safety and Health Administration [OSHA]. Both agencies have lots of tasks, but both place an emphasis on controlling exposures to carcinogens. Congress passed and amended law after law. The Clean Air Act, the Clean Water Act, the Safe Drinking Water Act, amendments to the Fungicide, Insecticide, and Rodenticide Act—the euphonious FIFRA—the Toxic Substances Control Act, and the Resource Recovery and Control Act poured forth from Capitol Hill.

And in return, EPA and OSHA, to justify their existence, generated scare after scare. They are aided by all kinds of people eager for explanations about their health problems or for government grants and contracts for research or other work or for money to compensate for health effects or other problems that could be blamed on chemicals.

In 1978, we had the occupational exposure scare. Astoundingly, according to a government report, six workplace substances caused 38 percent of all the cancers in the United States. It was nonsense, of course, and many scientists ridiculed the report, but the government never retracted it. The government scientists who contributed to it never repudiated it.

At about the same time, wastes disposed in Love Canal near Niagara Fall, NY, spewed liquids and gases into a residential community. The chemicals were blamed as the cause of cancers, birth defects, miscarriages, skin diseases, you name it. None of it was true, but waste sites around the country were routinely identified as "another Love Canal" or a "Love Canal in the making," and Congress gave the nation the Superfund Law. Since its passage, Superfund has enriched lawyers and provided secure employment to thousands who wear moon suits and dig up, burn, and rebury wastes, and done nothing for the nation's health. For those who doubt the importance of politics in the environmental health saga, it's worth recalling that every state had two waste sites on the first list of sites slated for priority cleanup under Superfund.

By the 1980s, EPA was chucking out the scares. We had the 2,4,5-T scare, the dioxin scare, the 2,4-D scare, the asbestos in schools scare, the radon in homes scare, the Alar scare, the EMF scare. I've left some out, but the common thread that linked the scares together was cancer. Each scare prompted investigations by affected industries and non-government scientists. Each scare fell apart, revealed as a house of cards jerry-rigged from bad science, worse interpretations of the science, and terrible policy.

In fact, by the late 1970s, there was ample evidence that the much-talked about "cancer epidemic" and the 90 percent statement were simply wrong. Cancer rates were not increasing and rates for some cancers were higher in industrial countries and rates for others were higher in non-industrial countries. The U.S. fell in the middle of countries when ranked by cancer rates. Sure, there are some carcinogenic substances in the nation's workplaces, but the best estimates are that they cause four percent or less of all cancers, and the percentage is decreasing because the biggest occupational threat, asbestos, is gone. Environmental exposures might cause two or three percent of cancer—on the outside—and they might cause much less.

The research into causes of cancer—not stories designed to bolster the chemicals cause cancer myth—did reveal that there are preventable causes of cancer. Not smoking is a good idea, as is eating lots of fruits and vegetables, not gaining too much weight, restricting the number of sexual partners, and, for those who are fair-skinned, being careful about sun exposures. It's not a lot different from what your mother or grandmother told you.

The government can take a nanny role in urging us to behave, but that's not where the big bucks are. The big bucks are in regulation, and regulation doesn't seem to have much to do with cancer.

In any case, cancer death rates began to fall in 1990, they've fallen since, and the fall is growing steeper. Maybe that information is blunting the cancer scare. I somehow doubt it. I think that the public has become numb to the cancer scare or that it fatalistically accepts the notion that "everything causes cancer." In any case, the environmentalists and the regulators needed a new scare.

The collapse of the cancer scare wasn't good news to everyone. Government bureaucrats and scientists in the anti-carcinogen offices and programs at EPA and elsewhere have secure jobs. Congress easily finds the will to write laws establishing environmental protection activities, but it lacks the will or patience to examine those activities to see if they've accomplished anything. And, let's face it, Congress doesn't eliminate established programs. But the growth of programs slows, and money can become scarce, and that can squeeze researchers who depend on EPA grants and contracts to fund their often senseless surveys and testing programs. Moreover, the fading of scares doesn't benefit environmental organizations that utter shrill cries about scares and coming calamities in their campaigns for contributions.

Here's an example of just how disappointed some people can be that cancer isn't on the rise. Dr. Theo Colborn, a wildlife biologist working for the Conservation Foundation in the late 1980s, was convinced that the chemicals in the Great Lakes were causing human cancer. She set out to prove it by reviewing the available literature about cancer rates in that region. She couldn't. In fact, she found that the rates for some cancers in the Great Lakes region were lower than the rates for the same cancers in other parts of the United States and Canada....

Failing to find cancer slowed her down but didn't stop her. She knew that those chemicals were causing something. All she had to do was find it.

And find it, she did. She collected every paper that described any abnormality in wildlife that live on or around the Great Lakes, and concluded that synthetic chemicals were mimicking the effects of hormones. They were causing every problem in the literature, whether it was homosexual behavior among gulls, crossed bills in other birds, cancer in fish, or increases or decreases in any wildlife population.

The chemicals that have those activities were called "environmental estrogens" or "endocrine disrupters." There was no more evidence to link them to every abnormality in wildlife than there had been in the 1960s to link every human cancer to chemicals. The absence of evidence wasn't much of a problem. Colborn and her colleagues believed that chemicals were the culprit, and the press and much of the public, nutured on the idea that chemicals were bad, didn't require evidence.

Even so, Colborn had a problem that EPA faced in its early days. Soon after EPA was established, the agency leaders realized that protecting wildlife and the environment might be a good thing, but that Congress might not decide to lavish funds on such activities. They were sure, however, that Congress would throw money at programs that were going to protect human health from environmental risks. Whether Colborn knew that history or not, she apparently realized that any real splash for endocrine disrupters depended on tying them to human health effects.

Using the same techniques she'd used to catalogue the adverse effects of endocrine disrupters on wildlife, she reviewed the literature about human health effects that some way or another might be related to disruption of hormone activity. The list was long, including cancers, birth defects, and learning disabilities, but the big hitter on the list was decreased sperm counts. According to Colborn and others' analyses of sperm counts made in different parts of the world under different conditions of nutrition and stress and at different time periods, sperm counts had decreased by 50 percent in the post–World War II period.

If there's anything that catches the attention of Congress, it's risks to males. Congress banned leaded gasoline after EPA released a report that said atmospheric lead was a cause of heart attacks in middle-aged men. The reported decrease in sperm counts leaped up for attention, and attention it got. Congressional hearings were held, magazine articles were written, experts opined about endocrine disrupters and sexual dysfunctions.

And then it fell apart. Scientists found large geographical variations in sperm counts that have not changed over time. Those geographical variations and poor study designs accounted for the reported decrease. That scare went away, but endocrine disrupters were here to stay.

Well-organized and affluent women's groups are convinced that breast cancer is unusually common on Long Island.... We know that obesity, estrogen replacement therapy, and late child-bearing or no child-bearing, all of which are more common in affluent women, are associated with breast cancer. Nevertheless, from the very beginning, environmental chemicals have been singled out as the cause of the breast cancer excess. The insecticide aldicarb, which is very resistant to degradation was blamed, but subsequent studies failed to confirm the link. A well-publicized study found a link between DDT and breast cancer, but larger, follow-up studies failed to confirm it. But there're lots of environmental chemicals, and no evidence is required to justify a suggestion of a link between the chemicals and cancer.

Senator Al D'Amato is from Long Island, and he shares his constituents' concerns. During a hearing about the Clean Water Act, Senator D'Amato heard testimony by Dr. Anna Soto from Tufts University about her "E-Screen." According to Dr. Soto, her quick laboratory test could identify chemicals that behave as environmental estrogens or endocrine disrupters for $500 a chemical. Since environmental estrogens seem to some people to be a likely cause of breast cancer, Soto's test appeared to be a real bargain.

Senator D'Amato pushed for an amendment to require E-Screen testing of chemicals that are regulated under the Clean Water Act, but he was unsuccessful. Later in 1995, a senior Senate staffer, Jimmy Powell, took the E-Screen amendment to a very junior Senate staffer and told her to incorporate it into the Safe Drinking Water Act as an "administrative amendment." She did, it passed the Senate, and, for the first time, there was a legislative requirement for endocrine testing.

In the spring of 1996, House Committees were considering legislation to amend the Safe Drinking Water Act and new legislation related to pesticides in food. Aware of the Senate's action, some members of the House committees were eager to include endocrine testing in their legislation, but there was resistance as well. Chemical companies viewed the imposition of yet another test as certain to be an expense, unlikely to cost as little as $500 a chemical, and bound to raise new concerns about chemicals that would require far more extensive tests and research to understand or discount.

Furthermore, so far as food was concerned, there was a general conviction that all the safety factors built into the testing of pesticides and other chemicals that might end up in food provided adequate margins of protection. That conviction was shattered by rumors that reached the House in May 1996. According to the rumors, Dr. John McLachlan and his colleagues at Tulane University had shown that mixtures of pesticides and other environmental chemicals such as PCBs were far more potent in activating estrogen receptors, the first step in estrogen modulation of biochemical pathways than were single chemicals. In the most extreme case, two chemicals at concentrations considered safe by all conventional toxicity tests were 1600 times as potent as estrogen receptor activators as either chemical by itself. The powerful synergy raised new alarms.

In May, everyone concerned about pesticides knew that EPA had a draft of the Tulane paper, and EPA staff were drifting around House offices, but they refused to answer questions about the Tulane results. The silence signaled the expected significance of the paper. A month later, in June, the paper appeared in *Science*. It was a big deal. *Science* ran a news article about the research with a picture of the Tulane researchers. It also ran an editorial by a scientist from the National Institutes of Health who offered some theoretical explanations for how combinations of pesticides at very low levels could affect cells and activate the estrogen receptors. *The New York Times, The Washington Post*, other major newspapers, and newsmagazines and TV reported the news. If there was ever any doubt that FQPA [Food Quality Protection Act] would require tests for endocrine activity, the flurry of news about the Tulane results erased them.

While the House was drafting the FQPA, Dr. Lynn Goldman, an assistant administrator at EPA, established a committee called the "Endocrine Disrupter Screening and Testing Advisory Committee" (EDSTAC) [which is now] considering tests for all of the 70,000 chemicals that it estimates are present in commerce, and it's not limiting its recommendations to tests for estrogenic activity.

It's adding tests for testosterone and thyroid hormone activity as well as for anti-estrogenic, anti-testosterone, and anti-thyroid activity. The relatively simple E-Screen, which is run on cultured cells, is to be supplemented by some whole animal tests. Tests on single compounds will have to be complemented by tests of mixtures of compounds. The FQPA requires that "valid" tests be used. None of the tests being considered by EDSTAC has been validated; many of them have never been done.

EDSTAC's estimate of 70,000 chemicals in commerce is on the high side—some of those chemicals are used in such small amounts and under such controlled conditions that there's no exposure to them. Dr. Dan Byrd has estimated that 50,000 is a more realistic number. He's also looked at the price lists from commercial testing laboratories to see how much they would charge for a battery of tests something like EDSTAC is considering. Some of the tests haven't been developed, but assuming they can be, Dr. Byrd estimates testing each chemical will cost between $100,000 and $200,000. The total cost would be between $5 and $10 billion....

The Tulane results played some major role in the passage of FQPA, the focus on endocrine disrupters, and Dr. Goldman's establishment of EDSTAC. The Tulane results are wrong. Several groups of scientists tried to replicate the Tulane results. None could. At first, Dr. McLachlan insisted his results were correct. He said that the experiments he reported required expertise and finesse and suggested that the scientists who couldn't repeat his findings were at fault, essentially incompetent. That changed. In July 1997, just 13 months after he published his report, he threw in the towel, acknowledging that neither his laboratory nor anyone else had been able to produce the results that had created such a stir.

Whether the initial results were caused by a series of mistakes or a willful desire to show, once and for all, that environmental chemicals, especially pesticides are bad, bad, bad, we don't know. We do know that the results were wrong.

No matter, EPA now assumes as a matter of policy that synergy occurs. Good science, repeatable science that showed the reported synergy didn't occur has been brushed aside. In its place, we have bad science or junk science. If the Tulane results were the products of honest mistakes, they're bad science; if they flowed from ideology, they're junk science. The effect is the same, but the reasons are different.

The estrogenic disrupter testing under FQPA is going to cost a lot of money and cause a lot of mischief. But the effects of that testing are off somewhere in the future. More immediately, a combination of ideology-driven science and congressional misreading of that science threatens to drive between 50 and 80 percent of all pesticides from the market.

In 1993, a committee of the National Research Council spun together the facts that childhood developmental takes place at specific times as an infant matures into a toddler and then a child, that infants, toddlers, and children eat, proportionally, far larger amounts of foods such as apple juice and apple sauce and orange juice than do adults, and that pesticides can be present in those processed foods. From those three observations, the committee concluded that an additional safety factor should be included in setting acceptable levels for pesticides in those foods. Left out from the analysis was any evidence that current exposures cause any harm to any infants, toddlers, or children. No matter.

Most people who worry about pesticides expected EPA and the Food and Drug Administration to react to the NRC recommendation by reducing the allowable levels of pesticides on foods that are destined for consumption by children. Maybe they would have. We'll never know. In the FQPA, Congress directed that a new ten-fold safety factor be incorporated into the evaluation of the risks from pesticides.

Safety factors are a fundamental part in the evaluation of pesticide risks. Pesticides are tested in laboratory animals to determine what concentrations to cause effects on the nervous, digestive, endocrine, and other systems. At some, sufficiently low dose that varies from pesticide to pesticide, the chemical does not cause those adverse effects. That dose, called the "No Observed Adverse Effect Level" (or NOAEL), is then divided by 100 to set the acceptable daily limit for human ingestion of the chemical. The FQPA requires division by another factor of 10, so the acceptable daily limit will be the NOAEL divided by 1000 instead of 100. Acceptable limits will be ten-fold less.

Dr. Byrd has estimated that up to 80 percent of all currently permitted uses of pesticides would be eliminated by an across the board application of the 1000-fold safety factor. He cites another toxicologist who estimates that 50 percent of all pesticides would be eliminated from the market. The extent to which these draconian reductions will be forced remains to be seen, but pesticide manufacturers and users can look forward to a period of even-greater limbo as EPA sorts through it new responsibilities and decides how to implement FQPA.

There's no convincing evidence that pesticides in food contribute to cancer causation and none that they cause other adverse health effects. Restrictions on pesticides in food will not have a demonstrable effect on human health. On the other hand, the estrogen testing program and the new safety factor will drive pesticide costs up and pesticide availability down.

Some manufacturers may lose profitable product lines; some may even lose their businesses. Farmers will pay more. They will pass those costs onto middlemen and processors, who, in turn, will pass them onto consumers. Increases in the costs of fruits and vegetables won't change the food purchasing habits of the middle class, but they may and probably will affect the purchases of the poor. The poor are already at greater risks because of poor diets, and the increased costs can be expected to further decrease their consumption of fresh fruits and vegetables.

POSTSCRIPT

Do Environmental Hormone Mimics Pose a Potentially Serious Health Threat?

Stephen H. Safe's "Environmental and Dietary Estrogens and Human Health: Is There a Problem?" *Environmental Health Perspectives* (April 1995) is often cited to support the contention that there is no causative link between environmental estrogens and human health problems. He draws a cautious conclusion, calling the link "implausible" and "unproven." Gough's belief that the battle against environmental estrogens is motivated by environmentalist ideology rather than facts is repeated by Angela Logomasini in "Chemical Warfare: Ideological Environmentalism's Quixotic Campaign Against Synthetic Chemicals," in Ronald Bailey, ed., *Global Warming and Other Eco-Myths: How the Environmental Movement Uses False Science to Scare Us to Death*, (Prima, 2002). Some caution is certainly warranted, for the complex and variable manner by which different compounds with estrogenic properties may affect organisms makes projections from animal effects to human effects risky.

Sheldon Krimsky, in "Hormone Disruptors: A Clue to Understanding the Environmental Cause of Disease," *Environment* (June 2001) summarizes the evidence that many chemicals released to the environment affect—both singly and in combination or synergistically—the endocrine systems of animals and humans and may threaten human health with cancers, reproductive anomalies, and neurological effects. He cautions that the regulatory machinery is likely to move very slowly, adding that we cannot wait for scientific certainty about the hazards before we act. See also M. Gochfeld, "Why Epidemiology of Endocrine Disruptors Warrants the Precautionary Principle," *Pure & Applied Chemistry* (December 1, 2003).

Theo Colborn, a senior scientist with the World Wildlife Fund, first drew public attention to the potential problems of environmental estrogens with the book *Our Stolen Future* (Dutton, 1996), coauthored by Dianne Dumanoski and John Peterson Myers. Colborn clearly believes that the problem is real; she finds the evidence that extensive damage is being done to wildlife by synthetic estrogenic chemicals convincing and thinks it likely that humans are experiencing similar health problems. Recent data reinforce her and Krimsky's points; see Rebecca Renner, "Human Estrogens Linked to Endocrine Disruption," *Environmental Science and Technology* (January 1, 1998) and Ted Schettler et al., *Generations at Risk: Reproductive Health and the Environment* (MIT Press, 1999). In 1999 the National Research Council published *Hormonally Active Agents in the Environment* (National Academy Press), in which the council's Committee on Hormonally Active Agents in the Environment reports on its evaluation of the scientific evidence pertaining to endocrine disruptors. The

National Environmental Health Association has called for more research and product testing; see Ginger L. Gist, "National Environmental Health Association Position on Endocrine Disrupters," *Journal of Environmental Health* (January–February 1998).

Elisabete Silva, Nissanka Rajapakse, and Andreas Kortenkamp, in "Something From 'Nothing'—Eight Weak Estrogenic Chemicals Combined at Concentrations Below NOECs Produce Significant Mixture Effects," *Environmental Science and Technology* (April 2002), find synergistic effects of exactly the kind dismissed by Gough. Also, after reviewing the evidence, the U.S. National Toxicology Program found that low-dose effects had been demonstrated in animals (see Ronald Melnick et al., "Summary of the National Toxicology Program's Report of the Endocrine Disruptors Low-Dose Peer Review," *Environmental Health Perspectives* [April 2002]). The evidence seems to favor the view that environmental hormone mimics have potentially serious effects.

ISSUE 18

Is the Superfund Program Successfully Protecting the Environment from Hazardous Wastes?

YES: Robert H. Harris, Jay Vandeven, and Mike Tilchin, from "Superfund Matures Gracefully," *Issues in Science & Technology* (Summer 2003)

NO: Margot Roosevelt, from "The Tragedy of Tar Creek," *Time* (April 26, 2004)

ISSUE SUMMARY

YES: Environmental consultants Robert H. Harris, Jay Vandeven, and Mike Tilchin argue that although the Superfund program still has room for improvement, it has made great progress in risk assessment and treatment technologies.

NO: Journalist Margot Roosevelt argues that because one-quarter of Americans live near Superfund sites, and sites such as Tar Creek, Oklahoma, remain hazardous, Superfund's work is clearly not getting done.

$\mathbf{T}$he potentially disastrous consequences of improper hazardous waste disposal burst upon the consciousness of the American public in the late 1970s. The problem was dramatized by the evacuation of dozens of residents of Niagara Falls, New York, whose health was being threatened by chemicals leaking from the abandoned Love Canal, which was used for many years as an industrial waste dump. Awakened to the dangers posed by chemical dumping, numerous communities bordering on industrial manufacturing areas across the country began to discover and report local sites where chemicals had been disposed of in open lagoons or were leaking from disintegrating steel drums. Such esoteric chemical names as dioxins and PCBs have become part of the common lexicon, and numerous local citizens' groups have been mobilized to prevent human exposure to these and other toxins.

The expansion of the industrial use of synthetic chemicals following World War II resulted in the need to dispose of vast quantities of wastes laden with organic and inorganic chemical toxins. For the most part, industry adopted a casual attitude toward this problem and, in the absence of regulatory restraint, chose the least expensive means of disposal available. Little attention was paid to the ultimate fate of chemicals that could seep into surface water or groundwater. Scientists have estimated that less than 10 percent of the waste was disposed of in an environmentally sound manner.

The magnitude of the problem is truly mind-boggling: Over 275 million tons of hazardous waste is produced in the United States each year; as many as 10,000 dump sites may pose a serious threat to public health, according to the federal Office of Technology Assessment; and other government estimates indicate that more than 350,000 waste sites may ultimately require corrective action at a cost that could easily exceed $500 billion.

Congressional response to the hazardous waste threat is embodied in two complex legislative initiatives. The Resource Conservation and Recovery Act (RCRA) of 1976 mandated action by the Environmental Protection Agency (EPA) to create "cradle to grave" oversight of newly generated waste, and the Comprehensive Environmental Response, Compensation, and Liability Act of 1980 (CERCLA), commonly called "Superfund," gave the EPA broad authority to clean up existing hazardous waste sites. The implementation of this legislation has been severely criticized by environmental organizations, citizens' groups, and members of Congress who have accused the EPA of foot-dragging and a variety of politically motivated improprieties. Less than 20 percent of the original $1.6 billion Superfund allocation was actually spent on waste cleanup.

Amendments designed to close RCRA loopholes were enacted in 1984, and the Superfund Amendments and Reauthorization Act (SARA) added $8.6 billion to a strengthened cleanup effort in 1986 and an additional $5.1 billion in 1990. While acknowledging some improvement, both environmental and industrial policy analysts remain very critical about the way that both RCRA and Superfund/SARA are being implemented. Efforts to reauthorize and modify both of these hazardous waste laws have been stalled in Congress since the early 1990s. But the work went on. The Superfund program continued to identify hazardous waste sites that warranted cleanup and to clean up sites; see http://www.epa.gov/superfund/ for the latest news. In 2004, it even declared that the infamous Love Canal site was finally safe.

In the following selections, Robert H. Harris, Jay Vandeven, and Mike Tilchin argue that the Superfund program has had to struggle with unexpectedly large cleanup tasks such as mines, harbors, and the ruins of the World Trade Center in New York; is subject to greater resource demands than ever before; and still has room for improvement. But, they say, the program has made great progress in risk assessment and treatment technologies. Margot Roosevelt argues that because one-quarter of Americans live near Superfund sites, and sites like Tar Creek remain hazardous, Superfund's work is not getting done.

YES

Robert H. Harris, Jay Vandeven,
and Mike Tilchin

Superfund Matures Gracefully

Superfund, one of the main programs used by the Environmental Protection Agency (EPA) to clean up serious, often abandoned, hazardous waste sites, has been improved considerably in recent years. Notably, progress has been made in two important areas: the development of risk assessments that are scientifically valid yet flexible, and the development and implementation of better treatment technologies.

The 1986 Superfund Amendments and Reauthorization Act (SARA) provided a broad refocus to the program. The act included an explicit preference for the selection of remediation technologies that "permanently and significantly reduce the volume, toxicity, or mobility of hazardous substances." SARA also required the revision of the National Contingency Plan (NCP) that sets out EPA's rules and guidance for site characterization, risk assessment, and remedy selection.

The NCP specifies the levels of risk to human health that are allowable at Superfund sites. However, "potentially responsible parties"—companies or other entities that may be forced to help pay for the cleanup—have often challenged the risk assessment methods used as scientifically flawed, resulting in remedies that are unnecessary and too costly. Since SARA was enacted, fundamental changes have evolved in the policies and science that EPA embraces in evaluating health risks at Superfund sites, and these changes have in turn affected which remedies are most often selected. Among the changes are three that collectively can have a profound impact on the selected remedy and attendant costs: EPA's development of land use guidance, its development of guidance on "principal threats," and the NCP requirement for the evaluation of "short-term effectiveness."

Before EPA's issuance in 1995 of land use guidance for evaluating the potential future public health risks at Superfund sites, its risk assessments usually would assume a future residential use scenario at a site, however unrealistic that assumption might be. This scenario would often result in the need for costly soil and waste removal remedies necessary to protect against hypothetical risks, such as those to children playing in contaminated soil or drinking contaminated ground water, even at sites where future residential use was highly improbable. The revised land use guidance provided a basis for selecting more realistic future use scenarios, with projected exposure patterns that may allow for less costly remedies.

Potentially responsible parties also complained that there was little room to tailor remedies to the magnitude of cancer risk at a site, and that the same costly remedies would be chosen for sites where the cancer risks may differ by several orders of magnitude. However, EPA's guidance on principal threats essentially established a risk-based hierarchy for remedy selection. For example, if cancer risks at a site exceed 1 in 1,000, then treatment or waste removal or both might be required. Sites that posed a lower lifetime cancer risk could be managed in other ways, such as by prohibiting the installation of drinking water wells, which likely would be far less expensive than intrusive remedies.

Revisions to the NCP in 1990 not only codified provisions required by the 1986 Superfund amendments, but also refined EPA's evolving remedy-selection criteria. For example, these revisions require an explicit consideration of the short-term effectiveness of a remedy, including the health and safety risks to the public and to workers associated with remedy implementation. EPA had learned by bitter experience that to ignore implementation risks, such as those associated with vapor and dust emissions during the excavation of wastes, could lead to the selection of remedies that proved costly and created unacceptable risks.

Although these changes in risk assessment procedures have brought greater rationality to the evaluation of Superfund sites, EPA still usually insists on the use of hypothetical exposure factors (for example, the length of time that someone may come in contact with the site) that may overstate risks. The agency has been slow in embracing other methodologies, such as probabilistic exposure analysis, that might offer more accurate assessments. Thus, some remedies are still fashioned on risk analyses that overstate risk.

Technological Evolution

Cleanup efforts in Superfund's early years were dominated by containment and excavation-and-disposal remedies. But over the years, cooperative work by government, industry, and academia have led to the development and implementation of improved treatment technologies.

The period from the mid-1980s to the early 1990s was marked by a dramatic increase in the use of source control treatment, reflecting the preference expressed in SARA for "permanent solutions and alternative treatment technologies or resource recovery technologies to the maximum extent practicable." Two types of source control technologies that have been widely used are incineration and soil vapor extraction. Although the use of incineration decreased during the 1990s because of cost and other factors, soil vapor extraction remains a proven technology at Superfund sites.

Just as early source control remedies relied on containment or excavation and disposal offsite, the presumptive remedy for groundwater contamination has historically been "pump and treat." It became widely recognized in the early 1990s that conventional pump-and-treat technologies had significant limitations, including relatively high costs. What emerged to fill the gap was an approach called "monitored natural attenuation" (MNA), which makes use of a variety of technologies, such as biodegradation, dispersion, dilution, absorption, and volatilization. As the name suggests, monitoring the effectiveness of the process is a key element of this technology. And although cleanup times still may be on the order of years, there is evidence that MNA can achieve comparable results in comparable periods and at significantly lower costs than conventional pump-and-treat systems. EPA has taken an active role in promoting this technology, and its use has increased dramatically in recent years.

As suggested by the MNA example, what may prove an even more formidable challenge than selecting specific remedies is the post-remedy implementation phase—that is, the monitoring and evaluation that will be required during coming decades to ensure that the remedy chosen is continuing to protect human health and the environment. Far too few resources have been devoted to this task, which will require not only monitoring and maintaining the physical integrity of the technology used and ensuring the continued viability of institutional controls, but also evaluating and responding to the developing science regarding chemical detection and toxicity.

Coming Challenges

In recent years, the rate at which waste sites are being added to the National Priorities List (NPL) has been decreasing dramatically as compared with earlier years. In fiscal years 1983 to 1991, EPA placed an average of 135 sites on the NPL annually. The rate dropped to an average of 27 sites per year between 1992 and 2001. Although many factors have contributed to this trend, three stand out:

- There was a finite group of truly troublesome sites before Superfund's passage, and after a few years most of those were identified.
- The program's enforcement authority has had a profound impact on how wastes are managed, significantly reducing, although not eliminating, the types of waste management practices that result in the creation of Superfund sites.
- A range of alternative cleanup programs, such as voluntary cleanup programs and those for brownfields, have evolved at both the federal and state levels. No longer is Superfund the only path for cleaning up sites.

But such programmatic changes are about more than just site numbers. In 1988, most NPL sites were in the investigation stage, and the program was widely criticized as being too much about studies and not enough about cleanup. Superfund is now a program predominantly focused on the design and construction of cleanup remedies.

This shift reflects the natural progress of sites through the Superfund pipeline, the changes in NPL listing activity, and a deliberate emphasis on achieving "construction completion," which is the primary measure of achievement for the program as established under the Government Performance and Results Act. It is a truism in regulatory matters that what gets done is what gets measured, and Superfund is no exception.

In the late 1990s, many observers believed that the demands on Superfund were declining and that it would be completed sometime in the middle of the first decade of the new century. But this is not proving to be true. Although expenditures have not been changing dramatically over time, the resource demands on the program are greater today than ever before.

Few people would have predicted, for example, that among the biggest technical and resource challenges facing Superfund at this date would be the cleanup of hard-rock mining sites and of large volumes of sediments from contaminated waterways and ports. These sites tend to be very costly to clean up, with the driver behind these great costs weighted more toward the protection of natural resources than of human health. In mapping the future course of the program, Congress and EPA must address the question of whether Superfund is the most appropriate program for cleaning up these types of sites.

There are other uncertainties, as well. The substantial role that Superfund has played in emergency response in the aftermath of 9/11, the response to the anthrax attacks of October 2001, and the program's role in the recovery of debris from the crash of the space shuttle Columbia were all totally unforeseeable. Although many valuable lessons have been learned over the past 20 years of the program, there remain substantial opportunities for improvement as well as considerable uncertainty about the kinds of environmental problems Superfund will tackle in the coming decade.

Margot Roosevelt

 **NO**

The Tragedy of Tar Creek

To get a better view of the situation, John Sparkman guns his flame-red truck up a massive pile of gravel. From the summit, a lifeless brown wasteland stretches to the horizon, like a scene from a science-fiction movie. Mountains of mine tailings, some as tall as 13-story buildings, others as wide as four football fields, loom over streets, homes, churches and schools. Dust, laced with lead, cadmium and other poisonous metals, blows off the man-made hills and 800 acres of dry settling ponds. "It gets in your teeth," says Sparkman, head of a local citizens' group. "It cakes in your ears and hair. It's like we've been environmentally raped."

Hyperbole? Drive through the desolate towns around Picher, Okla., and you might think differently. This is eco-assault on an epic scale. The prairie here in the northeast corner of the state is punctured with 480 open mine shafts and 30,000 drill holes. Little League fields have been built over an immense underground cavity that could collapse at any time. Acid mine waste flushes into drinking wells. When the water rises in Tar Creek, which runs through the site, a neon-orange scum oozes onto the roadside. Wild onions, a regional delicacy tossed into scrambled eggs, are saturated with cadmium—which may explain, local doctors say, why three different kidney dialysis centers have opened here to serve a population of only 30,000.

But the grimmest legacy of a century of intensive lead and zinc mining are the "lead heads," or "chat rats," as the kids who grew up around here are known. As toddlers, they played in sandboxes of chat—the powdery output of mills after ore is extracted from rock. As preteens, they rode their bikes across the gravel mounds and swam in lime-green sinkholes. Their parents used mine tailings to make driveways and foundations, never thinking that contaminated dust might blow through the heating ducts of their ranch houses. In the past decade, studies have shown that up to 38% of local children have had high levels of lead in their blood—an exposure that can cause permanent neurological damage and learning disabilities. "Our kids hit a brick wall," says Kim Pace, principal of the Picher-Cardin Elementary School. "Their eyes skip and jump. It takes them 100 repetitions to learn a sound."

At her kitchen table, Evona Moss helps her son Michael, 10, with his homework. Michael grew up across the street from a chat pile, and at one point the third-grader's lead levels measured 40% above the Centers for

Disease Control's danger level. He repeated kindergarten. "I used to think he was lazy," says his mother, "but he tries so hard. One minute he knows the words, and a half-hour later he doesn't. Every night he kneels down and prays to be a better reader."

It wasn't supposed to be like this. In 1980, Congress passed the Comprehensive Environmental Response, Compensation and Liability Act—commonly known as the Superfund law—one of the boldest environmental statutes in U.S. history. It was a law designed to fit all circumstances. It covered existing plants whose owners could be forced to clean up their dumps. It covered polluted sites long since abandoned by their owners: defunct factories, refineries and mines. Even when companies followed the standard, if dubious, practices of the day—dumping toxic waste in rivers, burying it in leaky drums or just leaving it, as in Oklahoma, to blow in the wind—they would be held accountable. And if they refused to clean up their messes, the Environmental Protection Agency (EPA) would do so for them and charge treble damages for its trouble. In the event that the perpetrators had disappeared or gone out of business, a general tax on polluting industries—a "Superfund"—would pay to fix the damage.

But today Superfund is a program under siege, plagued by partisan politics, industry stonewalling and bureaucratic inertia. The U.S. government has spent $27 billion on the effort and forced individual polluters to spend an additional $21 billion. Love Canal, the deadly dump in New York State that spurred the law's passage, has been capped with a layer of clay, and the EPA proposed last month to take it off the list. So far, 278 sites have been delisted. But there are thousands more out there. According to the General Accounting Office (GAO), 1 out of 4 Americans still lives within four miles of a Superfund site—many of them killing fields saturated with cancer-causing chemicals and other toxins.

The GAO reports that the program's budget fell 35% in inflation-adjusted dollars over the past decade. And environmentalists say that Bush appointees are slowing the pace of cleanups and failing to list potential new sites. According to the EPA's inspector general, 29 projects in 17 states were underfunded last year. The Administration, charges New Jersey Senator Frank Lautenberg, a Democrat, has "allowed—deliberately—these sites to rot where they are."

Tar Creek is a case in point. Two decades after it was targeted on the very first Superfund priority list, the 40-sq.-mi. site is worse off than ever. Early on, the government confined its effort to the polluted creek, without looking at chat piles, soil, air quality or the danger of subsidence. Was it a lack of knowledge of the danger, as EPA claims? Or industry influence, as environmentalists charge? Whatever the reason, federal attorneys settled with mining companies for pennies on the dollar. Now, after fruitless efforts to contain 28 billion gal. of acid mine water, contamination is spreading across a vast watershed. And although the EPA trucked out toxic dirt from about 2,000 homes and schools, Tar Creek's children still show elevated lead levels at six times the national average.

Administration officials say they are cleaning up the nation's 1,240 highest-priority sites as fast as they can. But that will be harder, since the multibillion-dollar industry-paid trust fund, set aside for abandoned sites such as Tar Creek, ran dry in October [2003]. The fund was supplied by taxes on the purchase of toxic chemicals and petroleum and on corporate profits above $2 million. But the Republican-led Congress allowed the fees to expire in 1995. Bush is the first President to oppose the levies, and last month Lautenberg and other Senate Democrats lost a narrow vote to reinstate them. In protest, the Sierra Club aired "Make Polluters Pay" TV ads in Pennsylvania, Florida and Michigan—all swing states. And on April 15, tax day, activists in 25 states picketed post offices to object. "We went from polluters paying to citizens paying," says Oklahoma environmentalist Earl Hatley. "Now EPA doesn't have the money for megasites like Tar Creek."

Meanwhile, Superfund defenders in Washington are bracing for a new battle: a Bush-appointed advisory committee, which they claim is heavily stacked with corporate members, issued a report [recently] that pushes for administrative changes. "It is a wonky thing," says Julie Wolk of the Public Interest Research Group. "But it could dramatically weaken the program." Companies want to limit liability and shift responsibility to the states, where rules are more flexible. Federal standards are "rigid and extreme," says Michael Steinberg of the Superfund Settlements Project, an industry group that includes General Electric, DuPont and IBM. "Groundwater must meet standards for tap water, even though at many of these sites no one drinks it. Soil at many sites must be clean enough so people could play in it. The costs exceed the benefits."

With the EPA's clout slackening, private attorneys are moving in. At Tar Creek, lawyers are suing seven mining companies on behalf of scores of lead-exposed children. A separate suit demanding a cleanup was filed by the Quapaw Indians, whose land was leased for the mines. And environmentalist Robert F. Kennedy Jr. has joined a class action to force companies to relocate the population of two polluted towns, Picher and Cardin. Court papers suggest that mining executives knew as early as the 1930s that the contaminated dust was dangerous but sought to, in their words, "dissuade" the government from intervening. A mining-company lawyer says the charge is based on "out-of-context reading" of historical documents.

Just how dangerous that dust might be is still a matter of dispute. Doctors at the Harvard School of Public Health have begun extensive studies in Tar Creek, not just of lead exposure but also of the cocktail mix of lead, manganese, cadmium and other metals that interact in unknown ways. "We're looking at four generations of poisoning," says Rebecca Jim of the L.E.A.D. agency, a local group. Meanwhile, parents like Evona Moss wonder what else the toxic brew might have done. Did it cause her obesity and bad teeth? Is it responsible for the malformation of her daughter's shins? Does her baby's asthma come from the chat? Her nephew's cancer? No one knows because no one has done careful, long-term studies.

Tar Creek is an extreme case. But like Tolstoy's unhappy families, every Superfund site is tragic and contentious in its own way. In Libby, Mont., a massive mine blanketed the town with asbestos dust, killing at least 215 people and sickening 1,100 more with cancer and lung disease—yet cleanup funds have been cut so sharply that it could take 10 to 15 years to finish the job. In Coeur d'Alene, Idaho, miners dumped 60 million tons of toxic metals into waterways, but state officials are fighting a Superfund cleanup, fearing a stigma that might hurt tourism. In New York, General Electric, which contaminated 40 miles of the Hudson River with cancer-causing PCBs, has hired high-profile attorney Laurence Tribe to convince federal courts that the Superfund law is unconstitutional. And in New Jersey, where the rabbits frolicking around the Chemical Insecticide Corp. plant once grew green-tinged fur, cleanup funds were restored only after locals sent green plush bunnies to members of Congress.

At Tar Creek, many residents have given up hope. Even the EPA, which has spent $107 million at the site, isn't sure if it can ever be repaired. "We don't have an off-the-shelf remedy," says EPA Superfund official Randy Deitz. "What do you do with the enormous chat piles? When does cleanup become impracticable? We have limited resources." In a show of no-confidence, the Oklahoma legislature last week passed a $5 million buyout for all families with children under 6. John Sparkman, who heads the Tar Creek Steering Committee, a group of buyout supporters, veers between cynicism and despair. "They think we're poor white trash," he says bitterly, driving past Picher's boarded-up storefronts. "The votes here don't affect any federal election—so why bother? We've agitated till we can't agitate anymore." Meanwhile, at Tar Creek, the toxic dust keeps blowing in the wind.

POSTSCRIPT

Is the Superfund Program Successfully Protecting the Environment from Hazardous Wastes?

Superfund cleanups, when those responsible for contaminated sites could not pay or could not be found, were to have been funded by taxes on industry (e.g., the Crude Oil Tax, the Chemical Feedstock Tax, the Toxic Chemicals Importation Tax, and the Corporate Environmental Income Tax). These taxes expired in 1995, and Congress has so far refused to reauthorize them; the Senate voted down the latest attempt in March 2004. The program exhausted its funds in September 2003 and is now running on government revenues. See Kara Sissell, "Senate Votes Down Bill to Reinstate Superfund Tax," *Chemical Week* (March 17, 2004) and "Superfund Program: Current Status and Future Fiscal Challenges," Report to the Chairman, Subcommittee on Oversight of Government Management, the Federal Workforce, and the District of Columbia, Committee on Governmental Affairs, U.S. Senate (GAO, July 2003). Activist groups, such as the Public Interest Research Group (PIRG), have issued calls for the Bush administration to reinstate funding without delay. But, says PIRG, "The Bush administration opposes reauthorization of the polluter pays taxes, supports a steep increase in the amount paid by taxpayers, and has dramatically slowed down the pace of cleanups at the nation's worst toxic waste sites." The Competitive Enterprise Institute objects that such taxes are an assault on consumer pocketbooks, as is the Comprehensive Environmental Response, Compensation, and Liability Act's (CERCLA's) "joint and several liability" clause, which can make minor contributors to toxics sites liable for large cleanup costs even when they acted according to all laws and regulations in force at the time.

Meanwhile, the hazardous waste problem takes new forms. Even in the 1990s, an increasingly popular method of disposing of hazardous wastes was to ship them from the United States to "dumping grounds" in developing countries. Iwonna Rummel-Bulska's "The Basel Convention: A Global Approach for the Management of Hazardous Wastes," *Environmental Policy and Law* (vol. 24, no. 1, 1994) describes an international treaty designed to prevent or at least limit such waste dumping. But eight years later, in February 2002, the Basel Action Network and the Silicon Valley Toxics Coalition (http://svtc.org) published *Exporting Harm: The High-Tech Trashing of Asia.* This lengthy report documents the shipping of electronics wastes, including defunct personal computers, monitors, and televisions, as well as circuit boards and other products rich in lead, beryllium, cadmium, mercury, and other toxic materials. Some 50 to 80 percent of the "e-waste" collected for recycling in the western United

States is shipped to destinations such as China, India, and Pakistan, where recycling and disposal methods lead to widespread human and environmental contamination. An updated version of this report, *Poison PCs and Toxic TVs: E-Waste Tsunami to Roll Across the US: Are We Prepared?* was released in February 2004. In August 2005, Greenpeace released K. Brigden, et al., *Recycling of Electronic Wastes in China and India: Workplace and Environmental Contamination* (http://www.e-takeback.org/press_open/greenpeace.pdf).

Among the solutions that have been urged to address the hazardous waste problem are "take-back" and "remanufacturing" practices. Gary A. Davis and Catherine A. Wilt, in "Extended Product Responsibility," *Environment* (September 1997), urge such solutions as crucial to the minimization of waste and describe how they are becoming more common in Europe. Brad Stone, "Tech Trash, E-Waste: By Any Name, It's an Issue," *Newsweek* (December 12, 2005), describes their appearance in the U.S., where some states are making take-back programs mandatory for computers, televisions, and other electronic devices. After *Exporting Harm* was published and drew considerable attention from the press, some industry representatives hastened to emphasize such practices as Hewlett Packard's recycling of printer ink cartridges. See Doug Bartholomew's "Beyond the Grave," *Industry Week* (March 1, 2002), which also stressed the need to minimize waste by intelligent design. The Institute of Industrial Engineers published in its journal *IIE Solutions* Brian K. Thorn's and Philip Rogerson's "Take It Back" (April 2002), which stressed the importance of designing for reuse or remanufacturing. Anthony Brabazon and Samuel Idowu, in "Costing the Earth" *Financial Management (CIMA)* (May 2001), note that "take-back schemes may [both] provide opportunities to build goodwill and [help] companies to use resources more efficiently."

ISSUE 19

Should the United States Reprocess Spent Nuclear Fuel?

YES: Phillip J. Finck, from Statement Before the House Committee on Science, Energy Subcommittee, Hearing on Nuclear Fuel Reprocessing (June 16, 2005)

NO: Matthew Bunn, from "The Case Against a Near-Term Decision to Reprocess Spent Nuclear Fuel in the United States," Testimony for the House Committee on Science, Energy Subcommittee, Hearing on Nuclear Fuel Reprocessing (June 16, 2005)

ISSUE SUMMARY

YES: Phillip J. Finck argues that by reprocessing spent nuclear fuel, the United States can enable nuclear power to expand its contribution to the nation's energy needs while reducing carbon emissions, nuclear waste, and the need for waste repositories such as Yucca Mountain.

NO: Matthew Bunn argues that there is no near-term need to embrace nuclear spent fuel reprocessing, costs are highly uncertain, and there is a worrisome risk that the increased availability of bomb-grade nuclear materials will increase the risk of nuclear war and terrorism.

INTRODUCTION

Nuclear waste is generated when uranium and plutonium atoms are split to make energy in nuclear power plants, when uranium and plutonium are purified to make nuclear weapons, and when radioactive isotopes useful in medical diagnosis and treatment are made and used. These wastes are radioactive, meaning that as they break down they emit radiation of several kinds. Those that break down fastest are most radioactive.

According to the U.S. Department of Energy, high-level waste includes spent reactor fuel (52,000 tons) and waste from weapons production (91 million gallons). Low- and mixed-level waste includes waste from hospitals and research labs, remnants of decommissioned nuclear plants, and air filters (472 million cubic feet). The high-level waste is the most hazardous and poses the most severe disposal problems and experts say such materials must be kept away from people and other living things.

The Nuclear Age began in the 1940s. As nuclear waste accumulated, there also developed a sense of urgency about finding a place to put it where it would not threaten humans or ecosystems for a quarter million years or more. Among the potential answers was reprocessing, which separates (and recycles) unused fuel from spent fuel and thereby reduces the quantity of waste while also extending the supply of fuel. After the Nuclear Nonproliferation Treaty went into force in 1970, it became United States policy not to reprocess spent nuclear fuel and thereby to limit the availability of bomb-grade material. As a consequence, spent fuel was not recycled, and the waste continued to accumulate.

In 1982, the Nuclear Waste Policy Act called for locating candidate disposal sites for high-level wastes and choosing one by 1998. Since no state chosen as a candidate site was happy, and many sites were for various reasons less than ideal, the schedule proved impossible to meet. In 1987, Congress attempted to settle the matter by designating Yucca Mountain, Nevada, as the one site to be intensively studied and developed. It would be opened for use in 2010. However, problems have plagued the project, as summarized by Chuck McCutcheon, "High-Level Acrimony In Nuclear Storage Standoff," *Congressional Quarterly Weekly Report* (September 25, 1999), and Sean Paige, "The Fight at the End of the Tunnel," *Insight on the News* (November 15, 1999).

In February 2002, U.S. Secretary of Energy Spencer Abraham recommended to the President that the nation go ahead with development of the Yucca Mountain site. His report made the points that a disposal site is necessary, that Yucca Mountain has been thoroughly studied, and that moving ahead with the site best serves "our energy future, our national security, our economy, our environment, and safety." Nevadans and activists of several kinds have objected strenuously; early in 2005, reports that researchers had lied about the data that showed Yucca Mountain to be a safe, long-term repository gave them hope that approval of the site would be delayed, perhaps indefinitely.

But the need to dispose of nuclear waste is not about to go away, especially if the United States expands its reliance on nuclear power (see Issue 12). In the following selections, Phillip J. Finck, Deputy Associate Laboratory Director, Applied Science and Technology and National Security, Argonne National Laboratory, argues that by reprocessing spent nuclear fuel the United States can enable nuclear power to expand its contribution to the nation's energy needs while reducing carbon emissions, nuclear waste, and the need for waste repositories such as Yucca Mountain. Matthew Bunn, a Senior Research Associate at Harvard University's John F. Kennedy School of Government, argues that there is no near-term need to embrace nuclear spent fuel reprocessing, costs are highly uncertain, and there is a worrisome risk that the increased availability of bomb-grade nuclear materials will increase the risk of nuclear war and terrorism.

YES

Phillip J. Finck

Statement Before the House Committee on Science, Energy Subcommittee, Hearing on Nuclear Fuel Reprocessing

Summary

Management of spent nuclear fuel from commercial nuclear reactors can be addressed in a comprehensive, integrated manner to enable safe, emissions-free, nuclear electricity to make a sustained and growing contribution to the nation's energy needs. Legislation limits the capacity of the Yucca Mountain repository to 70,000 metric tons from commercial spent fuel and DOE defense-related waste. It is estimated that this amount will be accumulated by approximately 2010 at current generation rates for spent nuclear fuel. To preserve nuclear energy as a significant part of our future energy generating capability, new technologies can be implemented that allow greater use of the repository space at Yucca Mountain. By processing spent nuclear fuel and recycling the hazardous radioactive materials, we can reduce the waste disposal requirements enough to delay the need for a second repository until the next century, even in a nuclear energy growth scenario. Recent studies indicate that such a closed fuel cycle may require only minimal increases in nuclear electricity costs, and are not a major factor in the economic competitiveness of nuclear power (The University of Chicago study, "The Economic Future of Nuclear Power," August 2004). However, the benefits of a closed fuel cycle can not be measured by economics alone; resource optimization and waste minimization are also important benefits. Moving forward in 2007 with an engineering-scale demonstration of an integrated system of proliferation-resistant, advanced separations and transmutation technologies would be an excellent first step in demonstrating all of the necessary technologies for a sustainable future for nuclear energy.

House Committee on Science, Energy Subcommittee, Hearing on Nuclear Fuel Reprocessing, June 16, 2005.

333

Nuclear Waste and Sustainability

World energy demand is increasing at a rapid pace. In order to satisfy the demand and protect the environment for future generations, energy sources must evolve from the current dominance of fossil fuels to a more balanced, sustainable approach. This new approach must be based on abundant, clean, and economical energy sources. Furthermore, because of the growing world-wide demand and competition for energy, the United States vitally needs to establish energy sources that allow for energy independence.

Nuclear energy is a carbon-free, secure, and reliable energy source for today and for the future. In addition to electricity production, nuclear energy has the promise to become a critical resource for process heat in the production of transportation fuels, such as hydrogen and synthetic fuels, and desalinated water. New nuclear plants are imperative to meet these vital needs.

To ensure a sustainable future for nuclear energy, several requirements must be met. These include safety and efficiency, proliferation resistance, sound nuclear materials management, and minimal environmental impacts. While some of these requirements are already being satisfied, the United States needs to adopt a more comprehensive approach to nuclear waste management. The environmental benefits of resource optimization and waste minimization for nuclear power must be pursued with targeted research and development to develop a successful integrated system with minimal economic impact. Alternative nuclear fuel cycle options that employ separations, transmutation, and refined disposal (e.g., conservation of geologic repository space) must be contrasted with the current planned approach of direct disposal, taking into account the complete set of potential benefits and penalties. In many ways, this is not unlike the premium homeowners pay to recycle municipal waste.

The spent nuclear fuel situation in the United States can be put in perspective with a few numbers. Currently, the country's 103 commercial nuclear reactors produce more than 2000 metric tons of spent nuclear fuel per year (masses are measured in heavy metal content of the fuel, including uranium and heavier elements). The Yucca Mountain repository has a legislative capacity of 70,000 metric tons, including spent nuclear fuel and DOE defense-related wastes. By approximately 2010 the accumulated spent nuclear fuel generated by these reactors and the defense-related waste will meet this capacity, even before the repository starts accepting any spent nuclear fuel. The ultimate technical capacity of Yucca Mountain is expected to be around 120,000 metric tons, using the current understanding of the Yucca Mountain site geologic and hydrologic characteristics. This limit will be reached by including the spent fuel from current reactors operating over their lifetime. Assuming nuclear growth at a rate of 1.8% per year after 2010, the 120,000 metric ton capacity will be reached around 2030. At that projected nuclear growth rate, the U.S. will need up to nine Yucca Mountain-type repositories by the end of this century. Until Yucca Mountain starts accepting waste, spent nuclear fuel must be stored in temporary facilities, either storage pools or above ground storage casks.

Today, many consider repository space a scarce resource that should be managed as such. While disposal costs in a geologic repository are currently quite affordable for U.S. electric utilities, accounting for only a few percent of the total cost of electricity, the availability of U.S. repository space will likely remain limited.

Only three options are available for the disposal of accumulating spent nuclear fuel:

- Build more ultimate disposal sites like Yucca Mountain.
- Use interim storage technologies as a temporary solution.
- Develop and implement advanced fuel cycles, consisting of separation technologies that separate the constituents of spent nuclear fuel into elemental streams, and transmutation technologies that destroy selected elements and greatly reduce repository needs.

A responsible approach to using nuclear power must always consider its whole life cycle, including final disposal. We consider that temporary solutions, while useful as a stockpile management tool, can never be considered as ultimate solutions. It seems prudent that the U.S. always have at least one set of technologies available to avoid expanding geologic disposal sites.

Spent Nuclear Fuel

The composition of spent nuclear fuel poses specific problems that make its ultimate disposal challenging. Fresh nuclear fuel is composed of uranium dioxide (about 96% U238, and 4% U235). During irradiation, most of the U235 is fissioned, and a small fraction of the U238 is transmuted into heavier elements (known as "transuranics"). The spent nuclear fuel contains about 93% uranium (mostly U238), about 1% plutonium, less than 1% minor actinides (neptunium, americium, and curium), and 5% fission products. Uranium, if separated from the other elements, is relatively benign, and could be disposed of as low-level waste or stored for later use. Some of the other elements raise significant concerns:

- The fissile isotopes of plutonium, americium, and neptunium are potentially usable in weapons and, therefore, raise proliferation concerns. Because spent nuclear fuel is protected from theft for about one hundred years by its intense radioactivity, it is difficult to separate these isotopes without remote handling facilities.
- Three isotopes, which are linked through a decay process (Pu241, Am241, and Np237), are the major contributors to the estimated dose for releases from the repository, typically occurring between 100,000 and 1 million years, and also to the long-term heat generation that limits the amount of waste that can be placed in the repository.
- Certain fission products (cesium, strontium) are major contributors to the repository's shortterm heat load, but their effects can be mitigated

by providing better ventilation to the repository or by providing a cooling-off period before placing them in the repository.
- Other fission products (Tc99 and I129) also contribute to the estimated dose.

The time scales required to mitigate these concerns are daunting: several of the isotopes of concern will not decay to safe levels for hundreds of thousands of years. Thus, the solutions to long-term disposal of spent nuclear fuel are limited to three options: the search for a geologic environment that will remain stable for that period; the search for waste forms that can contain these elements for that period; or the destruction of these isotopes. These three options underlie the major fuel cycle strategies that are currently being developed and deployed in the U.S. and other countries.

Options for Disposing of Spent Nuclear Fuel

Three options are being considered for disposing of spent nuclear fuel: the once-through cycle is the U.S. reference; limited recycle has been implemented in France and elsewhere and is being deployed in Japan; and full recycle (also known as the closed fuel cycle) is being researched in the U.S., France, Japan, and elsewhere.

1. Once-through Fuel Cycle

This is the U.S. reference option where spent nuclear fuel is sent to the geologic repository that must contain the constituents of the spent nuclear fuel for hundreds of thousands of years. Several countries have programs to develop these repositories, with the U.S. having the most advanced program. This approach is considered safe, provided suitable repository locations and space can be found. It should be noted that other ultimate disposal options have been researched (e.g., deep sea disposal; boreholes and disposal in the sun) and abandoned. The challenges of long-term geologic disposal of spent nuclear fuel are well recognized, and are related to the uncertainty about both the long-term behavior of spent nuclear fuel and the geologic media in which it is placed.

2. Limited Recycle

Limited recycle options are commercially available in France, Japan, and the United Kingdom. They use the PUREX process, which separates uranium and plutonium, and directs the remaining transuranics to vitrified waste, along with all the fission products. The uranium is stored for eventual reuse. The plutonium is used to fabricate mixed-oxide fuel that can be used in conventional reactors. Spent mixed-oxide fuel is currently not reprocessed, though the feasibility of mixed-oxide reprocessing has been demonstrated. It is typically stored or eventually sent to a geologic repository for disposal. Note that a reactor partially loaded with mixed-oxide fuel can destroy as much plutonium as it creates. Nevertheless, this approach always results in increased production of americium, a key contributor to the heat generation in a repository. This approach has two significant advantages:

- It can help manage the accumulation of plutonium.
- It can help significantly reduce the volume of spent nuclear fuel (the French examples indicate that volume decreases by a factor of 4).

Several disadvantages have been noted:

- It results in a small economic penalty by increasing the net cost of electricity a few percent.
- The separation of pure plutonium in the PUREX process is considered by some to be a proliferation risk; when mixed-oxide use is insufficient, this material is stored for future use as fuel.
- This process does not significantly improve the use of the repository space (the improvement is around 10%, as compared to a factor of 100 for closed fuel cycles).
- This process does not significantly improve the use of natural uranium (the improvement is around 15%, as compared to a factor of 100 for closed fuel cycles).

3. Full Recycle (the Closed Fuel Cycle)

Full recycle approaches are being researched in France, Japan, and the United States. This approach typically comprises three successive steps: an advanced separations step based on the UREX+ technology that mitigates the perceived disadvantages of PUREX, partial recycle in conventional reactors, and closure of the fuel cycle in fast reactors.

The first step, UREX+ technology, allows for the separations and subsequent management of highly pure product streams. These streams are:

- Uranium, which can be stored for future use or disposed of as low-level waste.
- A mixture of plutonium and neptunium, which is intended for partial recycle in conventional reactors followed by recycle in fast reactors.
- Separated fission products intended for short-term storage, possibly for transmutation, and for long-term storage in specialized waste forms.
- The minor actinides (americium and curium) for transmutation in fast reactors.

The UREX+ approach has several advantages:

- It produces minimal liquid waste forms, and eliminates the issue of the "waste tank farms."
- Through advanced monitoring, simulation and modeling, it provides significant opportunities to detect misuse and diversion of weapons-usable materials.
- It provides the opportunity for significant cost reduction.

• Finally and most importantly, it provides the critical first step in managing all hazardous elements present in the spent nuclear fuel.

The second step—partial recycle in conventional reactors—can expand the opportunities offered by the conventional mixed-oxide approach. In particular, it is expected that with significant R&D effort, new fuel forms can be developed that burn up to 50% of the plutonium and neptunium present in spent nuclear fuel. (Note that some studies also suggest that it might be possible to recycle fuel in these reactors many times—i.e., reprocess and recycle the irradiated advanced fuel—and further destroy plutonium and neptunium; other studies also suggest possibilities for transmuting americium in these reactors. Nevertheless, the practicality of these schemes is not yet established and requires additional scientific and engineering research.) The advantage of the second step is that it reduces the overall cost of the closed fuel cycle by burning plutonium in conventional reactors, thereby reducing the number of fast reactors needed to complete the transmutation mission of minimizing hazardous waste. This step can be entirely bypassed, and all transmutation performed in advanced fast reactors, if recycle in conventional reactors is judged to be undesirable.

The third step, closure of the fuel cycle using fast reactors to transmute the fuel constituents into much less hazardous elements, and pyroprocessing technologies to recycle the fast reactor fuel, constitutes the ultimate step in reaching sustainable nuclear energy. This process will effectively destroy the transuranic elements, resulting in waste forms that contain only a very small fraction of the transuranics (less than 1%) and all fission products. These technologies are being developed at Argonne National Laboratory and Idaho National Laboratory, with parallel development in Japan, France, and Russia.

The full recycle approach has significant benefits:

• It can effectively increase use of repository space by a factor of more than 100.
• It can effectively increase the use of natural uranium by a factor of 100.
• It eliminates the uncontrolled buildup of all isotopes that are a proliferation risk.
• The fast reactors and the processing plant can be deployed in small co-located facilities that minimize the risk of material diversion during transportation.
• The fast reactor does not require the use of very pure weapons usable materials, thus increasing their proliferation resistance.
• It finally can usher the way towards full sustainability to prepare for a time when uranium supplies will become increasingly difficult to ensure.
• These processes would have limited economic impact; the increase in the cost of electricity would be less than 10% (ref: OECD).
• Assuming that demonstrations of these processes are started by 2007, commercial operations are possible starting in 2025; this will require adequate funding for demonstrating the separations, recycle, and reactor technologies.

- The systems can be designed and implemented to ensure that the mass of accumulated spent nuclear fuel in the U.S. would always remain below 100,000 metric tons—less than the technical capacity of Yucca Mountain—thus delaying, or even avoiding, the need for a second repository in the U.S.

Conclusion

A well engineered recycling program for spent nuclear fuel will provide the United States with a long-term, affordable, carbon-free energy source with low environmental impact. This new paradigm for nuclear power will allow us to manage nuclear waste and reduce proliferation risks while creating a sustainable energy supply. It is possible that the cost of recycling will be slightly higher than direct disposal of spent nuclear fuel, but the nation will only need one geologic repository for the ultimate disposal of the residual waste.

Matthew Bunn  **NO**

The Case Against a Near-Term Decision to Reprocess Spent Nuclear Fuel in the United States

Madam chairwoman and members of the committee: It is an honor to be here today to discuss a subject that is very important to the future of nuclear energy and efforts to stem the spread of nuclear weapons—reprocessing of spent nuclear fuel.

I believe that, while research and development (R&D) on advanced concepts that may offer promise for the future should continue, a near-term decision to reprocess U.S. commercial spent nuclear fuel would be a serious mistake, with costs and risks far outweighing its potential benefits. Let me make seven points to support that view.

First, reprocessing by itself does not make any of the nuclear waste go away. Whatever course we choose, we will still need a nuclear waste repository such as Yucca Mountain. Reprocessing is simply a chemical process that separates the radioactive materials in spent fuel into different components. In the traditional process, known as PUREX, reprocessing produces separated plutonium (which is weapons-usable), recovered uranium, and high-level waste (containing all the other transuranic elements and fission products). In the process, intermediate and low-level wastes are also generated. More advanced processes now being examined, such as UREX+ and pyroprocessing, attempt to address some of the problems of the PUREX process, but whether they will do so successfully remains to be seen. Once the spent fuel has been reprocessed, the plutonium and uranium separated from the spent fuel can in principle be recycled into new fuel; in the more advanced processes, some other long-lived species would also be irradiated in reactors (or accelerator-driven assemblies) to transmute them into shorterlived species.

More Expensive

Second, reprocessing and recycling using current or near-term technologies would substantially increase the cost of nuclear waste management, even if the cost of both uranium and geologic repositories increase significantly. In a

House Committee on Science, Energy Subcommittee, Hearing on Nuclear Fuel Reprocessing, June 16, 2005.

recent Harvard study, we concluded, even making a number of assumptions that were quite favorable to reprocessing, that shifting to reprocessing and recycling would increase the costs of spent fuel management by more than 80% (after taking account of appropriate credits or charges for recovered plutonium and uranium from reprocessing). Reprocessing (at an optimistic reprocessing price) would not become economic until uranium reached a price of over $360 per kilogram—a price not likely to be seen for many decades, if then. Government studies even in countries such as France and Japan have reached similar conclusions. The UREX+ technology now being pursued adds a number of complex separation steps to the traditional PUREX process, in order to separate important radioactive isotopes for storage or transmutation, and there is little doubt that reprocessing and transmutation using this process would be even more expensive. Other processes might someday reduce the costs, but this remains to be demonstrated, and a number of recent official studies have estimated costs for reprocessing and transmutation that are far higher than the costs of traditional reprocessing and recycling, not lower.

To follow this course, either the current 1 mill/kilowatt-hour nuclear waste fee would have to be substantially increased, or billions of dollars in tax money would have to be used to subsidize the effort. Since facilities required for reprocessing and transmutation would not be economically attractive for private industry to build, the U.S. government would either have to build and operate these facilities itself, give private industry large subsidies to do so, or impose onerous regulations requiring private industry to do so with its own funds. All of these options would represent dramatic government intrusions into the nuclear fuel industry, and the implications of such intrusions have not been appropriately examined. I am pleased that the subcommittee plans a later hearing with representatives from the nuclear industry to discuss these economic and institutional issues.

Unnecessary Proliferation Risks

Third, traditional approaches to reprocessing and recycling pose significant and unnecessary proliferation risks, and even proposed new approaches are not as proliferation-resistant as they should be. It is crucial to understand that any state or group that could make a bomb from weapon-grade plutonium could make a bomb from the reactor-grade plutonium separated by reprocessing. Despite the remarkable progress of safeguards and security technology over the last few decades, processing, fabricating, and transporting tons of weaponsusable separated plutonium every year—when even a few kilograms is enough for a bomb—inevitably raises greater risks than not doing so. The dangers posed by these operations can be reduced with sufficient investment in security and safeguards, but they cannot be reduced to zero, and these additional risks are unnecessary.

Indeed, contrary to the assertion in the Energy and Water appropriations subcommittee report that plutonium reprocessing in other countries poses little risk because the plutonium is immediately recycled as fresh fuel—a

342 ISSUE 19 / Should the United States Reprocess Spent Nuclear Fuel?

conclusion that would not be correct even if the underlying assertion were true—the fact is that reprocessing is far outpacing the use of the resulting plutonium as fuel, with the result that over 240 tons of separated, weapons-usable civilian plutonium now exists in the world, a figure that will soon surpass the amount of plutonium in all the world's nuclear weapons arsenals combined. The British Royal Society, in a 1998 report, warned that even in an advanced industrial state like the United Kingdom, the possibility that plutonium stocks might be "accessed for illicit weapons production is of extreme concern."

Moreover, a near-term U.S. return to reprocessing could significantly undermine broader U.S. nuclear nonproliferation policies. President Bush has announced an effort to convince countries around the world to forego reprocessing and enrichment capabilities of their own; has continued the efforts of past administrations to convince other states to avoid the further accumulation of separated plutonium, because of the proliferation hazards it poses; and has continued to press states in regions of proliferation concern not to reprocess (including not only states such as North Korea and Iran, but also U.S. allies such South Korea and Taiwan, both of which had secret nuclear weapons programs closely associated with reprocessing efforts in the past). A U.S. decision to move toward reprocessing itself would make it more difficult to convince other states not to do the same.

Advocates argue that the more advanced approaches now being pursued would be more proliferation-resistant. Technologies such as pyroprocessing are undoubtedly better than PUREX in this respect. But the plutonium-bearing materials that would be separated in either the UREX+ process or by pyroprocessing would not be radioactive enough to meet international standards for being "self-protecting" against possible theft. Moreover, if these technologies were deployed widely in the developing world, where most of the future growth in electricity demand will be, this would contribute to potential proliferating states building up expertise, realworld experience, and facilities that could be readily turned to support a weapons program.

Proponents of reprocessing and recycling often argue that this approach will provide a nonproliferation benefit, by consuming the plutonium in spent fuel, which would otherwise turn geologic repositories into potential plutonium mines in the long term. But the proliferation risk posed by spent fuel buried in a safeguarded repository is already modest; if the world could be brought to a state in which such repositories were the most significant remaining proliferation risk, that would be cause for great celebration. Moreover, this risk will be occurring a century or more from now, and if there is one thing we know about the nuclear world a century hence, it is that its shape and contours are highly uncertain. We should not increase significant proliferation risks in the near term in order to reduce already small and highly uncertain proliferation risks in the distant future.

As-Yet-Unexamined Safety and Terrorism Risks

Fourth, reprocessing and recycling using technologies available in the near term would be likely to raise additional safety and terrorism risks. Until Chernobyl, the world's worst nuclear accident had been the explosion at the reprocessing plant at Khystym in 1957, and significant accidents at both Russian and Japanese reprocessing plants occurred as recently as the 1990s. No complete life-cycle study of the safety and terrorism risks of reprocessing and recycling compared to those of direct disposal has yet been done by disinterested parties. But it seems clear that extensive processing of intensely radioactive spent fuel using volatile chemicals presents more opportunities for release of radionuclides than does leaving spent fuel untouched in thick metal or concrete casks.

Limited Waste Management Benefits

Fifth, the waste management benefits that might be derived from reprocessing and transmutation are quite limited. Two such benefits are usually claimed: decreasing the repository volume needed per kilowatt-hour of electricity generated (potentially eliminating the need for a second repository after Yucca Mountain); and greatly reducing the radioactive dangers of the material to be disposed.

It is important to recognize that reprocessing and recycling as currently practiced (with only one round of recycling the plutonium as uranium-plutonium mixed oxide (MOX) fuel) does not have either of these benefits. The size of a repository needed for a given amount of waste is determined not by the volume of the waste but by its heat output. Because of the build-up of heat-emitting higher actinides when plutonium is recycled, the total heat output of the waste per kilowatt-hour generated is actually higher—and therefore the needed repositories larger and more expensive—with one round of reprocessing and recycling than it is for direct disposal. And the estimated long-term doses to humans and the environment from the repository are not noticeably reduced.

Newer approaches that might provide a substantial reduction in radiotoxic hazards and in repository volume are complex, likely to be expensive, and still in an early stage of development. Most important, even if they achieved their goals, the benefits would not be large. The projected long-term radioactive doses from a geologic repository are already low. No credible study has yet been done comparing the risk of increased doses in the near term from the extensive processing and operations required for reprocessing and transmutation to the reduction in doses thousands to hundreds of thousands of years in the future that might be achieved by this method.

With respect to reducing repository volume, while the Department of Energy (DOE) has not yet performed any detailed study of the maximum amount of spent fuel that could be emplaced at Yucca Mountain, there is little doubt that even without reprocessing, the mountain could hold far more than the current legislative limit. There are a variety of approaches to providing additional capacity at Yucca Mountain or elsewhere without recycling. Indeed, as a recent American Physical Society report noted, it is possible that even if all existing reactors receive license extensions allowing them to operate for 60 years, Yucca Mountain will be able to hold all the spent fuel they will generate in their lifetimes, without reprocessing. While proponents of reprocessing and transmutation point to the likely difficulty of licensing a second repository in the United States after Yucca Mountain's capacity is filled, it is likely to be at least as difficult to gain public acceptance and licenses for the facilities needed for reprocessing and transmutation—particularly as such facilities will likely pose more genuine hazards to their neighbors than would a nuclear waste repository.

Limited Energy Benefits

Sixth, the energy benefits of reprocessing and recycling would also be limited. Additional energy can indeed be generated from the plutonium and uranium in spent fuel. But in today's market, spent fuel is like oil shale: getting the energy out of it costs far more than the energy is worth. In the only approach to recycling that is commercially practiced today—which involves a single round of recycling as MOX fuel in existing light-water reactors—the amount of energy generated from each ton of uranium mined is increased by less than 20%. In principle, if, in the future, fast-neutron breeder reactors become economic, so that the 99.3% of natural uranium that is U-238 could be turned to plutonium and burned, the amount of energy that could be derived from each ton of uranium mined might be increased 50-fold.

But there is no near-term need for this extension of the uranium resource. World resources of uranium likely to be economically recoverable in future decades at prices far below the price at which reprocessing would be economic are sufficient to fuel a growing global nuclear enterprise for many decades, relying on direct disposal without recycling.

Nor does reprocessing serve the goal of energy security, even for countries such as Japan, which have very limited domestic energy resources. If energy security means anything, it means that a country's energy supplies will not be disrupted by events beyond that country's control. Yet events completely out of the control of any individual country—such as a theft of poorly guarded plutonium on the other side of the world—could transform the politics of plutonium overnight and make major planned programs virtually impossible to carry out. Japan's experience following the scandal over BNFL's falsification of safety data on MOX fuel, and following the accidents at Monju and Tokai, all of which have delayed Japan's plutonium programs by many years, makes this point clear. If anything, plutonium recycling is much *more* vulnerable to external events than reliance on once-through use of uranium, whose supplies are diverse, plentiful, and difficult to cut off.

Premature to Decide—and No Need to Rush

Seventh, there is no need to rush to make this decision in 2007, or in fact any time in the next few decades. Dry storage casks offer the option of storing spent fuel cheaply, safely, and securely for decades. During that time, technology will develop; interest will accumulate on fuel management funds set aside today, reducing the cost of whatever we choose to do in the long run; political and economic circumstances may change in ways that point clearly in one direction or the other; and the radioactivity of the spent fuel will decay, making it cheaper to process in the future, if need be. Our generation has an obligation to set aside sufficient funds so that we are not passing unfunded obligations on to our children and grandchildren, but it is not our responsibility to make and implement decisions prematurely, thereby depriving future generations of what might turn out to be better options developed later. Indeed, because the repository will remain open for 50–100 years, with the spent fuel readily retrievable, moving forward with direct disposal will still leave all options open for decades to come.

Similarly, there is no need to rush to set up new interim storage sites on DOE or military sites, and no possibility of performing the needed reviews and getting the needed licenses to do so by 2006, as the Energy and Water appropriations subcommittee proposed. There is a legitimate debate as to whether such interim spent fuel storage prior to emplacement in a geologic repository should be centralized at one or two sites, or whether in most cases the fuel should continue to be stored at existing reactor sites. In any case, the government should fulfill its obligations to the utilities by taking title to the fuel and paying the cost of storage. At the same time, we should continue to move toward opening a permanent geologic repository as quickly as we responsibly can—in part because public acceptance of interim spent fuel storage facilities is only likely to be forthcoming if the public is convinced that they will not become permanent waste dumps.

Nor is there any need to rush on deciding whether a second nuclear waste repository will be needed. While existing nuclear power plants will have discharged enough fuel to fill the current legislated capacity limit within a few years, the reality is that it will be decades before sufficient fuel to fill Yucca Mountain has in fact been emplaced. We can and should defer this decision, and take the time to consider the options in detail. Congress should consider amending current law and giving the Secretary of Energy another decade or more before reporting on the need for a second repository.

Proponents of deciding quickly on reprocessing sometimes argue that such decisions are necessary because no new nuclear reactors will be purchased unless sufficient geologic repository capacity for all the spent fuel they will generate throughout their lifetimes has already been provided. I do not believe this is correct. I believe that if the government is fulfilling its obligation to take title to spent fuel and pay the costs of managing it, and clear progress is being made toward opening and operating a nuclear waste repository, investors will have sufficient confidence that they will not be saddled with unexpected spent fuel obligations to move forward. By contrast, if the government were seriously considering drastic changes in spent fuel management approaches which might major increases in the nuclear waste fee, investors might well wish to wait to see the outcome of those decisions before investing in new nuclear plants.

It is a good thing there is no need to rush, as we simply do not have the information that would be needed to make a decision on reprocessing in 2007. The advanced reprocessing technologies now being pursued are in a very early stage of development. As of a year ago, UREX+ had been demonstrated on a total of one pin of real spent fuel, in a small facility—and had not met all of its processing goals in that test. Frankly, in my judgment there is little prospect that further development of complex multi-stage aqueous separations processes such as UREX+ will result in processes that will provide low costs, proliferation resistance, and waste management benefits sufficient to make them worth implementing in competition with direct disposal. Pyroprocessing has been tried on a somewhat larger scale over the years, but the process is designed for processing metals, and significant development is still needed to be confident in industrial-scale application to the oxide spent fuel from current reactors. Other, longer-term processes might offer more promise, but too little is known about them to know for sure.

So far, we do not have a credible life-cycle analysis of the cost of a reprocessing and transmutation system compared to that of direct disposal; DOE has yet to do any detailed estimate of how much spent fuel can be placed in Yucca Mountain, and of non-reprocessing approaches to extending that capacity; we do not have a realistic evaluation of the impact of a reprocessing and transmutation on the existing nuclear fuel industry; we do not have a serious evaluation of the licensing and public acceptance issues facing development and deployment of such a system; we do not have any serious assessment of the safety and terrorism risks of a reprocessing and transmutation system, compared to those of direct disposal; and we do not yet have assessments of the proliferation implications of the proposed systems that are detailed enough to support responsible decision-making. In short, now is the time for continued research and development, and additional systems analysis, not the time for committing to processing using any particular technology.

Recommendations

For the reasons just outlined, I recommend that we follow the advice of the bipartisan National Commission on Energy Policy, which reflected a broad spectrum of opinion on energy matters generally and on nuclear energy in particular, and recommended that the United States should:

1. "continue indefinitely the U.S. moratoria on commercial reprocessing of spent nuclear fuel and construction of commercial breeder reactors";
2. establish expanded interim spent fuel storage capacities "as a complement and interim backup" to Yucca Mountain;

3. proceed "with all deliberate speed" toward licensing and operating a permanent geologic waste repository; and
4. continue research and development on advanced fuel cycle approaches that might improve nuclear waste management and uranium utilization, without the huge disadvantages of traditional approaches to reprocessing.

At the same time, the U.S. government should redouble its efforts to: (a) limit the spread of reprocessing and enrichment technologies, as a critical element of a strengthened nonproliferation effort; (b) ensure that every nuclear warhead and every kilogram of separated plutonium and highly enriched uranium (HEU) worldwide are secure and accounted for, as the most critical step to prevent nuclear terrorism; and (c) convince other countries to end the accumulation of plutonium stockpiles, and work to reduce stockpiles of both plutonium and HEU around the world. The Bush administration should, in particular, resume the effort to negotiate a 20-year U.S.-Russian moratorium on separation of plutonium that was almost completed at the end of the Clinton administration.

Similar recommendations have been made in the MIT study on the future of nuclear energy, and in the American Physical Society study of nuclear energy and nuclear weapons proliferation.

It remains possible that someday approaches to reprocessing and recycling will be developed that make security, economic, political, and environmental sense. Research and development should explore such possibilities. Continued investment in R&D on advanced fuel cycle technologies is justified, in part to ensure that the United States will have the technological expertise and credibility to play a leading role in limiting the proliferation risks of the fuel cycle around the world. But the leverage of these technologies in meeting the most serious energy challenges of the 21st century is likely to be somewhat limited in comparison to the promise of other potential future energy technologies, and the emphasis that nuclear fuel cycle R&D should receive in the overall energy R&D portfolio should reflect that.

The global nuclear energy system would have to grow substantially if nuclear energy was to make a substantial contribution to meeting the world's 21st century needs for carbon-free energy. Building the support from governments, utilities, and publics needed to achieve that kind of growth will require making nuclear energy as cheap, as simple, as safe, as proliferation-resistant, and as terrorism-proof as possible. Reprocessing using any of the technologies likely to be available in the near term points in the wrong direction on every count. Those who hope for a bright future for nuclear energy, therefore, should oppose near-term reprocessing of spent nuclear fuel.

POSTSCRIPT

Should the United States Reprocess Spent Nuclear Fuel?

The nuclear waste disposal problem is real and it must be dealt with. If it is not, we may face the same kinds of problems created by the former Soviet Union, which disposed of some nuclear waste simply by dumping it at sea. For a recent summary of the nuclear waste problem and the disposal controversy, see Michael E. Long, "Half Life: The Lethal Legacy of America's Nuclear Waste," *National Geographic* (July 2002). The need for care in nuclear waste disposal is underlined by Tom Carpenter and Clare Gilbert, "Don't Breathe the Air," *Bulletin of the Atomic Scientists* (May/June 2004); they describe the Hanford Site in Hanford, Washington, where wastes from nuclear weapons production were stored in underground tanks. Leaks from the tanks have contaminated groundwater, and an extensive cleanup program is under way. But cleanup workers are being exposed to both radioactive materials and toxic chemicals, and they are falling ill. And in June 2004, the U.S. Senate voted to ease cleanup requirements.

In November 2005, President Bush signed the budget for the Department of Energy, which contained $50 million to start work toward a reprocessing plant; see Eli Kintisch, "Congress Tells DOE to Take Fresh Look at Recycling Spent Reactor Fuel," Science (December 2, 2005). Reprocessing spent nuclear fuel will be expensive, but the costs do not seem to be great enough to make nuclear power unacceptable; see "The Economic Future of Nuclear Power," University of Chicago (August 2004) (http://www.ne.doe.gov/reports/NuclIndustryStudy.pdf). Matthew L. Wald, "A New Vision for Nuclear Waste," *Technology Review* (December 2004), says that in the absence of Yucca Mountain, reprocessing, or other solutions to the problem of what to do with spent fuel, utilities have been storing the fuel on site. The latest techniques involve massive casks sitting on concrete pads and exposed to the air. The casks contain the waste, even against the threats of earthquakes and bombs, while the air carries away the heat generated by radioactive decay. Over time, the waste in the casks becomes less hazardous. Other potential solutions exist as well. Steven Ashley, in "Divide and Vitrify," *Scientific American* (June 2002), describes work on potential methods of separating the most hazardous components of nuclear waste. One such approach is to expose nuclear waste to neutrons from particle accelerators or special nuclear reactors and thereby greatly hasten the process of radioactive decay. William H. Hannum, Gerald E. Marsh, and George S. Stanford, "Smarter Use of Nuclear Waste," *Scientific American* (December 2005), discuss the use of fast-neutron reactors to accomplish this.

It is an unfortunate truth that the reprocessing of nuclear spent fuel does indeed increase the risks of nuclear proliferation. On February 28, 2004, *The Economist* ("The World Wide Web of Nuclear Danger") wrote that the risk that someone, somewhere, might detonate a bomb in anger is arguably greater than at any time since the 1962 Cuban missile crisis brought the cold-war world soberingly close to the brink." Both nations and terrorists itch to possess nuclear weapons, whose destructive potential makes present members of the "nuclear club" tremble. Can the risk be controlled? John Deutch, Arnold Kanter, Ernest Moniz, and Daniel Poneman, in "Making the World Safe for Nuclear Energy," *Survival* (Winter2004/2005), argue that present nuclear nations could supply fuel and reprocess spent fuel for other nations; nations that refuse to participate would be seen as suspect and subject to international action.

ISSUE 20

Is a Large-Scale Shift to Organic Farming the Best Way to Increase World Food Supply?

YES: **Brian Halweil**, from "Can Organic Farming Feed Us All?" *World Watch* (May/June 2006)

NO: **John J. Miller**, from "The Organic Myth: A Food Movement Makes a Pest of Itself," *National Review* (February 9, 2004)

ISSUE SUMMARY

YES: Brian Halweil, senior researcher at the Worldwatch Institute, argues that organic agriculture is potentially so productive that it could sustainably increase world food supply, although the future may be more likely to see a mix of organic and nonorganic techniques.

NO: John J. Miller argues that organic farming is not productive enough to feed today's population, much less larger future populations, it is prone to dangerous biological contamination, and it is not sustainable.

There was a time when all farming was organic. Fertilizer was compost and manure. Fields were periodically left fallow (unfarmed) to recover soil moisture and nutrients. Crops were rotated to prevent nutrient exhaustion. Pesticides were nonexistent. And farmers were at the mercy of periodic droughts (despite irrigation) and insect infestations.

As population grew, so did the demand for food. In Europe and America, the concomitant demand for fertilizer led in the nineteenth century to a booming trade in guano mined from Caribbean and Pacific islands where deposits of seabird dung could be a hundred and fifty feet thick. When the guano deposits were exhausted, there was an agricultural crisis that was relieved only by the invention of synthetic nitrogen-containing fertilizers early in the twentieth century. See Jimmy Skaggs, *The Great Guano Rush: Entrepreneurs and American Overseas Expansion* (St. Martin's, 1994), and G. J. Leigh, *The World's Greatest Fix: A History of Nitrogen and Agriculture* (Oxford, 2004). Unfortunately, synthetic fertilizers do not maintain the soil's content of organic matter (humus). This deficit can be amended by tilling in sewage sludge, but the public is not usually very receptive to the idea, partly because of the "yuck factor," but also because sewage sludge may contain human pathogens and chemical contaminants.

Synthetic pesticides, beginning with DDT, came into use in the 1940s. Their history is nicely outlined in Keith S. Delaplane, *Pesticide Usage in the United States: History, Benefits, Risks, and Trends* (University of Georgia Extension Bulletin 1121) (http://pubs.caes.uga.edu/caespubs/pubcd/B1121.htm). When they turned out to have problems—target species quickly became resistant, and when the chemicals reached human beings and wildlife on food and in water, they proved to be toxic—some people sought alternatives. These alternatives to synthetic fertilizers and pesticides (among other things) are what is usually meant by "organic farming." Proponents of organic farming have called it holistic, biodynamic, ecological, and natural and claimed a number of advantages for its practice. They say it both preserves the health of the soil and provides healthier food for people. They also argue that it should be used more, even to the point of replacing chemical-based "industrial" agriculture. Because proponents of chemicals hold that fertilizers and pesticides are essential to produce food in the quantities that a world population of over six billion people requires, and to hold food prices down to affordable levels, one strand of debate has been over whether organic agriculture can do the job.

In the following selections, Worldwatch researcher Brian Halweil argues that organic agriculture is potentially so productive that it could sustainably increase world food supply beyond current levels while strengthening the economic position of small farmers. However, the realities of the marketplace are such that many farmers are likely to use a mixture of organic and nonorganic techniques known as "low-input" farming. John J. Miller argues that organic farming is not productive enough to feed today's population, much less larger future populations, it is prone to dangerous biological contamination, and it is not sustainable. "Wishful thinking is at the heart of the organic-food movement."

YES

Brian Halweil

Can Organic Farming Feed Us All?

The only people who think organic farming can feed the world are delusional hippies, hysterical moms, and self-righteous organic farmers. Right?

Actually, no. A fair number of agribusiness executives, agricultural and ecological scientists, and international agriculture experts believe that a large-scale shift to organic farming would not only *increase* the world's food supply, but might be the only way to eradicate hunger.

This probably comes as a surprise. After all, organic farmers scorn the pesticides, synthetic fertilizers, and other tools that have become synonymous with high-yield agriculture. Instead, organic farmers depend on raising animals for manure, growing beans, clover, or other nitrogen-fixing legumes, or making compost and other sources of fertilizer that cannot be manufactured in a chemical plant but are instead grown—which consumes land, water, and other resources. (In contrast, producing synthetic fertilizers consumes massive amounts of petroleum.) Since organic farmers can't use synthetic pesticides, one can imagine that their fields suffer from a scourge of crop-munching bugs, fruitrotting blights, and plant-choking weeds. And because organic farmers depend on rotating crops to help control pest problems, the same field won't grow corn or wheat or some other staple as often.

As a result, the argument goes, a world dependent on organic farming would have to farm more land than it does today—even if it meant less pollution, fewer abused farm animals, and fewer carcinogenic residues on our vegetables. "We aren't going to feed 6 billion people with organic fertilizer," said Nobel Prize-winning plant breeder Norman Borlaug at a 2002 conference. "If we tried to do it, we would level most of our forest and many of those lands would be productive only for a short period of time." Cambridge chemist John Emsley put it more bluntly: "The greatest catastrophe that the human race could face this century is not global warming but a global conversion to 'organic firming'—an estimated 2 billion people would perish."

In recent years, organic farming has attracted new scrutiny, not just from critics who fear that a large-scale shift in its direction would cause billions to starve, but also from farmers and development agencies who actually suspect that such a shift could *better* satisfy hungry populations. Unfortunately, no one had ever systematically analyzed whether in fact a widespread shift to organic farming would run up against a shortage of nutrients and a lack of yields—until recently. The results are striking.

From *World Watch Magazine*, May/June 2006, pp. 18–24. Copyright © 2006 by Worldwatch Institute. Reprinted by permission. www.worldwatch.org

High-Tech, Low-Impact

There are actually myriad studies from around the world showing that organic farms can produce about as much, and in some settings much more, than conventional farms. Where there is a yield gap, it tends to be widest in wealthy nations, where farmers use copious amounts of synthetic fertilizers and pesticides in a perennial attempt to maximize yields. It is true that farmers converting to organic production often encounter lower yields in the first few years, as the soil and surrounding biodiversity recover from years of assault with chemicals. And it may take several seasons for farmers to refine the new approach.

But the long-standing argument that organic farming would yield just one-third or one-half of conventional farming was based on biased assumptions and lack of data. For example, the often-cited statistic that switching to organic farming in the United States would only yield one-quarter of the food currently produced there is based on a U.S. Department of Agriculture study showing that all the manure in the United States could only meet one-quarter of the nation's fertilizer needs—even though organic farmers depend on much more than just manure.

More up-to-date research refutes these arguments. For example, a recent study by scientists at the Research Institute for Organic Agriculture in Switzerland showed that organic farms were only 20 percent less productive than conventional plots over a 21-year period. Looking at more than 200 studies in North America and Europe, Per Pinstrup Andersen (a Cornell professor and winner of the World Food Prize) and colleagues recently concluded that organic yields were about 80 percent of conventional yields. And many studies show an even narrower gap. Reviewing 154 growing seasons' worth of data on various crops grown on rain-fed and irrigated land in the United States, University of California–Davis agricultural scientist Bill Liebhardt found that organic corn yields were 94 percent of conventional yields, organic wheat yields were 97 percent, and organic soybean yields were 94 percent. Organic tomatoes showed no yield difference.

More importantly, in the world's poorer nations where most of the world's hungry live, the yield gaps completely disappear. University of Essex researchers Jules Pretty and Rachel Hine looked at over 200 agricultural projects in the developing world that converted to organic and ecological approaches, and found that for all the projects—involving 9 million farms on nearly 30 million hectares—yields increased an average of 93 percent. A seven-year study from Maikaal District in central India involving 1,000 farmers cultivating 3,200 hectares found that average yields for cotton, wheat, chili, and soy were as much as 20 percent higher on the organic farms than on nearby conventionally managed ones. Farmers and agricultural scientists attributed the higher yields in this dry region to the emphasis on cover crops, compost, manure, and other practices that increased organic matter (which helps retain water) in the soils. A study from Kenya found that while organic farmers in "high-potential areas" (those with above-average rainfall and high soil quality) had lower maize yields than nonorganic farmers, organic farmers in areas with poorer resource endowments consistently outyielded conventional growers. (In both regions, organic farmers had higher net profits, return on capital, and return on labor.)

Contrary to critics who jibe that it's going back to farming like our grandfathers did or that most of Africa already farms organically and it can't do the job, organic farming is a sophisticated combination of old wisdom and modern ecological innovations that help harness the yield-boosting effects of nutrient cycles, beneficial insects, and crop synergies. It's heavily dependent on technology—just not the technology that comes out of a chemical plant.

High-Calorie Farms

So could we make do without the chemical plants? Inspired by a field trip to a nearby organic farm where the farmer reported that he raised an amazing 27 tons of vegetables on six-tenths of a hectare in a relatively short growing season, a team of scientists from the University of Michigan tried to estimate how much food could be raised following a global shift to organic farming. The team combed through the literature for any and all studies comparing crop yields on organic farms with those on nonorganic farms. Based on 293 examples, they came up with a global dataset of yield ratios for the world's major crops for the developed and the developing world. As expected, organic farming yielded less than conventional farming in the developed world for most food categories, while studies from the developing world showed organic farming boosting yields. The team then ran two models. The first was conservative in the sense that it applied the yield ratio for the developed world to the entire planet, i.e., they assumed that every farm regardless of location would get only the lower developed-country yields. The second applied the yield ratio for the developed world to wealthy nations and the yield ratio for the developing world to those countries.

"We were all surprised by what we found," said Catherine Badgley, a Michigan paleoecologist who was one of the lead researchers. The first model yielded 2,641 kilocalories ("calories") per person per day, just under the world's current production of 2,786 calories but significantly higher than the average caloric requirement for a healthy person of between 2,200 and 2,500. The second model yielded 4,381 calories per person per day, 75 percent greater than current availability—and a quantity that could theoretically sustain a much larger human population than is currently supported on the world's farmland. (It also laid to rest another concern about organic agriculture; see sidebar at left.)

The team's interest in this subject was partly inspired by the concern that a large-scale shift to organic farming would require clearing additional wild areas to compensate for lower yields—an obvious worry for scientists like Badgley, who studies present and past biodiversity. The only problem with the argument, she said, is that much of the world's biodiversity exists in close proximity to farmland, and that's not likely to change anytime soon. "If we simply try to maintain biodiversity in islands around the world, we will lose most of it," she said. "It's very important to make areas between those islands friendly to biodiversity. The idea of those areas being pesticide-drenched fields is just going to

ENOUGH NITROGEN TO GO AROUND?

In addition to looking at raw yields, the University of Michigan scientists also examined the common concern that there aren't enough available sources of non-synthetic nitrogen—compost, manure, and plant residues—in the world to support large-scale organic farming. For instance, in his book *Enriching the Earth: Fritz Haber, Carl Bosch, and the Transformation of World Food Production,* Vaclav Smil argues that roughly two-thirds of the world's food harvest depends on the Harber-Bosch process, the technique developed in the early 20th century to synthesize ammonia fertilizer from fossil fuels. (Smil admits that he largely ignored the contribution of nitrogen-fixing crops and assumed that some of them, like soybeans, are net users of nitrogen, although he himself points out that on average half of all the fertilizer applied globally is wasted and not taken up by plants.) Most critics of organic farming as a means to feed the world focus on how much manure—and how much related pastureland and how many head of livestock—would be needed to fertilize the world's organic farms. "The issue of nitrogen is different in different regions," says Don Lotter, an agricultural consultant who has published widely on organic farming and nutrient requirements. "But lots more nitrogen comes in as green manure than animal manure."

Looking at 77 studies from the temperate areas and tropics, the Michigan team found that greater use of nitrogen-fixing crops in the world's major agricultural regions could result in 58 million metric tons more nitrogen than the amount of synthetic nitrogen currently used every year. Research at the Rodale Institute in Pennsylvania showed that red clover used as a winter cover in an oat/wheat–corn–soy rotation, with no additional fertilizer inputs, achieved yields comparable to those in conventional control fields. Even in arid and semi-arid tropical regions like East Africa, where water availability is limited between periods of crop production, drought-resistant green manures such as pigeon peas or groundnuts could be used to fix nitrogen. In Washington state, organic wheat growers have matched their non-organic neighbor's wheat yields using the same field pea rotation for nitrogen. In Kenya, farmers using leguminous tree crops have doubled or tripled corn yields as well as suppressing certain stubborn weeds and generating additional animal fodder.

The Michigan results imply that no additional land area is required to obtain enough biologically available nitrogen, even without including the potential for intercropping (several crops grown in the same field at the same time), rotation of livestock with annual crops, and inoculation of soil with *Azobacter, Azospirillum,* and other free-living nitrogen-fixing bacteria.

be a disaster for biodiversity, especially in the tropics. The world would be able to sustain high levels of biodiversity much better if we could change agriculture on a large scale."

Badgley's team went out of the way to make its assumerions as conservative as possible: most of the studies they used looked at the yields of a single crop, even though many organic farms grow more than one crop in a field at the same time, yielding more total food even if the yield of any given crop may be lower. Skeptics may doubt the team's conclusions—as ecologists, they are likely to be sympathetic to organic farming—but a second recent study of the potential of a global shift to organic farming, led by Niels Halberg of the Danish Institute of Agricultural Sciences, came to very similar conclusions, even though the authors were economists, agronomists, and international development experts.

Like the Michigan team, Halberg's group made an assumption about the differences in yields with organic farming for a range of crops and then plugged those numbers into a model developed by the World Bank's International Food Policy Research Institute (IFPRI). This model is considered the definitive algorithm for predicting food output, farm income, and the number of hungry people throughout the world. Given the growing interest in organic farming among consumers, government officials, and agricultural scientists, the researchers wanted to assess whether a large-scale conversion to organic farming in Europe and North America (the world's primary food exporting regions) would reduce yields, increase world food prices, or worsen hunger in poorer nations that depend on imports, particularly those people living in the Third World's swelling megacities. Although the group found that total food production declined in Europe and North America, the model didn't show a substantial impact on world food prices. And because the model assumed, like the Michigan study, that organic farming would boost yields in Africa, Asia, and Latin America, the most optimistic scenario even had hunger-plagued sub-Saharan Africa exporting food surpluses.

"Modern non-certified organic farming is a potentially sustainable approach to agricultural development in areas with low yields due to poor access to inputs or low yield potential because it involves lower economic risk than comparative interventions based on purchased inputs and may increase farm level resilience against climatic fluctuations," Halberg's team concluded. In other words, studies from the field show that the yield increases from shifting to organic farming are highest and most consistent in exactly those poor, dry, remote areas where hunger is most severe. "Organic agriculture could be an important part of increased food security in sub-Saharan Africa," says Halberg.

That is, if other problems can be overcome. "A lot of research is to try to kill prejudices," Halberg says—like the notion that organic farming is only a luxury, and one that poorer nations cannot afford. "I'd like to kill this once and for all. The two sides are simply too far from each other and they ignore the realities of the global food system." Even if a shift toward organic farming boosted yields in hungry African and Asian nations, the model found that nearly a billion people remained hungry, because any surpluses were simply exported to areas that could best afford it.

Wrong Question?

These conclusions about yields won't come as a surprise to many organic farmers. They have seen with their own eyes and felt with their own hands how productive they can be. But some supporters of organic farming shy away from even asking whether it can feed the world, simply because they don't think it's the most useful question. There is good reason to believe that a global conversion to organic farming would not proceed as seamlessly as plugging some yield ratios into a spreadsheet.

To begin with, organic farming isn't as easy as farming with chemicals. Instead of choosing a pesticide to prevent a pest outbreak, for example, a particular organic farmer might consider altering his crop rotation, planting a crop that will repel the pest or one that will attract its predators—decisions that require some experimentation and long-term planning. Moreover, the IFPRI study suggested that a large-scale conversion to organic farming might require that most dairy and beef production eventually "be better integrated in cereal and other cash crop rotations" to optimize use of the manure. Bringing cows back to one or two farms to build up soil fertility may seem like a no-brainer, but doing it wholesale would be a challenge—and dumping ammonia on depleted soils still makes for a quicker fix.

Again, these are just theoretical assumptions, since a global shift to organic farming could take decades. But farmers are ingenious and industrious people and they tend to cope with whatever problems are at hand. Eliminate nitrogen fertilizer and many farmers will probably graze cows on their fields to compensate. Eliminate fungicides and farmers will look for fungus-resistant crop varieties. As more and more farmers begin to farm organically, everyone will get better at it. Agricultural research centers, universities, and agriculture ministries will throw their resources into this type of farming—in sharp contrast to their current neglect of organic agriculture, which partly stems from the assumption that organic farmers will never play a major role in the global food supply.

So the problems of adopting organic techniques do not seen insurmountable. But those problems may not deserve most of our attention; even if a mass conversion over, say, the next two decades, dramatically increased food production, there's little guarantee it would eradicate hunger. The global food system can be a complex and unpredictable beast. It's hard to anticipate how China's rise as a major importer of soybeans for its feedlots, for instance, might affect food supplies elsewhere. (It's likely to drive up food prices.) Or how elimination of agricultural subsidies in wealthy nations might affect poorer countries. (It's likely to boost farm incomes and reduce hunger.) And would less meat eating around the world free up food for the hungry? (It would, but could the hungry afford it?) In other words, "Can organic farming feed the world?" is probably not even the right question, since feeding the world depends more on politics and economics than any technological innovations.

'"Can organic farming feed the world' is indeed a bogus question," says Gene Kahn, a long-time organic farmer who founded Cascadian Farms organic foods and is now vice president of sustainable development for General Mills. "The real question is, can we feed the world? Period. Can we fix the disparities in human nutrition?" Kahn notes that the marginal difference in today's organic yields and the yields of conventional agriculture wouldn't matter if food surpluses were redistributed.

But organic farming will yield other benefits that are too numerous to name. Studies have shown, for example, that the "external" costs of organic farming—erosion, chemical pollution to drinking water, death of birds and other wildlife—are just one-third those of conventional farming. Surveys from every continent show that organic farms support many more species of birds, wild plants, insects, and other wildlife than conventional farms. And tests by several governments have shown that organic foods carry just a tiny fraction of the pesticide residues of the nonorganic alternatives, while completely banning growth hormones, antibiotics, and many additives allowed in many conventional foods. There is even some evidence that crops grown organically have considerably higher levels of health-promoting antioxidants.

There are social benefits as well. Because organic farming doesn't depend on expensive inputs, it might help shift the balance towards smaller farmers in hungry nations. A 2002 report from the UN Food and Agriculture Organization noted that "organic systems can double or triple the productivity of traditional systems" in developing nations but suggested that yield comparisons offer a "limited, narrow, and often misleading picture" since farmers in these countries often adopt organic farming techniques to save water, save money, and reduce the variability of yields in extreme conditions. A more recent study by the International Fund for Agricultural Development found that the higher labor requirements often mean that "organic agriculture can prove particularly effective in bringing redistribution of resources in areas where the labour force is underemployed. This can help contribute to rural stability."

Middle Earth

These benefits will come even without a complete conversion to a sort of organic Utopia. In fact, some experts think that a more hopeful, and reasonable, way forward is a sort of middle ground, where more and more farmers adopt the principles of organic farming even if they don't follow the approach religiously. In this scenario, both poor farmers and the environment come out way ahead. "Organic agriculture is not going to do the trick," says Roland Bunch, an agricultural extensionist who has worked for decades in Africa and the Americas and is now with COSECHA (Association of Consultants for a Sustainable, Ecological, and People-Centered Agriculture) in Honduras. Bunch knows first-hand that organic agriculture can produce more than conventional farming among poorer farmers. But he also knows that these farmers cannot get the premium prices paid for organic produce elsewhere, and that they are often unable, and unwilling, to shoulder some of the costs and risks associated with going completely organic.

Instead, Bunch points to "a middle path," of eco-agriculture, or low-input agriculture that uses many of the principles of organic farming and depends on just a small fraction of the chemicals. "These systems can immediately produce two or three times what smallholder farmers are presently producing" Bunch says. "And furthermore, it is attractive to smallholder farmers because it is less costly per unit produced." In addition to the immediate

FOOD VERSUS FUEL

$\mathbf{S}$ometimes, when humans try to solve one problem, they end up creating another. The global food supply is already under serious strain: more than 800 million people go hungry every day, the world's population continues to expand, and a growing number of people in the developing world are changing to a more Western, meat-intensive diet that requires more grain and water per calorie than traditional diets do. Now comes another potential stressor: concern about climate change means that more nations are interested in converting crops into biofuels as an alternative to fossil fuels. But could this transition remove land from food production and further intensify problems of world hunger?

For several reasons, some analysts say no, at least not in the near future. First, they emphasize that nearly 40 percent of global cereal crops are fed to livestock, not humans, and that global prices of grains and oil seeds do not always affect the cost of food for the hungry, who generally cannot participate in formal markets anyway.

Second, at least to date, hunger has been due primarily to inadequate income and distribution rather than absolute food scarcity. In this regard, a biofuels economy may actually help to reduce hunger and poverty. A recent UN Food and Agriculture Organization report argued that increased use of biofuels could diversify agricultural and forestry activities, attract investment in new small and medium-sized enterprises, and increase investment in agricultural production, thereby increasing the incomes of the world's poorest people.

Third, biofuel refineries in the future will depend less on food crops and increasingly on organic wastes and residues. Producing biofuels from corn stalks, rice hulls, sawdust, or waste paper is unlikely to affect food production directly. And there are drought-resistant grasses, fast-growing trees, and other energy crops that will grow on marginal lands unsuitable for raising food.

Nonetheless, with growing human appetites for both food and fuel, biofuels' long-run potential may be limited by the priority given to food production if bioenergy systems are not harmonized with food systems. The most optimistic assessments of the long-term potential of biofuels have assumed that agricultural yields will continue to improve and that world population growth and food consumption will stabilize. But the assumption about population may prove to be wrong. And yields, organic or otherwise, may not improve enough if agriculture in the future is threatened by declining water tables or poor soil maintenance.

gains in food production, Bunch suggests that the benefits for the environment of this middle path will be far greater than going "totally organic," because "something like five to ten times as many smallholder farmers will adopt it per unit of extension and training expense, because it behooves them economically. They aren't taking food out of their kids' mouths. If five farmers eliminate half their use of chemicals, the effect on the environment will be two and one-half times as great as if one farmer goes totally organic."

And farmers who focus on building their soils, increasing biodiversity, or bringing livestock into their rotation aren't precluded from occasionally turning to biotech crops or synthetic nitrogen or any other yield-enhancing innovations in the future, particularly in places where the soils are heavily depleted, "In the end, if we do things right, we'll build a lot of organic into conventional systems," says Don Lotter, the agricultural consultant. Like Bunch, Lotter notes that such an "integrated" approach often out-performs both a strictly organic and chemical-in tensive approach in terms of yield, economics, and environmental benefits. Still, Lotter's not sure we'll get there tomorrow, since the world's farming is hardly pointed in the organic direction—which could be the real problem for the world's poor and hungry. "There is such a huge area in sub-Saharan Africa and South America where the Green Revolution has never made an impact and it's unlikely that it will for the next generation of poor farmers," argues Niels Halberg, the Danish scientist who lead the IFPR] study. "It seems that agro-ecological measures for some of these areas have a beneficial impact on yields and food insecurity. So why not seriously try it out?"

John J. Miller

 **NO**

The Organic Myth: A Food Movement Makes a Pest of Itself

Somewhere in the cornfields of Britain, a hungry insect settled on a tall green stalk and decided to have a feast. It chewed into a single kernel of corn, filled its little belly, and buzzed off—leaving behind a tiny hole that was big enough to invite a slow decay. The agent of the decomposition was a fungus known to biologists as *Fusarium*. Farmers have a much simpler name for it: corn ear rot.

As the mold spread inside the corn, it left behind a cancer causing residue called fumonisin. This sequence repeated itself thousands and thousands of times until the infested corn was harvested and sold last year as Fresh and Wild Organic Maize Meal, Infinity Foods Organic Maize Meal, and several other products.

Consuming trace amounts of fumonisin is harmless, but large doses can be deadly. Last fall, the United Kingdom's Food Standards Agency detected alarming concentrations of the toxin in all six brands of organic corn meal subjected to testing—for a failure rate of 100 percent. The average level of contamination was almost 20 times higher than the safety threshold Europeans have set for fumonisin. The tainted products were immediately recalled from the food chain. In contrast, inspectors determined that 20 of the 24 non-organic corn meal products they examined were unquestionably safe to eat.

Despite this, millions of people continue to assume that organic foods are healthier than non-organic ones, presumably because they grow in pristine settings free from icky chemicals and creepy biotechnology. This has given birth to an energetic political movement. In 2002, activists in Oregon sponsored a ballot initiative that essentially would have required the state to slap biohazard labels on anything that wasn't produced in ways deemed fit by anti-biotech agitators. Voters rejected it, but the cause continues to percolate. Hawaiian legislators are giving serious thought to banning biotech crop tests in their state. In March, California's Mendocino County may outlaw biotech plantings altogether.

Beneath it all lurks the belief that organic food is somehow better for us. In one poll, two-thirds of Americans said that organic food is healthier. But they're wrong: It's no more nutritious than food fueled by industrial fertilizers, sprayed with synthetic pesticides, and genetically altered in science labs. And the problem isn't limited to the fungal infections that recently cursed

From *National Review*, February 9, 2004, pp. 35–37. Copyright © 2004 by National Review, Inc., 215 Lexington Avenue, New York, NY 10016. Reprinted by permission.

organic corn meal in Britain; bacteria are a major source of disease in organic food as well. To complicate matters further, organic farming is incredibly inefficient. If its appeal ever grew beyond the boutique, it would pose serious threats to the environment. Consumers who go shopping for products emblazoned with the USDA's "organic" seal of approval aren't really helping themselves or the planet—and they're arguably hurting both.

No Fear

Here's the good news: At no point in human history has food been safer than it is today, despite occasional furors like the recent one over an isolated case of mad-cow disease here in the U.S. People still get sick from food—each year, about 76 million Americans pick up at least a mild illness from what they put in their mouths—but modern agricultural methods have sanitized our fare to the point where we may eat without fear. This is true for all food, organic or otherwise.

And that raises a semantic question: What is it about organic food that makes it "organic"? The food we think of as nonorganic isn't really *in*organic, as if it were composed of rocks and minerals. In truth, everything we eat is organic—it's just not "organic" the way the organic-food movement has come to define the word.

About a decade ago, the federal government decided to wade into this semantic swamp. There was no compelling reason for this, but Congress nonetheless called for the invention of a National Organic Rule. It became official in 2002. Organic food, said the bureaucrats, is produced without synthetic fertilizers, conventional pesticides, growth hormones, genetic engineering, or germ-killing radiation. There are also varying levels of organic-ness: Special labels are available for products that are "made with organic ingredients" (which means the food is 70 percent organic), "organic" (which means 95 percent organic), and "100 percent organic." It's not at all clear what consumers are supposed to do with this information. As the Department of Agriculture explains on its website, the "USDA makes no claims that organically produced food is safer or more nutritious than conventionally produced food."

It doesn't because it can't: There's no scientific evidence whatsoever showing that organic food is healthier. So why bother with a National Organic Rule? When the thing was in development, the Clinton administration's secretary of agriculture, Dan Glickman, offered an answer: "The organic label is a marketing tool. It is not a statement about food safety." In other words, those USDA labels are intended to give people warm fuzzies for buying pricey food.

And herein lies one of the dirty secrets of organic farming: It's big business. Although the organic movement has humble origins, today most of its food isn't produced on family farms in quaint villages or even on hippie communes in Vermont. Instead, the industry has come to be dominated by large corporations that are normally the dreaded bogeymen in the minds of many organic consumers. A single company currently controls about 70 percent of the market in organic milk. California grows about $400 million per year in organic produce—and about half of it comes from just five farms. The membership list of the Organic Trade Association includes the biggest names in agribusiness, such as Archer Daniels Midland, Gerber, and Heinz. Even Nike is a member. When its capitalist slavedrivers aren't exploiting child labor in Third World sweatshops (as they do in the fevered imaginations of campus protesters), they're promoting Nike Organics, a clothing line made from organic cotton.

The Yum Factor

There are, in fact, good reasons to eat organic food. Often it's yummier—though this has nothing to do with the fact that it's "organic." If an organic tomato tastes better than a non-organic one, the reason is usually that it has been grown locally, where it has ripened on the vine and taken only a day or two to get from the picking to the selling. Large-scale farming operations that ship fruits and vegetables across the country or the world can't compete with this kind of homegrown quality, even though they do make it possible for people in Minnesota to avoid scurvy by eating oranges in February. Conventional produce is also a good bargain because organic foods can be expensive—the profit margins are quite high, relative to the rest of the food industry.

Unfortunately, money isn't always the sole cost. Although the overwhelming majority of organic foods are safe to eat, they aren't totally risk-free. Think of it this way: Organic foods may be fresh, but they're also fresh from the manure fields.

Organic farmers aren't allowed to enrich their soils the way most non-organic farmers do, which is with nitrogen fertilizers produced through an industrial process. In their place, many farmers rely on composted manure. When they spread the stuff in their fields, they create luscious breeding grounds for all kinds of nasty microbes. Take the dreaded *E. coli*, which is capable of killing people who ingest it. A study by the Center for Global Food Issues found that although organic foods make up about 1 percent of America's diet, they also account for about 8 percent of confirmed *E. coli* cases. Organic food products also suffer from more than eight times as many recalls as conventional ones.

Some of this problem would go away if organic farmers used synthetic sprays—but this, too, is off limits. Conventional wisdom says that we should avoid food that's been drenched in herbicides, pesticides, and fungicides. Half a century ago, there was some truth in this: Sprays were primitive and left behind chemical deposits that often survived all the way to the dinner table. Today's sprays, however, are largely biodegradable. They do their job in the field and quickly break down into harmless molecules. What's more, advances in biotechnology have reduced the need to spray. About one-third of America's corn crop is now genetically modified. This corn includes a special gene that produces a natural toxin that's safe for every living creature to eat except caterpillars with alkaline guts, such as the European corn borer, a moth larva that can ravage whole harvests. This kind of biotech innovation has helped farmers reduce their reliance on pesticides by about 50 million pounds per year.

Organic farmers, of course, don't benefit from any of this. But they do have some recourse against the bugs, weeds, and fungi that can devastate a crop: They spray their plants with "natural" pesticides. These are less effective than synthetic ones and they're certainly no safer. In rat tests, rotenone—an insecticide extracted from the roots of tropical plants—has been shown to cause the symptoms of Parkinson's disease. The Environmental Protection Agency has described pyrethrum, another natural bug killer, as a human carcinogen. Everything is lethal in massive quantities, of course, and it takes huge doses of pyrethrum to pose a health hazard. Still, the typical organic farmer has to douse his crops with it as many as seven times to have the same effect as one or two applications of a synthetic compound based on the same ingredients. Then there's one of the natural fungicides preferred by organic coffee growers in Guatemala: fermented urine. Think about that the next time you're tempted to order the "special brew" at your local organic java hut.

St. Anthony's Fire, etc.

Fungicides are worth taking seriously—and not just because they might have prevented Britain's corn meal problem a few months ago. Before the advent of modern farming, when all agriculture was necessarily "organic," food-borne fungi were a major problem. One of the worst kinds was ergot, which affects rye, wheat, and other grains. During the Middle Ages, ergot poisoning caused St. Anthony's Fire, a painful contraction of blood vessels that often led to gangrene in limb extremities and sometimes death. Hallucinations also were common, as ergot contains lysergic acid, which is the crucial component of LSD. Historians of medieval Europe have documented several episodes of towns eating large batches of ergot-polluted food and falling into mass hysteria. There is some circumstantial evidence suggesting that ergot was behind the madness of the Salem witches: The warm and damp weather just prior to the infamous events of 1692 would have been ideal for an outbreak. Today, however, chemical sprays have virtually eradicated this affliction.

The very worst thing about organic farming requires the use of a word that doomsaying environmentalists have practically trademarked: It's not *sustainable*. Few activities are as wasteful as organic farming. Its yields are about half of what conventional farmers expect at harvest time. Norman Borlaug, who won the Nobel Peace Prize in 1970 for his agricultural innovations, has said, "You couldn't feed more than 4 billion people" on an all-organic diet.

If organic-food consumers think they're making a political statement when they eat, they're correct: They're declaring themselves to be not only friends of population control, but also enemies of environmental conservation. About half the world's land area that isn't covered with ice or sand is devoted to food production. Modern farming techniques have enabled this limited supply to produce increasing quantities of food. Yields have fattened so much in the last few decades that people refer to this phenomenon as the "Green Revolution," a term that has nothing to do with enviro-greenies and everything to do with improvements in breeding, fertilization, and irrigation. Yet even greater challenges lie ahead, because demographers predict that world population will rise to 9 billion by 2050. "The key is to produce more food," says Alex Avery of CGFI. "Growing more per acre leaves more land for nature." The alternative is to chop down rainforests so that we may dine on organic soybeans.

There's one more important reason that organics can't feed the world: There just isn't enough cow poop to go around. For fun, pretend that U.N. secretary-general Kofi Annan chowed on some ergot rye, decreed that all of humanity must eat nothing but organic food, and that all of humanity responded by saying, "What the heck, we'll give it a try." Forget about the population boom ahead. The immediate problem would be generating enough manure to fertilize all the brand-new, low-yield organic crop fields. There are a little more than a billion cattle in the world today, and each bovine needs between 3 and 30 acres to support it. Conservative estimates say it would take around 7 or 8 billion cattle to produce sufficient heaps of manure to sustain our all-organic diets. The United States alone would need about a billion head (or rear, to be precise). The country would be made up of nothing but cities and manure fields—and the experiment would give a whole new meaning to the term "fruited plains."

This is the sort of future the organic-food movement envisions—and its most fanatical advocates aren't planning to win any arguments on the merits or any consumers on the quality of organic food. In December, when a single U.S. animal was diagnosed with mad-cow disease, nobody was more pleased than Ronnie Cummins of the Organic Consumers Association, who has openly hoped for a public scare that would spark a "crisis of confidence" in American food. No such thing happened, but Cummins should be careful about what he wishes for: Germany's first case of mad-cow disease surfaced at a slaughterhouse that specializes in organic beef.

But then wishful thinking is at the heart of the organic-food movement. Its whole market rationale depends on the misperception that organic foods are somehow healthier for both consumers and Mother Earth. Just remember: Nature's Valley can't be found on any map. It's a state of mind.

POSTSCRIPT

Is a Large-Scale Shift to Organic Farming the Best Way to Increase World Food Supply?

Is organic food better or safer for the consumer than non-organic food? Faidon Magkos, et al., "Organic Food: Buying More Safety or Just Peace of Mind? A Critical Review of the Literature," *Critical Reviews in Food Science & Nutrition* (January 2006), report that the quality and safety of organic food are largely a matter of perception. "Relevant scientific evidence . . . is scarce, while anecdotal reports abound." Pesticide and herbicide residues may be lower, but even on non-organic food they are low. Environmental contaminants are likely to affect both organic and non-organic foods. "'Organic' does not automatically equal 'safe.'"

Is organic farming better for the environment? Soil fertility is in decline in many parts of the world; see Alfred E. Herteminck, "Assessing Soil Fertility Decline in the Tropics Using Soil Chemical Data," *Advances in Agronomy* (2006). In Africa, the situation is extraordinarily serious. According to the International Center for Soil Fertility and Agricultural Development (http://www.ifdc. org/New_Design/Whats_New/africasfailingagriculture033006. pdf), "About 75 percent of the farmland in sub-Saharan Africa is plagued by severe degradation, losing basic soil nutrients needed to grow the crops that feed Africa, according to a new report . . . on the precipitous decline in African soil health from 1980 to 2004. Africa's crisis in food production and battle with hunger are largely rooted in this 'soil health crisis.'" Proponents of organic farming argue that it is essential to relieving the crisis, but one study of changes in soil fertility, as indicated by crop yield, earthworm numbers, and soil properties, after converting from conventional to organic practices, found that different soils responded differently, with some improving and some not; see Anne Kjersti Bakken, et al., "Soil Fertility in Three Cropping Systems after Conversion from Conventional to Organic Farming, "*Acta Agriculturae Scandinavica: Section B, Soil & Plant Science* (June 2006). Richard Wood, et al., "A Comparative Study of Some Environmental Impacts of Conventional and Organic Farming in Australia," *Agricultural Systems* (September 2006), find in a comparison of organic and conventional farms "that direct energy use, energy related emissions, and greenhouse gas emissions are higher for the" former.

According to Catherine M. Cooney, "Sustainable Agriculture Delivers the Crops," *Environmental Science & Technology* (February 15, 2006), "Sustainable agriculture, such as crop rotation, organic farming, and genetically modified seeds, increased crop yields by an average of 79 percent" while also improving the lives of farmers in developing countries. Cong Tu, et al., "Responses of Soil Microbial Biomass and N Availability to Transition Strategies from Conventional to Organic Farming Systems," *Agriculture, Ecosystems & Environment* (April 2006), note that a serious barrier to changing from conventional to organic farming, despite soil improvements, is an initial reduction in yield and increase in pests. David Pimentel, et al., "Environmental, Energetic, and Economic Comparisons of Organic and Conventional Farming Systems," *Bioscience* (July 2005), find that organic farming uses less energy, improves soil, and has yields comparable to those of conventional farming but crops probably cannot be grown as often (because of fallowing), which then reduces long-term yields. However, "because organic foods frequently bring higher prices in the marketplace, the net economic return per [hectare] is often equal to or higher than that of conventionally produced crops."

That last point means that organic farming is currently good for the organic farmer. However, the advantage would disappear if the world converted to organic farming. Initial declines in yield mean the conversion would be difficult, but the longer we wait and the more population grows, the more difficult it will be. Whether the conversion is essential depends on the availability of alternative solutions to the problem, and it is worth noting that high energy prices make chemical fertilizers increasingly expensive. Sadly, Stacey Irwin, "Battle High Fertilizer Costs," *Farm Industry News* (January 2006), does not mention the possibility of using organic methods.

ISSUE 21

Should the Endangered Species Act Be Strengthened?

YES: John Kostyack, from Testimony before the Oversight Hearing on the Endangered Species Act, U.S. Senate Committee on Environment & Public Works (May 19, 2005)

NO: Monita Fontaine, from Testimony before the Oversight Hearing on the Endangered Species Act, U.S. Senate Committee on Environment & Public Works (May 19, 2005)

ISSUE SUMMARY

YES: Representing the National Wildlife Foundation, John Kostyack argues that the Endangered Species Act has been so successful that it should not be weakened but strengthened.

NO: Speaking for the National Endangered Species Act Reform Coalition, a group that represents those affected by the Endangered Species Act, Monita Fontaine argues that federal regulation under the ESA should be replaced by a system that relies more on voluntary and state species conservation efforts.

Today, human activities are an important cause of species loss mostly because humans destroy or alter habitat but also because of hunting (including commercial fishing), the introduction of foreign species as novel competitors, and the introduction of diseases. According to Martin Jenkins, "Prospects for Biodiversity," *Science* (November 14, 2003), says species losses may lead to a "biologically impoverished" world, with consequences. Jenkins states that the consequences for human life are "unforeseeable but potentially catastrophic."

Awareness of the problem has been growing. When the United States adopted the Endangered Species Act (ESA) in 1973, the goal was to protect species that were so reduced in numbers or restricted in habitat that a single untoward event could wipe them out. Both environmental groups and politicians were concerned over declining populations of some birds and plants. According to Ted Williams, "Law of Salvation," *Audubon* (November/December 2005), "Protecting the planet's genetic wealth made sense morally and economically. It was considered, rightly enough, what decent, civilized people do." The ESA therefore barred construction projects that would further threaten endangered species. One famous case involved the spotted owl, which was threatened by logging in the Northwest. Those in favor of the dam or the timber industry felt that the endangered species was trivial compared to the human benefits at stake. Those in favor of the act argued that the loss of a single species might not matter to the world, but where one species went, others would follow. Protecting one species also protects others.

The ESA has had some notable successes even though it has been called a political lightning rod: see "Endangered Species: Time and Costs Required to Recover Species Are Largely Unknown," *GAO Reports* (GAO-06-463R) (April 6, 2006). Some species have recovered to the point of being removed from the list of threatened and endangered species. Others have simply been kept from dying out. But to achieve these successes, the ESA has restricted the activities—and the opportunities to make money—of real estate developers, farmers, ranchers, loggers, miners, and other private interests. The result has been lengthy legal battles and pressures to ease restrictions on activities that might damage species or their habitat. These pressures have made it difficult to maintain the ESA. It was last authorized in 1988. Since that authorization expired in 1992, it has been kept in play only because Congress has kept funding it. Efforts to reauthorize it have repeatedly stalled. According to Peter Uimonen and John Kostyack, "Unsound Economics: The Bush Administration's New Strategy for Undermining the Endangered Species Act," National Wildlife Federation (June 2004) (http://www.nwf.org/wildlife/pdfs/UnsoundEconomics.pdf), the current administration has reduced critical habitat protection, suppressed and distorted information on the economic benefits to local economies of habitat conservation, exaggerated costs, and reduced funding. Erik Stokstad, "What's Wrong with the Endangered Species Act?" *Science* (September 30, 2005), describes the Republican-sponsored attempt to "reform" the ESA with the Threatened and Endangered Species Recovery Act as further restricting the ESA budget by requiring that landowners be compensated "for the fair market value of any development or other activity that the government vetoes because it would impact endangered species." It would also insulate landowners against the hazards of regulation as long as they have a plan in place to mitigate the effects of their actions and drastically limit the time federal agencies have to object to projects that might harm endangered species. Critics charge that the legislation, which passed the House late in 2005, is clearly more favorable to private interests than to endangered species. Proponents of reform argue that improving protection of private property rights is crucial.

In the following selections, John Kostyack argues that the Endangered Species Act has been so successful that it should not be weakened. What is needed is increased funding, earlier, speedier, and more effective recovery planning, reinforced habitat protection, and financial assistance for cooperative landowners. Speaking for the National Endangered Species Act Reform Coalition, a group that represents those affected by the Endangered Species Act, Monita Fontaine argues that federal regulation under the ESA should be replaced by a system that relies more on voluntary and state species conservation efforts.

YES

John Kostyack

Testimony

Good morning, Senator Chafee and members of the subcommittee. My name is John Kostyack, and I am Senior Counsel and Director of Wildlife Conservation Campaigns with the National Wildlife Federation. I appreciate your invitation for me to testify here today on the Endangered Species Act. I have been working on Endangered Species Act law and policy, both here in Washington, D.C., and in various regions around the country, for the past 12 years. Over this time my appreciation for the value and wisdom of this law has grown continuously.

I'd like to talk today about how Congress could update the law to deal with the wildlife conservation challenges of the coming decades. The challenges are many. Consider, for example, the following threats, each of which is accelerating over time:

- *Invasive Species.* According to the USDA, 133 million acres of land in the U.S. are already covered by invasive plants, and each year another 1.7 million acres are invaded. Invasive species threaten the survival of nearly half of all listed species.
- *Sprawling Development Patterns.* The amount of land covered by urban and suburban development in the U.S. has quadrupled since 1950, with the rate of land consumption greatly outpacing population growth and increasing every decade. According to Endangered by Sprawl (2005), a study recently completed by National Wildlife Federation, Smart Growth America, and Nature Serve, over 1,200 plant and animal species will be threatened with extinction by sprawl in just the next two decades.
- *Global Warming.* According to the U.S. State Department's recent Climate Action Report (2002), global warming poses serious risks to species and habitat types throughout the United States, threatening, among other things, alpine meadows across the West, prairie potholes in the Great Plains, and salmon spawning habitats in the Pacific Northwest.

If we truly want to pass on this nation's wildlife heritage to our children and grandchildren, we are going to need a strong Endangered Species Act to address these threats.

Before moving to some suggested updates to the Endangered Species Act, I would first like to talk about what kind of law we already have. It is crucial that Congress understands the benefits the law is already providing, and the

United States Senate Committee on Environment & Public Works Hearing on the National Wildlife Federation Oversight on the Endangered Species Act, May 19, 2005.

law's many on-the-ground success stories, before it proceeds to reauthorization. The positive accomplishments of the past 32 years are the foundation that future changes to the Act must be built upon.

The Benefits of the Endangered Species Act

The Endangered Species Act represents the only effort by this nation to grapple in a comprehensive way with the problem of human-caused extinctions. For the many animal and plant species at risk of extinction, it is the only safety net that our nation provides.

Fortunately, the Endangered Species Act has been quite successful in rescuing plants and animals from extinction.

- Over 98% of species ever protected by the Act remain on the planet today.
- Of the listed species whose condition is known, 68% are stable or improving and 32% are declining.
- The longer a species enjoys the ESA's protection, the more likely its condition will stabilize or improve.

This is the most important thing for Congress to understand about the Endangered Species Act. **It has worked to keep species from disappearing forever into extinction and, over time, it has generally stabilized and improved the condition of species.** As a result, we have a fighting chance of achieving recovery, and more importantly, we are passing on to future generations the practical and aesthetic benefits of wildlife diversity that we have enjoyed.

The other key benefit provided by the Endangered Species Act, besides stopping extinction, is that it **protects the habitats that species depend upon for their survival.** The habitats protected by the Act are not only essential for wildlife, they are oftentimes the very natural areas that people count on to filter drinking water, prevent flooding, provide healthy conditions for hunting, fishing and other outdoor recreation, and provide a quiet and peaceful respite from our noisy and frenetic everyday lives.

To this date, no one has come up with a better way to protect our wildlife and wild places for future generations. So, when our children peer into the eyes of a manatee swimming by their canoe in a clear cool Florida river, or listen to a wolf howl in Yellowstone, or watch a condor soar majestically over the Grand Canyon, our generation and the one before ours should take pride in what we have done for them in the past 32 years. As a result of the commitment Congress made in enacting the Endangered Species Act in 1973, and as a result of the efforts of many people working with the law ever since, we still have a rich and wonderful wildlife legacy to pass along.

Measuring Success—A Lesson from the Ivory-Billed Woodpecker

In the past few years, opponents of the Endangered Species Act have repeatedly tried to persuade the American people that despite the law's success in stopping extinction, the law is broken and needs a radical overhaul. Their argument boils down to a single statistic: only 13 or so species have been removed from the endangered species list due to recovery.

Recovery and delisting are certainly goals that the National Wildlife Federation shares, and I will speak in a moment about how to improve the odds of achieving them. However, I must first challenge the premise of the ESA's opponents that recovery and delisting should be the only measure of the success of the Endangered Species Act. Because it is not the only measure of success—it is not even the best measure—the entire case for a radical overhaul of the Act evaporates.

The story of the ivory-billed woodpecker highlights three reasons why the Endangered Species Act cannot be evaluated based upon the number of species fully recovered and delisted. Although the ESA has not yet been applied to the ivory bill, this species symbolizes the challenges facing wildlife agencies today. It shows that some of the biggest obstacles to recovery and delisting are largely beyond the influence of the Endangered Species Act.

First, Restoring Species and Habitats Requires Funding

Although the ivory-billed woodpecker has been listed as endangered under the ESA and predecessor laws since 1967, it has been presumed extinct since the 1940s. In perhaps one of the most exciting wildlife stories in our nation's history, a single bird was recently sighted in the Cache River National Wildlife Refuge in eastern Arkansas. We hope and expect that there are more birds in that area, but in any case, the bird's numbers are extremely low.

The ivory bill historically inhabited swampy bottomland hardwood forests. It prefers older trees, where it finds its primary food source, beetle larvae, living under the bark. In the southeastern U.S. where the bird once ranged, the vast majority of these old-growth forests are now gone, cleared for farms and pine plantations, and it will take decades to grow them back.

Restoring the habitats that the ivory bill needs to recover is going to take a lot more than the Endangered Species Act. Although safe harbor agreements under the ESA can remove disincentives, substantial public and private dollars will be needed to create positive incentives for private landowners to plant bottomland hardwood trees and protect them until they reach the stage where they are suitable habitat for the ivory bill. The fact that the ivory bill is listed as endangered under the Endangered Species Act will help concentrate everyone's attention on this task. However, if sufficient restoration dollars are not raised, it will not be a failure of the Endangered Species Act. Congress and other key actors need to provide funding to make this large-scale restoration project happen.

Second, as a Matter of Biology, Achieving Full Recovery Often Takes a Long Time

The average period of time in which species have been listed under the ESA is 15.5 years. In that amount of time, our best-case scenario is that we will have discovered and begun protecting a few more ivory bills and developed a strategy for accommodating range expansion. As a matter of simple biology—there aren't currently enough old trees around that could sustain a viable meta-population—full recovery of the ivory bill will take many decades.

Although the condition of most other listed species is not as dire as the ivory bill, many have severely depleted population numbers and habitats. As with the ivory bill, bringing their population numbers back and restoring their habitats often takes a long time for reasons of biology alone. Add in economic and political obstacles—such as the fact that many areas that need to be restored as habitat have potentially competing uses—and you can reasonably expect that recovery will not be completed for many species for a long while.

Third, Delisting Requires Putting in Place Non-ESA Regulatory Measures

Once a species' numbers and habitats are restored to the point of long-term viability, delisting still may not be feasible. Under the ESA, the Fish and Wildlife Service or NOAA Fisheries must first ensure that adequate regulatory measures are in place to prevent immediate backsliding after delisting.

For the ivory bill and many other listed species, there are no protections in place to prevent immediate habitat losses after the Endangered Species Act's protections are removed. In addition, many species require continuing management even after their population sizes and habitats have been restored to targeted levels. Conservation agreements with funding, monitoring and enforcement mechanisms must be negotiated with land managers to ensure that this management is carried out over the long run.

In summary, those who claim the ESA is broken due to the absence of a sizable number of delistings are ignoring the facts. The realities that impede quick recovery and delisting—inadequate funding, slow biological processes, and the absence of any alternative safety net—are not the fault of the Endangered Species Act.

The Endangered Species Act is making an essential contribution to recovery by stabilizing and improving the condition of species over time. Thanks to the Act, the ivory bill has a real chance of making it into the next century. But Congress needs to look outside the four corners of the Act to fully understand and address the reasons why so few species are removed from the threatened and endangered list due to recovery each year.

In addition, members of Congress should stop relying a single statistic about delistings as the measure of the Act's success, and instead encourage the wildlife agencies to develop new and better mechanisms for tracking progress. As authors Michael Scott and Dale Goble point out in the April 2005 issue of BioScience, the wildlife agencies currently do not maintain a database enabling policymakers and the public to track Endangered Species Act actions. A database that identifies, among other things, how much habitat is being conserved and how much is being authorized for destruction as a result of ESA consultation processes, would greatly inform the debate over the effectiveness of the law.

On-the-Ground Success Stories to Build Upon

The Endangered Species Act has produced numerous on-the-ground successes. The small list of examples below is designed simply to highlight the variety and creativity of the conservation actions that the law has fostered. These examples show that the Endangered Species Act is empowering people to find a place for wildlife in a country that is increasingly crowded with extractive industries, real estate developments, and other human uses of natural resources. Because of the Act's safety net features and its recovery programs, native wildlife still has a place on the American landscape.

1. **Whooping Crane** The whooping crane is a dynamic and charismatic bird that, if it were not for the Endangered Species Act and its predecessors, would probably no longer exist in the wild today. As a result of a recovery program developed under the Act, birds have been bred in captivity, released into the wild, and trained with the help of an aircraft to fly and migrate. Endangered Species Act enforcement action to protect the bird's designated critical habitat led to the creation of the Platte River Critical Habitat Maintenance Trust, which has acquired over 10,000 acres of riparian habitat along the crane's migratory route. Prior to the Endangered Species Act, a mere 16 birds existed in the wild. Today, nearly 200 birds thrive in the wild, attracting birdwatchers from around the world.

2. **Florida Panther** The Florida panther is one of the most endangered large mammals in the world. As recently as fifteen years ago, its numbers had been reduced to somewhere between 30 and 50. Due to the Endangered Species Act, a number of innovative conservation measures have been taken to bring the animal back from the brink. The U.S. Fish and Wildlife Service successfully addressed the panther's inbreeding problem by bringing Texas cougars (a closely related subspecies) into south Florida. Vehicle mortality, one of the leading causes of panther deaths, has been greatly reduced with the construction of highway underpasses. The underpasses created for the Florida panther now serve as a world model for facilitating movement of wildlife in an urbanizing landscape. Today, the number of cats living in the wild approaches 100. The Florida panther is still a long way from full recovery, but it has a fighting chance.

3. **Gray Wolf** Although the gray wolf once ranged across much of the continental United States, several centuries of hunting and predator control programs, reduction of prey, and habitat loss greatly reduced the species' numbers. By the mid-1960s, the only gray wolves in the lower 48 states were the 200 to 500 animals in Minnesota and roughly 20 on Isle Royale, Michigan. Today, thanks to the Endangered Species Act, there are thriving gray wolf populations in the Western Great Lakes and Northern Rockies, a small population in the Southwest, and occasional wolf sightings in the Northeast and Pacific Northwest. The dramatic recovery of the gray wolf in the Northern Rockies was jump-started by an historic reintroduction of wolves to Yellowstone National Park and the central Idaho wilderness—one of the most successful wildlife reintroductions in the nation's history.

4. **Bald Eagle** In the 1960s, the bald eagle, our Nation's symbol, had fewer than 500 breeding pairs remaining in the continental U.S. Widespread use of the pesticide DDT in the post-World War II period had contaminated the majestic bird's food supply, causing its populations across the country to plummet. Although the federal ban on DDT in 1972 was a major factor in turning around the bald eagle's decline, the Endangered Species Act also played an essential role in its recovery. The Act protected the bird's key habitat and facilitated translocations of eaglets from areas where the bird was numerous to states where it had been eliminated or severely depleted. Today, the number of bald eagles in the lower 48 states exceeds 7,600 breeding pairs.

5. **Puget Sound Chinook Salmon** Chinook salmon have long been a symbol of the Pacific Northwest, providing important cultural values for Native American tribes and sustenance and recreation for all residents. The Puget Sound population of the Chinook was listed in 1999 after declining steadily due to logging, mining, dam-building and suburban development in its habitat, and interbreeding of hatchery fish. Recently, in response to the Endangered Species Act, Seattle City Light improved prospects for the fish by modifying its dam operations on the Skagit, the Puget Sound's largest river. Prospects for the fish and habitats also have improved due to the emergence of Shared Strategy, a groundbreaking collaborative effort by a diverse array of citizens and organizations to build an ESA recovery plan for the Puget Sound chinook from the ground up, watershed by watershed. This effort will ensure broad public support for the array of recovery actions that will ultimately be needed to bring the chinook back to full recovery.

6. **Robbins' Cinquefoil** The Robbins' cinquefoil is a species of the rose family, found at just two locations on the slopes of the White Mountains in New Hampshire. In the 1970s, its numbers were reduced to roughly 1,800 plants due to trampling by horses and hikers and harvesting by commercial plant collectors. After listing and critical habitat designation pursuant to the Endangered Species Act, the Appalachian Mountain Club and New England Wild Flower Society teamed up with federal agencies to relocate a hiking trail, educate the public and reestablish healthy populations. By 2002, the species' numbers had rebounded to over 14,000 plants in two populations, and the species was removed from the endangered list. A cooperative agreement with the U.S. Forest Service helps ensure the continuation of the Robbins' cinquefoil's success story through management and monitoring.

Opportunities for Updating and Improving the Act

Many lessons can be learned from the successes described above and from the numerous other positive experiences implementing the Endangered Species Act. The following are some ideas for updating and improving the Act that are drawn from these experiences.

- **Implement Recovery Plans and Encourage Proactive Conservation.** Any effort to update the Endangered Species Act must begin with steps to promote greater and earlier progress toward recovery. As discussed above, due to Act's flexibility the Nation has benefited in recent years from numerous collaborative initiatives to restore species and habitats. Wildlife agencies should build recovery plans around these proactive recovery initiatives, and Congress should support them with funding so long as they are consistent with recovery plans. If such an approach were taken, ESA conflicts would be reduced because there would be greater buy-in to the Act's implementation. Because greater amounts of habitats would be restored, wildlife agencies would have greater management flexibility.

The Endangered Species Act already provides a solid foundation for this approach. Section 4(f) calls for one of the two wildlife agencies to develop a recovery plan with objective measurable criteria for success and to implement it. However, recovery plans oftentimes are not completed for many years after listing, and thus there is no early blueprint to guide management and restoration actions. A simple solution to this problem would be to require that recovery plans be finalized within a specified time after listing (e.g., 3 years).

A related problem is that the two wildlife agencies are typically not in the position to carry out many of the actions that are needed to bring about recovery. Section 7(a)(1) of the Act requires all federal agencies to utilize their authorities in furtherance of species recovery, but it does not link this duty to the recovery plan. As a result, agencies have often chosen recovery actions in an arbitrary manner.

A solution to this problem would be for federal agencies to be required to develop and implement Recovery Implementation Plans to set forth the specific actions, timetables, and funding needed for that agency to help achieve the recovery goals set forth in the Recovery Plan. The Western Governors Association developed a variation of this idea when it adopted its ESA legislative proposal in the 1990s. "Implementation agreements" for federal and state agencies to help carry out recovery plans remains part of WGA policy to this day.

Another problem related to implementation of recovery plans is that federal agencies oftentimes carry out actions that are at odds with those plans. For example, the Corps of Engineers has issued dredge-and-fill permits for development in Florida panther habitat despite the fact that the habitat is deemed essential for the species in the recovery plan. Congress could easily fix this problem by clarifying that federal agencies must ensure that their actions do not undermine the recovery needs of listed species. The recovery needs of the species would be identified in the recovery plan, and updated by the latest scientific data. If Congress were to adopt this approach, agency decisions would more likely to contribute to the Act's recovery goal. They would also be easier to defend in court, and less likely to attract litigation, because they would be tied to a larger strategic framework, the recovery plan.

- **Provide incentives for private landowners to contribute to recovery.**
  According to the GAO, roughly 80 percent of all listed species have at least some of their habitat on non-federal land; about 50 percent have the majority of their habitat on non-federal land. Much of this non-federal land is private land, and yet the current Endangered Species Act

does not provide many incentives for private landowners to carry out the management measures that are often needed for listed species to thrive. Although ESA regulatory programs such as Safe Harbor remove disincentives, they do not provide incentives. Technical assistance programs can help, but by far the most meaningful incentive that Congress can provide is financial assistance. To ensure a reliable source of funding, this assistance should be provided through the tax code. In return for conservation agreements in which private landowners commit to actively manage habitats for the benefit of listed species, Congress should defer indefinitely federal estate taxes or provide immediate income tax credits for expenses incurred.

- **Protect critical habitat.** The Administration has attempted to justify its efforts to weaken the Act's critical habitat protections by claiming that these protections are redundant with other ESA protections and therefore without value to listed species. At the same time, the Administration contradicts itself by generating cost-benefit analyses claiming that critical habitat protections are imposing enormous costs on the private sector. None of this rhetoric is supported by any meaningful analysis of data. The only quantitative studies on critical habitat have shown that critical habitat indeed provides benefits to many listed species. Species with critical habitat designations tend to do better than species without such designations.

  Critical habitat is particularly important when it comes to protecting unoccupied habitat, because the other protections in the Endangered Species Act generally do not adequately protect such habitat. Most species will never recover unless they can return to some part of their historic range that is currently unoccupied.

  Because of the hostility shown by the current Administration toward critical habitat, it will be essential for Congress, when it reauthorizes the ESA, to strongly reaffirm the importance of critical habitat protection. Congress should push back the deadlines to three years after listing, thereby giving the wildlife agencies the time they need to get the science right. It also should encourage the wildlife agencies to integrate recovery plan and critical habitat designation decisions. Congress also should develop a schedule, and authorize the funding, for cleaning up the backlog of species awaiting critical habitat designations. When the late Senator Chafee took these steps in S. 1100 back in 1999, they attracted broad public support.

- **Provide adequate funding.** Finally, there perhaps can be no more important step that Congress can take to improve implementation of the Endangered Species Act than to increase funding to reasonable levels. At a bare minimum, Congress must provide the funding that the wildlife agencies need to carry out their mandatory duties. For example, the U.S. Fish and Wildlife Service has estimated that it would take approximately $153 million over 10 years to eliminate the current backlog of listings and critical habitat designations. Congress could immediately eliminate dozens of lawsuits simply by providing these funds and other funds needed for the basic implementation steps of the Act. In addition, many of the concerns about the Act's impact on states, local governments and private landowners could be alleviated if Congress were to expand its Section 6 and other grant funding for recovery actions.

Testimony

T he Endangered Species Act (ESA) was enacted in 1973 with the promise that we can do better in the job of protecting and conserving our nation's resident species and the ecosystems that support them. Today, over thirty years later, I bring that same message back to this Committee—*we can, and must, do better.* We have learned many lessons over the past three decades about how and what can be done to protect endangered and threatened species, and it is time to update and improve the ESA to reflect those lessons.

I am here before you today on behalf of the National Endangered Species Act Reform Coalition (NESARC), an organization of 110 national associations, businesses and individuals that are working to develop bipartisan legislation that updates and improves the ESA. Personally, my organization, the National Marine Manufacturers Association (NMMA), joined NESARC in 2003 largely due to our members' experiences with listed marine species such as the manatee population in Florida, as well for as the opportunity to join a diverse group of interests working on this matter. I have the pleasure of sitting on the NESARC Board of Directors. On behalf of the NESARC Board of Directors and, all of the NESARC members, I want to commend the efforts being undertaken by members of this Committee, other members of the Senate and in the House of Representatives to develop a bipartisan bill that updates and improves the ESA. We look forward to working with the Committee, its able staff, and other members of the Senate to find common ground.

NESARC members come from a wide range of backgrounds. Among our ranks are farmers, ranchers, cities and counties, rural irrigators, electric utilities, forest and paper operators, mining, homebuilders and other businesses and individuals throughout the United States. What our members have in common is that they have been impacted by the operation of the ESA. Frankly speaking, the burdens and rewards of protecting listed species are borne, in a very large part, by the members of NESARC. NESARC members are actively involved in a broad range of species conservation efforts including:

- The development of State management plans for wolf populations in the Rocky Mountains and in Minnesota, Michigan and Wisconsin.
- Recovery implementation programs such as the Upper Colorado and San Juan Rivers Endangered Fish Recovery Implementation Program and Platte River Endangered Species Recovery Program;

United States Senate Committee on Environment & Public Works Hearing on the National Wildlife Federation Oversight on the Endangered Species Act, May 19, 2005.

- Numerous habitat conservation plans ranging from county-wide HCPs in Southern California to single parcel plans for covering agricultural operations; and
- Observation, research and monitoring programs for listed and candidate species.

Many environmental groups (including some of those who are testifying today) have recognized the need for on-the-ground partnerships. The reality is that, without the support and active commitment to the protection of listed species by the private landowners, businesses and communities where the species reside, the chances of success are slim. We need to learn from the experiences of those who are faced with the real-world decisions on how to make a living and still protect species if we are to make the Act work better.

If we are to do a better job protecting endangered and threatened species, we need an ESA that can fully accommodate the range of efforts that are necessary. As detailed later in my testimony, NESARC has developed a number of recommendations for ways to improve the ESA. These recommendations are the product of an extensive reassessment by NESARC members as to what improvements to the ESA would be useful for the future implementation of the Act.

At the end of 2003, NESARC decided to look inward, to reassess the state of the ESA's implementation on the ground and to identify the success stories of its members in protecting endangered and threatened species as well as those roadblocks that had to be overcome. What we learned was that, more often than not, our members have succeeded in protecting endangered and threatened species *in spite of, rather than because of,* the ESA.

When we asked our members to share their success stories and positive experiences, what we received were very personal observations from the ground reporting that success is occurring—but not easily.

"Our HCP process has had some very beneficial elements, but it's been painfully slow and costly to get there. Given the experience, [it is] hard to endorse it for others to pursue. Yet an HCP embodies concepts for species protection which are very good and could be more effective. [We] advocate moving to a system with more incentives and much greater penalties for abuses." **Carol Rische, Humboldt Bay Metropolitan Water District.**

"Some of the regulators that we deal with are very results-oriented. Their practical approach has been beneficial to our operations and beneficial to species recovery. Working together with practical regulators to the benefit of the species has been a positive experience." **Tom Squeri, Granite Rock Company.**

The experience of my own members within NMMA is similar—with the hope of cooperative efforts between federal and state agencies limited by the realities of working within an Act that was enacted more than thirty years ago and does not provide the necessary flexibility and tools to effectively and efficiently develop workable solutions. As many of you know, Florida has a long history of protecting its endangered manatee population—in which NMMA members have actively participated. As a result of efforts led by the State of Florida and stakeholders, the manatee population has grown from an estimated 1,465 manatees in 1991 to at least 3,142 (as documented by a 2005 aerial survey)—more than a doubling of the population in approximately 14 years. Further, the U.S. Fish & Wildlife Service has joined with the Florida Fish & Wildlife Conservation Commission to begin a "manatee forum" which is aimed at developing a consensus, science-based approach to continuing to protect and enhance manatee populations in balance with marine activities. However, such cooperative efforts remain the exception, not the norm.

As I am here today representing NESARC, I do not wish to dwell on the particular problems facing boaters, marina operators and other marine services; however, to the extent that the Committee wishes to hear more about the personal experiences of any of our individual NESARC members, including NMMA, we are happy to provide that information and brief you or your staff on particular issues of interest.

Drawing from our members' experiences and observations, NESARC identified a series of guideposts from which to consider future improvements to the ESA, which include the following:

- Encourage Sound Decision-making
- Promote Innovation
- Promote Certainty
- Increase Funding
- Reduce Economic Impacts
- Increase Roles for State, Local Governments
- Provide Greater Public Participation
- Limit Litigation

After developing these initial guideposts, over the latter half of 2004, NESARC worked to draft a white paper which was publicly released in November 2004. This white paper is attached to my testimony and provides an outline of a new approach to ESA legislation that we hope the Members of this Committee will take into consideration.

In sum, a new approach is needed to change the focus of the debate from a clash over existing terms and programs to the development of new tools that improve the Act. We need *new provisions* of the Act that encourage recovery of listed species through voluntary species conservation efforts and the active involvement of States. This new approach can and should maintain the goal of species conservation. Simultaneously, we must recognize that species conservation and recovery will only be accomplished if we can find ways to provide stakeholders the tools and flexibility to take action and, most importantly, certainty that quantifiable success will be rewarded by the lifting of the ESA restrictions.

As this Committee reviews ways to improve the ESA, we would ask that you take into consideration the following proposals:

- **Expand and Encourage Voluntary Conservation Efforts**—A universal concern with the Act is that it does not fully promote and accommodate voluntary conservation efforts. Many landowners want to help listed species, but the ESA doesn't let them. A critical element of updating and improving the Act must be the development of additional voluntary conservation programs. These efforts should include: (1) creating a habitat reserve program, (2) tax incentives, (3) loan or grant programs and (4) other initiatives that encourage landowners to voluntarily participate in species conservation efforts. Further, existing programs like the Safe Harbor Agreements should be codified.

- **Give the States the Option of Being On the Front Line of Species Conservation**—In 1973, the National Wildlife Federation testified before Congress that "[s]tates should continue to exercise the prime responsibility for endangered species" and "should be given the opportunity to prepare and manage recovery plans and retain jurisdiction over resident species." Thirty-plus years later, the Western Governors' Association, in a February 25, 2005 letter (attached) noted that "[t]he [ESA] can be effectively implemented only through a full partnership between the states and the federal government" and asked Congress to "give us the tools and authority to make state and local conservation efforts meaningful."

    NESARC agrees that States should have a wider role in facilitating landowner/operator compliance with the Act and, ultimately, the recovery of species. States have significant resources, research capabilities and coordination abilities that can allow for better planning of species management activities. Further, States know their lands and are often better situated to work with stakeholders to protect and manage the local resources and species.

- **Increase Funding of Voluntary and State Programs for Species Conservation**—A significant amount of federal funding for ESA activities is presently tied up in addressing multiple lawsuits and the review of existing and new listing and critical habitat proposals. In contrast, actual funding for on-the ground projects that will recover species is limited.

    Federal funding priorities need to be re-focused to active conservation measures that ultimately serve to achieve the objectives of the Act. Further, we need to financially support the voluntary, community-based programs that are critical to ensuring species recovery.

- **Encourage Prelisting Measures**—Recently, a nationwide coalition of state and local governments, stakeholders and conservation organizations worked together to develop a comprehensive sage grouse conservation program that has been able to stand in the place of a listing of that species under the ESA. Those efforts were supported by many members of this committee including Senator Harry Reid of Nevada who stated that, ". . . I have advocated using the Farm Security and Rural Investment Act of 2002 (Farm Bill) conservation programs to help local communities like Elko, Nevada, engage in voluntary conservation efforts for species like sage grouse. In fact, the Farm Bill's Wildlife Habitat Incentives Program (WHIP) encourages private and public agencies to develop wildlife habitat on their properties, and specifically has directed funds to enhance habitats for sage grouse. I know more can be done, and I am committed to improving local conservation efforts." *Statement of Senator Harry Reid, September 24, 2004.*

    Private landowners, State and local governmental agencies should be encouraged to develop and implement programs for species that are being considered for listing. The protections afforded by all such programs (including existing activities) should be considered in deter- mining whether a listing is warranted or whether such voluntary pro- grams, other federal agency programs and State/local conservation efforts already provide sufficient protections and enhance species popu- lations so that application of the ESA is not necessary.

- **Establish Recovery Objectives**—We need to be able to identify and establish recovery objectives. Knowing what ultimately must be achieved is a critical first step in understanding what must be done. Since the goal of the ESA is to assure recovery of endangered and threat- ened species, implementation of the ESA should reward progress when it is made toward recovery. There must be a determination of specific recovery goals necessary to reach the point where a species can and will be downlisted or delisted—and there must be certainty in such a goal so that the goal is not continually shifted to perpetuate a listing.

- **Strengthen the Critical Habitat Designation Process**—We need to strengthen the critical habitat designation process by ensuring that these designations are supported by sound decision-making pro- cedures, do not overlap with existing habitat protection measures (such as habitat conservation plans, safe harbor agreements or candi- date conservation agreements, and other state and federal land conser- vation or species management programs) and rely on timely field survey data.

- **Improve Habitat Conservation Planning Procedures and Codify "No Surprises"**—The HCP process has the potential to be a success story, but too often private property owners are stymied by the delays and costs of getting HCP approval. HCP approval should be stream- lined, and the HCP process must be adapted so that it is practical for the smaller landowner. Further, landowners involved in conservation efforts need to be certain that a "deal is a deal." The "No Surprises" policy must be codified under the Act and cover all commitments by private parties to voluntary protection and enhancement of species and habitat—not just HCPs.

- **Ensure an Open and Sound Decision-Making Process**—The ESA must be open to new ideas and data. A good example of this principle is the emerging data regarding the effect of boat speeds on manatees and their avoidance mechanisms. Because the principal threat to manatees is impact from boat propellers, federal and state manatee-protection poli- cies historically have focused on slowing boats passing through manatee habitats. However, research by Dr. Edmund Gerstein of Florida Atlantic University and Joseph E. Blue, retired director of the Naval Undersea Warfare Center and the Naval Research Laboratory's Underwater Sound Reference Detachment challenges some of the existing protection mea- sures. This new research shows that while manatees have good hearing abilities at high frequencies, they have relatively poor sensitivity in the low frequency ranges associated with boat noise, which means that manatees may be least able to hear the propellers of boats that have slowed down in compliance with boat speed regulations designed to reduce collisions. My point is not to suggest that there should not be speed limits in areas occupied by manatees, but rather that we need to make sure that our policy decisions (like setting boat speed limits) are informed by up-to-date research. By providing for better data collection and independent scientific review, we can ensure that the necessary and appropriate data is available.

In addition to making sure we have better information upon which to act, we need a decision-making process that allows for full public participation in the listing, critical habitat and recovery decisions. It has been my experience that providing full and open access to the decision-making processes—beyond simply the submission of letter comments—through mechanisms like stakeholder representatives and data collection programs provides a much more diverse and ultimately stronger record from which to act.

For more than a decade, Congress has struggled with the question of what, if any, changes to the ESA should be made. In the interim, stakeholders like NESARC members, have had to take the existing Act and make it work. It has been time-consuming, expensive and often frustrating--and the successes have been limited. Today, less than 1% of all listed species in the United States have been recovered.

The Congressional history on ESA legislation has had its ebbs and flows over the past thirteen years with at least two distinct sets of legislative efforts— both of which ultimately failed. NESARC is not interested in going down that same path again where stakeholders (on both sides) re-open old battles and try to right perceived wrongs from past court decisions. NESARC urges this Committee to take stock of the lessons we have learned and successes that have been achieved in order to identify the improvements that are necessary to make this Act work better in the future.

POSTSCRIPT

Should the Endangered Species Act Be Strengthened?

After the Threatened and Endangered Species Recovery Act passed the House of Representatives in September 2005 (see "House Passes Bill Reforming Endangered Species Act," *Human Events* [October 31, 2005]), Senator Mike Enzi (R-Wyo) introduced in December similar legislation in the Senate, calling it a Christmas present for "Wyoming farmers and ranchers. . . . We have a bill that will both help recover species and preserve landowner livelihood." Six U.S. senators, both Republican and Democrat, asked the Keystone Center to prepare a report on changes needed in the Endangered Species Act. The Keystone Center is a Colorado think tank that helps "leaders from governmental, non-governmental, industrial, and academic organizations to find productive solutions to controversial and complex public policy issues." Its report, released in April 2006 (http://www.keystone.org/spp/env-esa.html), concluded that the ESA's effectiveness for species recovery could be improved in several ways and that the burdens imposed by the ESA on private interests could be relieved with incentives, but reconciling the two goals is difficult and the House bill is not the way to do it. Senator Lincoln Chafee (R-RI), one of the six who requested the Keystone report, said in March 2006 that Senate action was unlikely for that year. The fall 2006 shift in political domination of Congress from Republicans to Democrats underscores Chafee's prediction.

Michael J. Bean, "The Endangered Species Act under Threat," *Bioscience* (February 2006), cautioned that the House act undermines "the government's long-standing trust responsibility to safeguard wildlife. The Senate should think long and hard before embracing the House's radical proposals."

As a contribution to that thinking process, the Ecological Society of America, together with other scientific societies, published the "Scientific Societies's Statement on the Endangered Species Act" (http://www.esa.org/pao/policyStatements/pdfDocuments/2-2006_finalStatement_Scientific%20Societies%20ESA.pdf) on February 27, 2006. Among other things, the statement objects to bureaucratic attempts to limit the definition of useful scientific data (included in the House legislation), calls for eliminating delays in evaluating rare species for listing as threatened or endangered, improving funding for research, and restoring protections for critical habitat. It does recognize that "parties experiencing economic and social impacts from recovery activities should be included" in planning.

Contributors to This Volume

EDITOR

THOMAS A. EASTON is a professor of science at Thomas College in Waterville, Maine, where he has been teaching since 1983. He received a B.A. in biology from Colby College in 1966 and a Ph.D. in theoretical biology from the University of Chicago in 1971. He is a prolific writer on scientific and futuristic issues. His books include *Focus on Human Biology,* 2nd ed., coauthored with Carl E. Rischer (HarperCollins, 1995), and *Careers in Science,* 4th ed. (VGM Career Horizons, 2004). Dr. Easton is also a well-known writer and critic of science fiction.

STAFF

Larry Loeppke	Managing Editor
Jill Peter	Senior Developmental Editor
Susan Brusch	Senior Developmental Editor
Beth Kundert	Production Manager
Jane Mohr	Project Manager
Tara McDermott	Design Specialist
Nancy Meissner	Editorial Assistant
Julie Keck	Senior Marketing Manager
Mary Klein	Marketing Communications Specialist
Alice Link	Marketing Coordinator
Tracie Kammerude	Senior Marketing Assistant
Lori Church	Pemissions Coordinator

AUTHORS

JULIAN AGYEMAN is assistant professor of Urban and Environmental Policy and Planning at Tufts University and the author of *Sustainable Communities and the Challenge of Environmental Justice* (New York University Press, 2005).

RONALD BAILEY is a science correspondent for *Reason* magazine. A member of the Society of Environmental Journalists, his articles have appeared in many popular publications, including the *Wall Street Journal, The Public Interest,* and *National Review.* He has produced several series and documentaries for PBS television and *ABC News,* and he was the Warren T. Brookes Fellow in Environmental Journalism at the Competitive Enterprise Institute in 1993. He is the editor of *Earth Report 2000: Revisiting the True State of the Planet* (McGraw-Hill, 1999) and the author of *Global Warming and Other Eco-Myths: How the Environmental Movement Uses False Science to Scare Us to Death* (Prima, 2002).

MICHAEL BEHAR is a freelance writer and editor based in Washington, D.C. His beat includes environmental issues and scientific innovations.

DAVID L. BODDE is professor and senior fellow at the International Center for Automotive Research, Arthur M. Spiro Center for Entrepreneurial Leadership, at Clemson University.

MATTHEW BUNN is a senior research associate in the Project on Managing the Atom in the Belfer Center for Science and International Affairs at Harvard University's John F. Kennedy School of Government. His current research interests include nuclear theft and terrorism; disposition of excess plutonium; and nuclear waste storage, disposal, and reprocessing.

JAMIE CLARK is senior vice president for conservation programs, National Wildlife Federation.

BENEDICT S. COHEN is deputy general counsel for environment and installations, Department of Defense.

GERALD D. COLEMAN is the former rector of St. Patrick's Seminary and University in Menlo Park, California. His books include several on Catholic views of sexuality.

GIULIO A. De LEO is an associate professor of applied ecology and environmental impact assessment in the Dipartimento di Scienze Ambientali at the Universit degli Studi di Parma in Parma, Italy.

PHILLIP J. FINCK is deputy associate laboratory director, applied science and technology and national security, Argonne National Laboratory.

MONITA FONTAINE is a member of the board of directors of the National Endangered Species Act Reform Coalition and the National Marine Manufacturers Association's vice president for government relations.

DAVID FRIEDMAN is a writer, an international consultant, and a fellow in the MIT Japan program.

MARINO GATTO is a professor of applied ecology in the Dipartimento di Elettronica e Informazione at Politecnico di Milano in Milan, Italy. His main research interests include ecological models and the management of renewable resources. Gato is associate editor of *Theoretical Population Biology*.

MICHAEL GOUGH, a biologist and expert on risk assessment and environmental policy, has participated in science policy issues at the Congressional Office of Technology Assessment, in Washington think tanks, and on various advisory panels. He most recently edited *Politicizing Science: The Alchemy of PolicyMaking* (Hoover Institution Press, 2003).

JOHN D. GRAHAM, administrator of the office of information and regulatory affairs at the Office of Management and Budget, was appointed in October 2005 to be Dean of the Frederick S. Pardee RAND Graduate School.

BRIAN HALWEIL is a senior researcher at the Worldwatch Institute. His latest book is *Eat Here: Reclaiming Homegrown Pleasures in a Global Supermarket* (W. W. Norton, 2004).

ROBERT H. HARRIS is a principal with ENVIRON International Corporation. He has over 25 years of experience in the area of environmental health and toxic chemicals, with particular emphasis on water and air pollution and hazardous waste issues. He is recognized nationally as an expert consultant on the treatment and disposal of municipal solid and hazardous waste, as well as on air, soil, and groundwater contamination.

MICHEL J. KAISER is professor of marine ecology at the School of Ocean Sciences, University of Wales, Bangor. He was recently appointed to the editorial board of the *Journal of Sea Research* and the journal *Fish and Fisheries*.

JOHN KOSTYACK is senior counsel and director of Wildlife Conservation Campaigns with the National Wildlife Federation.

DAVID N. LABAND is a professor of economics and policy at the Forest Policy Center in the School of Forestry and Wildlife Sciences at Auburn University in Alabama. He is the author, with George McClintock, of *The Transfer Society: Economic Expenditures on Transfer Activity* (National Book Network, 2001). Labandís research interests include forest economics, causes and consequences of environmental policy, and land use planning.

DWIGHT R. LEE is the Ramsey Professor of Economics and Private Enterprise in the Terry College of Business at the University of Georgia. He received his Ph.D. from the University of California at San Diego in 1972 and his research has covered a variety of areas, including personal finance, public finance, the economics of political decision making, and the economics of the environment and natural resources. Lee has published over 100 articles and commentaries in academic journals, magazines, and newspapers. He is coauthor, with Richard B. McKenzie, of *Getting Rich in America: Eight Simple Rules for Building a Fortune and a Satisfying Life* (HarperBusiness, 2000).

SEAN McDONAGH is a Columban priest and ex-missionary to the Philippines. His latest book is *The Death of Life: The Horror of Extinction* (Columba Press, Dublin, 2004).

ANNE PLATT McGINN is a senior researcher at the Worldwatch Institute and the author of "Why Poison Ourselves? A Precautionary Approach to Synthetic Chemicals," Worldwatch Paper 153 (November 2000).

EDWARD J. MARKEY has represented the Seventh Congressional District of Massachusetts in the U.S. House of Rrepresentatives since 1976. He is a member of the Energy and Commerce Committee and the U.S. House Select Committee on Homeland Security.

MICHAEL MEYER, the European editor for *Newsweek International*, is a member of the New York Council on Foreign Relations and was an inaugural fellow at the American Academy in Berlin. He won the Overseas Press Club's Morton Frank Award for business/economic reporting from abroad in 1986 and 1988. He is the author of *The Alexander Complex* (Times Books, 1989), an examination of the psychology of American empire builders.

JOHN J. MILLER is a political reporter for the *National Review,* a contributing editor for *Reason* magazine, a former vice president of the Center for Equal Opportunity, and a Bradley Fellow at the Heritage Foundation. His most recent book is *A Gift of Freedom: How the John M. Olin Foundation Changed America* (Encounter, 2005).

JIM MORRISON is a Virginia journalist who writes frequently on environmental topics.

NANCY MYERS is communications director for the Science and Environmental Health Network. She is the co-editor, with Carolyn Raffensperger, of *Precautionary Tools for Reshaping Environmental Policy* (MIT Press, 2005).

DAVID NICHOLSON-LORD is an environmental writer, formerly with *The Times, The Independent,* and *The Independent on Sunday,* where he was environment editor. He is the author of *The Greening of the Cities* (Routledge, 1987) and of *Green Cities—And Why We Need Them* (New Economics Foundation, 2003). He is a member of UNESCO's UK Man and the Biosphere Urban Forum, an executive of the Urban Wildlife Network, and a trustee of the National Wildflower Centre. He also teaches environment in the journalism faculty at City University, London.

PUBLIC CITIZEN is a national, nonprofit organization founded in 1971 to represent consumer interests in Congress, the executive branch, and the courts.

JEREMY RIFKIN is the president of the Foundation on Economic Trends in Washington, D.C., and has written many books on the impact of scientific and technological changes on the economy, the workforce, society, and the environment. Among his latest books is *The European Dream: How Europe's Vision of the Future Is Quietly Eclipsing the American Dream* (Tarcher/Penguin, 2004).

DONALD R. ROBERTS is a professor in the Division of Tropical Public Health, Department of Preventive Medicine and Biometrics, Uniformed Services University of the Health Sciences.

MARGOT ROOSEVELT is a national correspondent for *Time* magazine. Since moving to Los Angeles, California, in 1994, she has specialized in social issues, covering immigration, education, crime, trade, energy, and environmental stories, including controversies over genetically modified food.

DONALD F. SANTA, JR., is the president of the Interstate Natural Gas Association of America.

CHARLES W. SCHMIDT is a freelance science writer based in Portland, Maine. He received an M.S. in public health from the University of Massachusetts at Amherst, with a concentration in toxicology. He has written for *Environmental Health Perspectives, Technology Review, New Scientist, Popular Science,* the *Washington Post, Child* magazine, *Environmental Science and Technology,* and more.

JERALD L. SCHNOOR occupies the Allen S. Henry Chair in engineering at the University of Iowa, where he is a professor of both civil and environmental engineering and occupation and environmental health. He is also the co-director of the Center for Global and Regional Environmental Research.

MIKE TILCHIN is a vice president of CH2M HILL.

BRIAN TOKAR is an associate faculty member at Goddard College in Plainfield, Vermont. A regular correspondent for Z magazine, he has been an activist for over 20 years in the peace, antinuclear, environmental, and green politics movements. He is the author of *The Green Alternative: Creating an Ecological Future,* 2d. ed. (R & E Miles, 1987).

MICHELE L. TRANKINA is a professor of biological sciences at St. Maryís University and an adjunct associate professor of physiology at the University of Texas Health Science Center, both in San Antonio, Texas.

JAY VANDEVEN is a principal with ENVIRON International Corporation. He has 16 years of experience in the assessment and remediation of soil and groundwater contamination, contaminant fate and transport, environmental cost allocation, and environmental insurance claims.

MICHAEL J. WALLACE is executive vice president of constellation energy, a leading supplier of electricity to large commercial and industrial customers.

ROBERT R. WARNER is a professor of marine ecology at the University of California at Santa Barbara. He has served on the Science Advisory Panel to the Marine Reserves Working Group for the Channel Islands National Marine Sanctuary, and as the research chair on the Sanctuary Advisory Council.

HOWARD YOUTH is a researcher and writer on wildlife conservation issues.

Index